MW01011801

Student's Study Guide
Part 1

Thomas/Finney

Calculus and Analytic Geometry

9TH Edition

Maurice D. Weir
U.S. Naval Postgraduate School

Addison-Wesley Publishing Company

Reading, Massachusetts • Menlo Park, California • New York
Don Mills, Ontario • Wokingham, England • Amsterdam • Bonn
Sydney • Singapore • Tokyo • Madrid • San Juan • Milan • Paris

Reproduced by Addison-Wesley from camera ready copy supplied by the author.

Copyright © 1996 by Addison-Wesley Publishing Company, Inc.

All rights reserved. No part of this publication may be reproduced, stored in a retrieval system, or transmitted, in any form or by any means, electronic, mechanical, photocopying, recording, or otherwise, without the prior written permission of the publisher. Printed in the United States of America.

ISBN 0-201-53181-X
2 3 4 5 6 7 8 9 10 ML 989796

PREFACE TO THE STUDENT

This study guide has been designed especially for you, the student. It conforms with the ninth edition of CALCULUS AND ANALYTIC GEOMETRY by George B. Thomas and Ross L. Finney. It is intended as a self-study workbook to assist you in mastering the basic ideas in calculus. Although this manual was written to conform to the Thomas/Finney CALCULUS AND ANALYTIC GEOMETRY it can be used to accompany any standard calculus textbook and course.

ORGANIZATION AND LEARNING OBJECTIVES

The study manual is organized section by section to correspond with the Thomas/Finney text. For each section we specify its main ideas by stating appropriate learning OBJECTIVES. Each objective states a particular task for you to perform in order to master that objective. Usually the task requires you to solve a certain type of problem related to the discussion in the text; sometimes the task requires you to demonstrate proficiency with certain key terms or concepts. In every case the objective is highly specific and states exactly what you must do.

One or more examples follows each objective and illustrates its requirements. Each example is written in a semi-programmed format; that is, the example is only partially worked out, so you must supply some of the intermediate results yourself. Correct answers to each intermediate result are supplied at the bottom of the page. Thus, each example is broken down into a sequence of steps to guide you through the procedures and techniques associated with its solution. Each example has been carefully selected not to repeat examples or problems in the Thomas/Finney text; thus you retain the full array of the text problems for practice and further learning.

SELF-TESTS

At the end of each chapter there is a SELF-TEST. Each test is followed by complete solutions to all the test problems. The test problems cover the objectives and are similar in scope and difficulty to the examples in this manual and the examples and problems in the Thomas/Finney text. The test should be useful in preparing for class examinations.

HOW TO USE THIS STUDY GUIDE

We recommend that this manual be used in the following way:

1. Read the textbook: Carefully read the section of Thomas/Finney assigned you by your calculus instructor.

2. Study the learning objectives and examples: Read each objective and work through the associated example(s) in the corresponding section of this manual. You should conceal the answers to the examples at the bottom of the page. Work with pencil and scratch paper as you are guided through each solution, writing in the intermediate results in the blanks provided.

3. Check your answers: After all the blanks for a given problem are filled in, compare your answers with the correct answers at the bottom of the page. If you have difficulty or do not fully understand the answers given, review the material in the textbook or consult with your instructor.

4. Do the chapter self-test: After you complete a chapter in Thomas/Finney, review the objectives for that chapter in this manual. Then take the chapter self-test and compare your solutions with those provided. Problems in the self-test sometimes bring together several ideas from the chapter.

GUIDANCE FROM YOUR INSTRUCTOR

We caution you that the learning objectives given in this manual by no means exhaust all the possible objectives that could be written for a careful study of calculus; we have tried to identify the main ones. However, your instructor may have additional requirements. For instance, he or she may want you to be able to prove certain theorems or derive results in the text. We have not stated objectives of this sort. Also, your instructor may consider some objectives far more important than others and not require that you master some objectives at all. Moreover your instructor may require critical thinking and writing exercises that ask you to explain the reasoning behind your answers. So it is imperative that you find out specifically what your instructor considers essential, and study accordingly. This manual should be helpful to you both in identifying the tasks and successfully mastering them. The problems assigned by your instructor should help you discover those concepts and applications of calculus that your instructor wishes to stress.

MAURICE D. WEIR
Monterey, California

TABLE OF CONTENTS

PRELIMINARIES

P.1 REAL NUMBERS AND THE REAL LINE

OBJECTIVE A: Know the special subsets of real numbers.

1. The numbers 1, 2, 3, 4, ... are called the natural numbers.

2. The integers are the numbers $0, \pm 1, \pm 2, \pm 3, \ldots$.

3. The rational numbers are the numbers that can be expressed in the form $\frac{m}{n}$ where m and n are integers and n $\neq 0$.

4. Rational numbers have decimal expansions that are either terminating or repeating .

5. Irrational numbers are real numbers that are not ~~integers~~ rational . They are characterized by having ~~a radical sign~~ nonterminating and nonrepeating decimal expansions.

6. An interval is a subset of the real line that contains at least two numbers and all the real numbers lying between any two of its elements.

OBJECTIVE B: Find an interval or intervals of numbers that satisfy an inequality in x.

7. Solve the inequality $\frac{5 - x}{3} < \frac{2x + 3}{4}$.

   Solution. Multiplying both sides of the inequality by 12 gives

   $20 - 4x <$ 6x + 9

   20 $< 10x + 9$ (Add 4x to both sides)

   $11 < 10x$ (Subtract 9 from both sides)

   The solution set is the interval $\left(\frac{11}{10}, \infty\right)$.

---

1. natural
2. $0, \pm 1, \pm 2, \pm 3, \ldots$
3. $\frac{m}{n}$, integers, n
4. terminating or repeating
5. rational, nonterminating
6. two, all, two
7. $6x + 9$, 20, 9, $\left(\frac{11}{10}, \infty\right)$

OBJECTIVE C: Define <u>absolute value</u> and know its basic properties.

8. The absolute value of a number x is denoted by ______, and is defined by ______ if $x \geq 0$ and ______ if $x < 0$. Thus, $|-3| = -($______$) =$ ______ and $\left|\frac{2}{3}\right| =$ ______ .

9. Geometrically, $|x|$ represents the distance from ______ to the ______ on the real line. More generally, $|x - y|$ is the ______ .

10. $\sqrt{x^2} =$ ______ . If you already know $x \geq 0$ you can write $\sqrt{x^2} =$ ______ .

11. The basic absolute value properties are

$$|a + b| \leq \underline{\qquad\qquad}$$
$$|ab| = \underline{\qquad\qquad}$$
$$|-a| = \underline{\qquad\qquad}$$
$$\left|\frac{a}{b}\right| = \underline{\qquad\qquad} .$$

OBJECTIVE D: Solve elementary equations or inequalities involving absolute values.

12. The equation $\left|4 - \frac{3}{x}\right| = 1$ says that $4 - \frac{3}{x} =$ ______ or $4 - \frac{3}{x} =$ ______ . Thus there are two possibilities: $4x - 3 = x$ or ______ . Solution of the first of these equations gives $x =$ ______ . The second equation is equivalent to $5x =$ ______ or $x = \frac{3}{5}$. The equation $\left|4 - \frac{3}{x}\right| = 1$ has two solutions: $x =$ ______ and $x =$ ______ .

13. If $|x+2| \leq 7$, then $|x-(-2)| \leq$ ______ . This is equivalent to the inequality ______ . Therefore x must lie within the closed interval ______ .

14. Solve the inequality $|4 - 3x| \geq 2$.
<u>Solution</u>.

$$4 - 3x \geq 2 \qquad \text{or} \qquad \underline{\qquad\qquad} \geq 2$$
$$2 \geq \underline{\qquad\qquad} \qquad \text{or} \qquad 4 - 3x \leq \underline{\qquad\qquad}$$

---

8. $|x|$, x, $-x$, -3, 3, $\frac{2}{3}$

9. x, origin, distance between x and y

10. $|x|$, x

11. $|a| + |b|$, $|a|\,|b|$, $|a|$, $\frac{|a|}{|b|}$

12. 1 or -1, $4x - 3 = -x$, 1, 3, 1, $\frac{3}{5}$

13. 7, $-7 \leq x+2 \leq 7$, $[-9,5]$

$\underline{2/3} \geq x$ or $\underline{2} \leq x$

The solution set is $\underline{(-\infty, \frac{2}{3}] \cup [2, \infty)}$.

P.2 COORDINATES, LINES, AND INCREMENTS

OBJECTIVE A: Draw a rectangular coordinate system and plot or locate points within it.

15. Finish labeling the coordinate system at the right and plot the point

$$P = P(-3, 2).$$

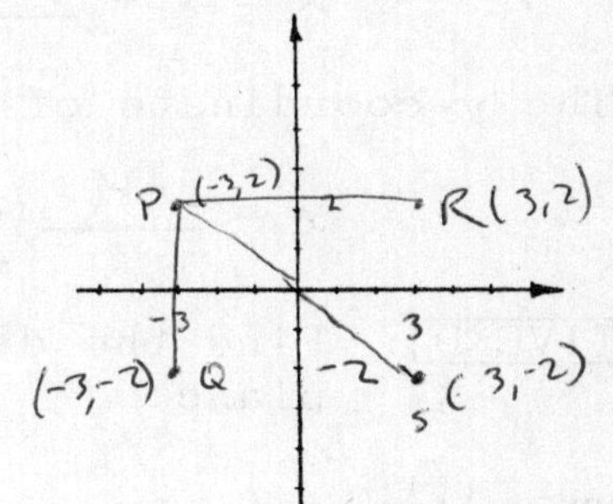

16. Use the same diagram to locate the point Q such that PQ is perpendicular to the x-axis and bisected by it. The coordinates of Q are $\underline{(-3,-2)}$.

17. Use the same diagram to locate the point R such that PR is perpendicular to the y-axis and bisected by it. The coordinates of R are $\underline{(3,2)}$.

18. Use the same diagram to locate the point S such that PS is bisected by the origin. The coordinates of S are $\underline{(3,-2)}$.

OBJECTIVE B: Find the net changes Δx and Δy in a particle's coordinates as it moves from a point P to a point Q.

19. If a particle starts at P(2, -1) and goes to Q(-7, -3), then its x-coordinate changes by

$$\Delta x = -7 - \underline{2} = \underline{-9} .$$

20. Its y-coordinate changes by

$$\Delta y = \underline{-3} - (-1) = \underline{-2} .$$

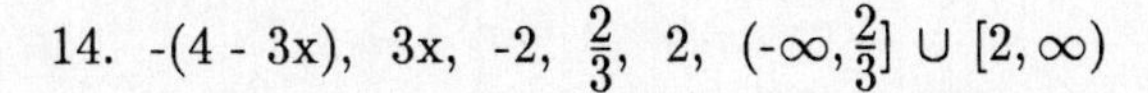
14. -(4 - 3x), 3x, -2, $\frac{2}{3}$, 2, $(-\infty, \frac{2}{3}] \cup [2, \infty)$

15. P(-3, 2) 16. Q(-3, -2)

17. R(3, 2) 18. S(3, -2)

19. 2, -9 20. -3, -2

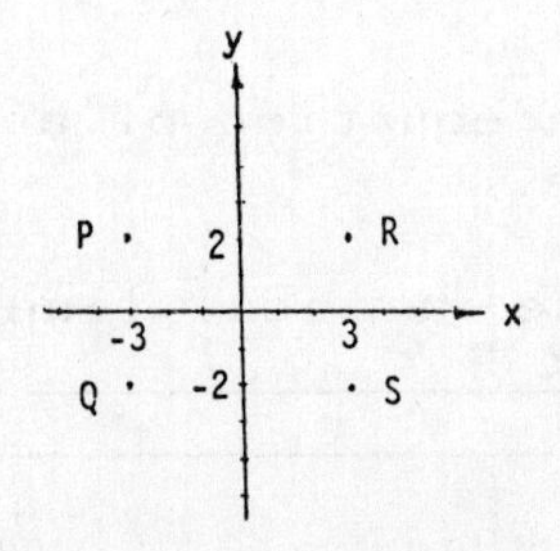

OBJECTIVE C: Given the increments from the point P to the point Q and the coordinates of one of these points, determine the coordinates of the other point.

21. The coordinates of a particle change by $\Delta x = -3$ and $\Delta y = 5$ in moving from P(1, -4) to Q(x, y). The x-coordinate of Q is given by

$$x = 1 + \underline{\Delta x \;(-3)} = \underline{-2}\ .$$

22. The y-coordinate of Q is given by

$$y = \underline{-4} + 5 = \underline{1}\ .$$

OBJECTIVE D: Find the distance between two given points in the plane.

23. The distance between the points P(-1, 3) and Q(2, -5) is $d = \sqrt{(-1 - \underline{2})^2 + (3 - \underline{-5})^2} = \underline{\sqrt{73}}$.

9 + 64

OBJECTIVE E: Define the slope of a straight line and calculate the slope (if any) of the line determined by two given points.

24. The slope of the line through the points $P_1(x_1, y_1)$ and $P_2(x_2, y_2)$ is given by

$$m = \frac{\text{rise}}{\text{run}} = \underline{\frac{\Delta y}{\Delta x}} = \underline{\frac{y_2 - y_1}{x_2 - x_1}},\ \text{provided that } x_1 \neq x_2\ .$$

25. If $x_1 = x_2$, then the line through the points $P_1(x_1, y_1)$ and $P_2(x_2, y_2)$ is a __vertical__ line. For vertical lines, the __slope__ is not defined.

26. The slope of the line through the points $A(-\frac{1}{2}, 1)$, $B(0, -2)$ is $m = \underline{-6}$.

$$\frac{-2-1}{0-\frac{-1}{2}} = \frac{-3}{\frac{1}{2}} = -3 \cdot 2 = -6$$

OBJECTIVE F: Find the slope (if any) of a line perpendicular to a line determined by two given points.

27. The slope of the line perpendicular to AB in Problem 26 is $m = \underline{1/6}$.

OBJECTIVE G: Write an equation of any vertical line given a point on the line.

28. An equation of the vertical line passing through the point P(4, -7) is __x = 4__ .

---

21. -3, -2
22. -4, 1
23. 2, -5, $\sqrt{73}$
24. $\frac{y_2 - y_1}{x_2 - x_1}$, $\frac{y_1 - y_2}{x_1 - x_2}$
25. vertical, slope
26. -6
27. $\frac{1}{6}$
28. x = 4

OBJECTIVE H: Write an equation of any line with given slope and passing through a given point.

29. Using the point-slope equation of the line, we have $y = y_1 + m(x - x_1)$. Thus an equation of the line with slope $m = -2$ through the point $(1,3)$ is given by ____________.

30. The line perpendicular to the line in Problem 29 has slope $m = -\left(\frac{1}{-2}\right) = \frac{1}{2}$. Thus an equation of the perpendicular through $(1,3)$ is ____________.

OBJECTIVE I: Write an equation of any line given two points on the line.

31. Let $P_1(-3,0)$ and $P_2(2,-1)$ be two points on the line L. The slope of L is $m =$ ________. Thus, an equation of L using P_1 is ____________, using P_2 an equation is ____________. In either case, solving for y we obtain the equation $y =$ ____________.

32. Let $P_1(1,-3)$ and $P_2(1,5)$ be two points on the line L. Since the x-coordinates of the points are the same, we conclude that L is a ____________ line and hence has no ____________. An equation for L is ____________.

OBJECTIVE J: Recognize an equation as representing a line and determine the slope (if any), the x-intercept (if any), and the y-intercept (if any).

33. The equation $3x - 2y = 6$ represents a straight line because it contains only ____________ powers of x and y. When $x = 0$, $y =$ ____________ which gives the value where the line crosses the y-axis. This is called the ____________. When $y = 0$, $x =$ ____________ giving the value where the line crosses the ____________. This is called the x-intercept.

34. The equation $y = 3$ represents a straight line that is parallel to the ____________. It is called a ____________ line and has slope $m =$ ____________.

35. The equation $xy = 1$ does not represent a straight line because it is not a ____________ equation when the variables x and y are multiplied together.

---

29. $y = 3 + (-2)(x - 1)$ or $y = 5 - 2x$

30. $y = 3 + \frac{1}{2}(x - 1)$ or $y = \frac{5}{2} + \frac{1}{2}x$

31. $-\frac{1}{5}$, $y = 0 + \left(-\frac{1}{5}\right)(x + 3)$, $y = -1 + \left(-\frac{1}{5}\right)(x - 2)$, $y = -\frac{x + 3}{5}$

32. vertical, slope, $x = 1$

33. first, -3, y-intercept, 2, x-axis

34. x-axis, horizontal, 0

35. linear

OBJECTIVE K: Graph any equation representing a line.

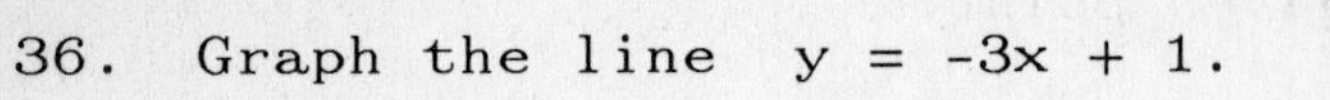

36. Graph the line $y = -3x + 1$.

37. Graph the line $\frac{x}{2} - \frac{y}{3} = \frac{1}{2}$.

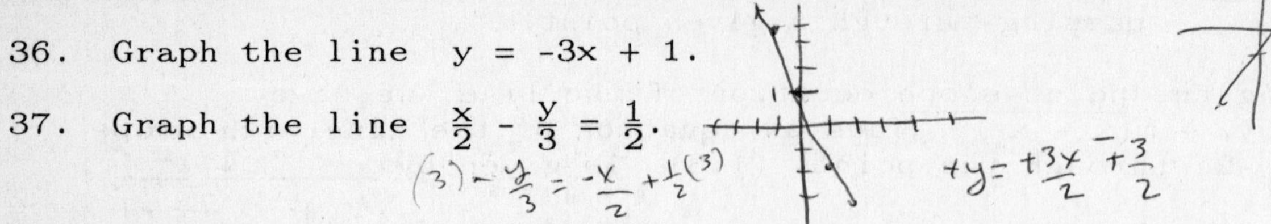

OBJECTIVE L: Find an equation of the line passing through a given point and parallel or perpendicular to a given line.

38. The line containing the point $(-1,2)$ that is parallel to the line $3x - y - 1 = 0$ has slope $m =$ ________ . Since the line contains the point $(-1,2)$, its equation in point-slope form is ________ .

39. The line containing the point $(4,1)$ that is perpendicular to the line $2y - 3x = 5$ has slope $m =$ ________ . Since the line contains the point $(4,1)$, its equation in point-slope form is ________ .

P.3 FUNCTIONS

In this section of the textbook, there are several terms associated with the concept of a function with which you will need to become familiar. The following items are designed to assist you in learning the precise mathematical meanings of these various terms.

40. Calculus is concerned with how variables are related. If to each value of the variable x there corresponds a unique value of the variable y, then y is said to be a ________ of x. The key word in this definition of function is ________ : we do not want to input a single value for the variable x with two or more possible outcomes for y. Every function is determined by two things: (1) the ________ of the first variable x and (2) the ________ or condition describing how y is obtained from x. The variable x is called the ________ variable of the function; the second variable y is called the ________ variable. The set of

---

36.

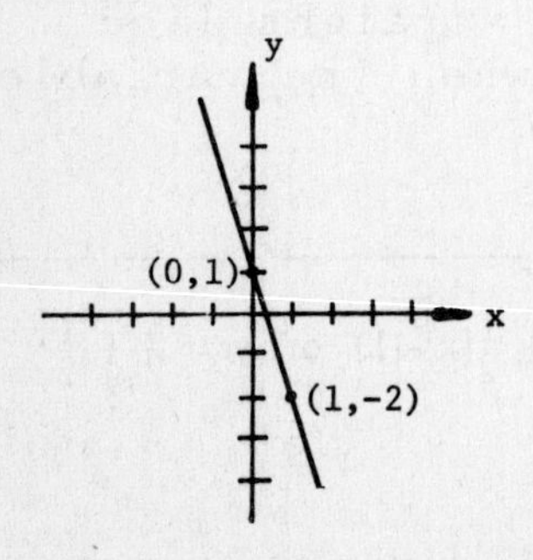

37.

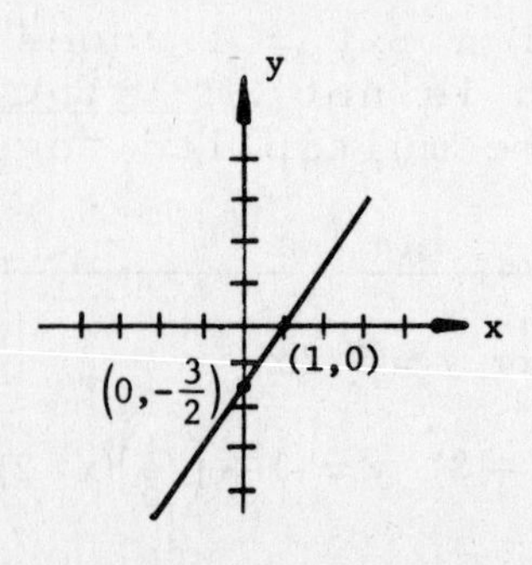

38. 3, $y = 2 + 3(x + 1)$

39. $-\frac{2}{3}$, $y = 1 - \frac{2}{3}(x - 4)$

values taken on by the dependent variable y is called the range of the function.

41. Two important restrictions on the domain of a real-valued function are: (a) never divide by 0; (b) never take square roots of negative numbers.

42. Convention: If the domain of a function is not stated explicitly, then the domain is automatically the largest set of x values for which the formula for the function gives real y-values. This is the function's natural domain.

43. The graph of the function $y = f(x)$ is the set of points (x, y) in the plane whose coordinates are the input output pairs of the function.

OBJECTIVE A: Given an equation for a function $y = f(x)$ calculate the value of f at a specified point, find the domain and range of f, and graph f by making a table of pairs.

$3x = 2 - y \qquad x = 2/3 - y/3$

44. Consider the function $y = -3x + 2$. The domain of the function is the interval $(-\infty, \infty)$. Solving the equation for x, gives $x =$ $2/3 - y/3$ so that the variable y may take on any value whatsoever. Thus, the range of the function is the interval $(-\infty, \infty)$. Sketch the graph.

45. Consider the function $y = \frac{x^2 - 1}{x + 1}$. The function is defined for all values of x except $x = -1$; hence the domain consists of the union of the intervals $(-\infty, -1)$ and $(-1, \infty)$. When $x \neq -1$, $y = \frac{x^2 - 1}{x + 1} = \frac{(x - 1)(x + 1)}{x + 1} =$ $x - 1$. Therefore, the range of the function is all real numbers except for

---

40. function, unique, domain, rule, independent, dependent, range

41. (a) by zero (b) negative numbers

42. the largest set of x-values, real y-values, natural

43. (x, y) in the plane, input-output

44. $(-\infty, \infty)$, $x = \frac{2 - y}{3}$, $(-\infty, \infty)$

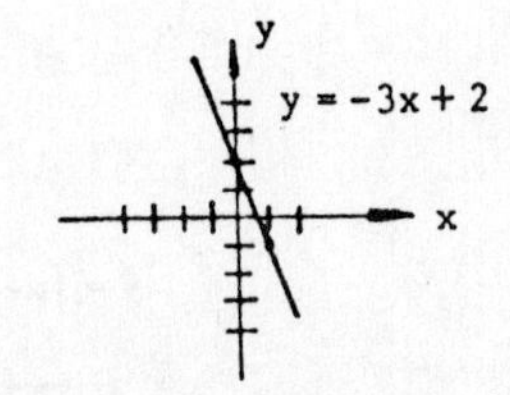

$y =$ ________ (because $x \neq -1$), so that the range is the union of the two intervals ________ and ________ .

46. The domain of the function $y = -\sqrt{1 - x}$ is the interval ________ , since $\sqrt{1 - x}$ is defined whenever $1 - x \geq 0$. Squaring both sides and solving the resultant equation for x, we obtain $x =$ ________ . We see from this last equation that y can take on any value. However, since y is the negative square root, the range is the interval ________ .

47. If $g(x) = \dfrac{1}{\sqrt{x - 2}}$, the domain of g is ________ . The value $g(3)$ is ________ ; $g(11)$ is ________ ; $g(a)$ is ________ ; $g(b + 2)$ is ________ .

48. Consider the function $y = [x - 1] + 2$, where $[x - 1]$ denotes the greatest integer in ________ . The domain of this function is the interval ________ . A table of some of the values for this function is given by (complete the table):

x	-2.0	-1.5	-1.0	-.5	0	.5	1.0	1.5	2.0	2.5
y	-1.0									

Sketch a graph of the function using the table. The range of this function is not an interval, but the set of numbers ________ .

OBJECTIVE B: Given two functions f and g, find formulas for $f+g$, $f-g$, $f \cdot g$, f/g, and g/f.

49. If $f(x) = x + 1$ and $g(x) = x^2 - 1$, then
$(f+g)(x) =$ ________ ;
$(f-g)(x) =$ ________ ;
$(f \cdot g)(x) = (x + 1)(x^2 - 1) =$ ________ ;

---

45. $x = -1$,
$(-\infty,-1)$ and $(-1,\infty)$,
$x - 1$, -2,
$(-\infty,-2)$ and $(-2,\infty)$

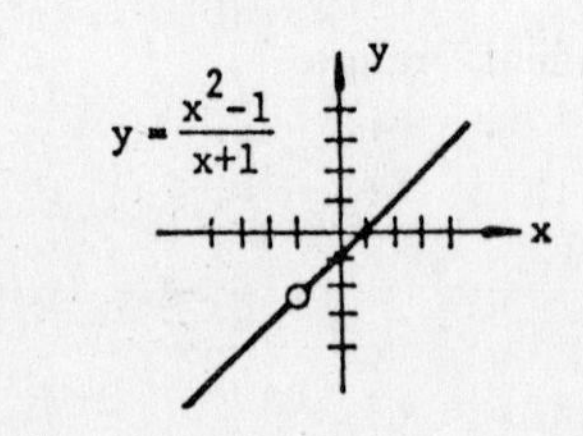

46. $(-\infty, 1]$,
$x = 1 - y^2$,
$(-\infty, 0]$

47. $(2,\infty)$, 1, $\frac{1}{3}$, $\frac{1}{\sqrt{a-2}}$, $\frac{1}{\sqrt{b}}$

48. $x - 1$, $(-\infty,\infty)$,

x	-2.0	-1.5	-1.0	-.5	0
y	-1.0	-1.0	0	0	1.0

x	.5	1.0	1.5	2.0	2.5
y	1.0	2.0	2.0	3.0	3.0

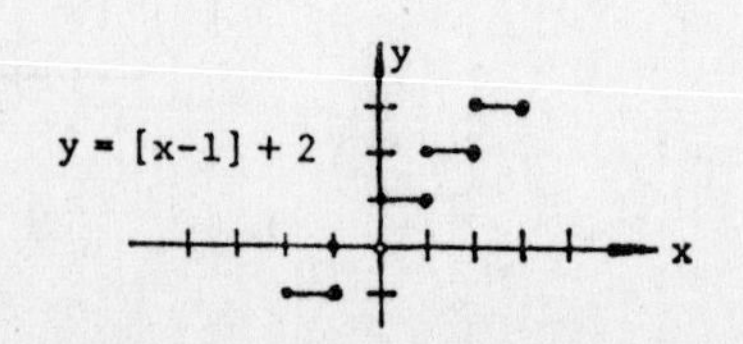

range: $\{\ldots,-2,-1,0,1,2,3,\ldots\}$

$\left(\frac{f}{g}\right)(x) = \frac{x + 1}{x^2 - 1} =$ ________________ provided $x \neq$ ________ .

$\left(\frac{g}{f}\right)(x) =$ ________________ provided $x \neq$ ________ .

OBJECTIVE C: Given two functions f and g, write an expression for their composite $f(g(x))$.

50. If $f(x) = 5x + 2$ and $g(x) = x^2$, then a formula for $f(g(x))$ is obtained as follows:

$$f(g(x)) = f(x^2) = \text{\_\_\_\_\_\_\_\_} .$$

The domain of $y = f(g(x))$ is all values of x in the domain of g such that $f(g(x))$ is defined. This is the interval ________ .

51. If $f(x) = \sqrt{x - 1}$ and $g(x) = x + 1$, then

$$f(g(x)) = \text{\_\_\_\_\_\_\_\_} .$$

The domain of the composite is all values of x in the domain of g such that $f(g(x))$ is defined. This is the interval ________ .

52. Let $f(x) = x^2$ and $g(x) = \sqrt{x - 1}$. Then $f(g(x)) =$ ________ . The domain of g is the set of all real numbers x satisfying ________ . Thus, the domain of the composite $y = f(g(x))$ is the interval ________ .

OBJECTIVE D: Find two functions f and g that will produce a given composite function h such that $h(x) = f(g(x))$.

53. Consider $h(x) = \sin(x^2 - 1)$. If we let $f(x) = \sin x$ and $u = g(x) =$ ________, then $h(x) =$ ________ $= f(x^2 - 1) =$ ________ .

54. If $h(x) = \sqrt{x^5 + 2x^3 - 1}$, then for $f(x) = \sqrt{x}$ and $u = g(x) =$ ________, it is true that $h(x) = f(g(x))$.

OBJECTIVE E: Test a given function to find what symmetries its graph has.

55. If whenever the point (x, y) lies on the graph, then the point $(-x, -y)$ also lies on the graph, we say that the graph is symmetric about the ________ .

---

49. $x + x^2$, $x - x^2 + 2$, $x^3 + x^2 - x + 1$, $\frac{1}{x - 1}$, $x \neq 1, -1$

50. $5x^2 + 2$, $(-\infty, \infty)$

51. $\sqrt{x}$, $[0, \infty)$

52. $x - 1$, $x \geq 1$, $[1, \infty)$

53. $x^2 - 1$, $f(g(x))$, $\sin(x^2 - 1)$

54. $x^5 + 2x^3 - 1$

55. origin

56. If whenever the point (x,y) lies on the graph, then the point ________ also lies on the graph, we say that the graph is symmetric about the x-axis.

57. If whenever the point (x,y) lies on the graph, then the point $(-x,y)$ also lies on the graph, we say that the graph is symmetric about the ________ .

58. The graph of the function $y = 3 - x^2$ is symmetric about the ________ . The reason is that (x,y) on the graph implies

$$y = 3 - x^2 \quad \Rightarrow \quad y = 3 - (-x)^2 .$$

Therefore, the point ________ also lies on the graph. The graph has no other symmetries.

OBJECTIVE F: For a given function $y = f(x)$, identify if the function is even, odd, or neither.

59. The function $y = f(x)$ is even if $f(-x) =$ ________ . The graph of an even function is symmetric about ________ .

60. If $f(-x) = -f(x)$, the function $y = f(x)$ is said to be an ________ function. The graph of an odd function is symmetric about the ________ .

61. For the function $f(x) = x^2 + x^4$ we find $f(-x) = (-x)^2 + (-x)^4 =$ ________ . Thus the function is ________ .

62. For the function $f(x) = x^2 - x^5$ we find $f(-x) = (-x)^2 - (-x)^5 =$ ________ Is $f(-x) = f(x)$? Is $f(-x) = -f(x) = -x^2 + x^5$? Thus, this function is neither an even nor an odd function.

OBJECTIVE G: Make a table of values and graph a piecewise defined function.

63. Complete the table of values and graph the function defined by

$$y = \begin{cases} 1 - x, & x < 0, \\ x^2 - 1, & x \geq 0. \end{cases}$$

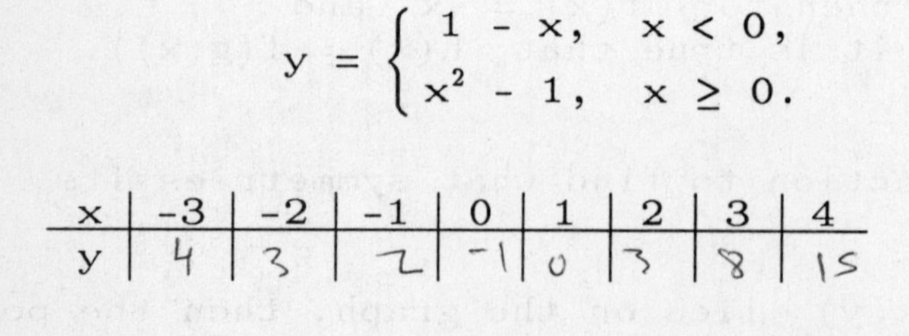

x	-3	-2	-1	0	1	2	3	4
y								

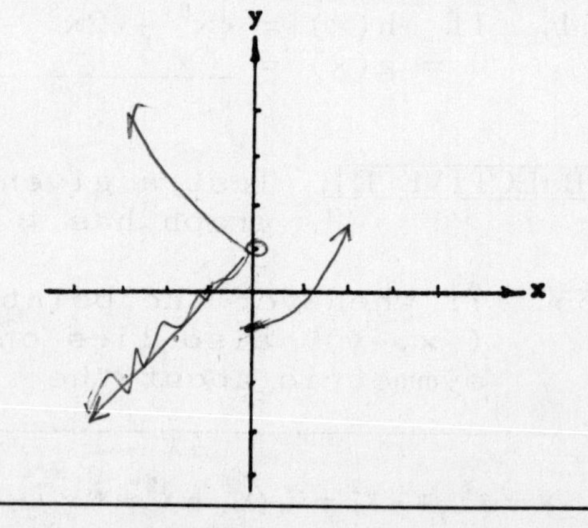

56. $(x,-y)$ 57. y-axis 58. y-axis, $(-x,y)$ 59. $f(x)$, the y-axis

60. odd, origin 61. $x^2 + x^4$ 62. $x^2 + x^5$, no, no

P.4 SHIFTING GRAPHS

64. The equation $y = f(x) + k$ shifts the graph of f ________ k units if $k < 0$.

65. The equation $y = f(x-h)$ shifts the graph of f ________ h units if $h > 0$.

66. The equation $y = f(x) + k$ shifts the graph of f ________ k units if $k > 0$.

67. The equation $y = f(x-h)$ shifts the graph of f ________ h units if $h < 0$.

OBJECTIVE A: Find an equation for a circle whose center and radius are known.

68. An equation of the circle with center at the point (h,k) and radius a is given by ________ .

69. An equation of the circle with center $(-1,3)$ and radius $a = \sqrt{11}$ is ________ or

$$x^2 + y^2 + 2x - 6y + (\_\_\_\_\_\_) = 0 .$$

70. The inequality $x^2 + y^2 \leq 4$ represents a circle of radius $a =$ ________ centered at the ________ plus its ________ .

OBJECTIVE B: Given an equation representing a circle, find the coordinates of its center and the radius.

71. Consider the circle $2x^2 + 2y^2 - 8x + 5y + 8 = 0$. Transposing the constant term and dividing by 2, we find $x^2 - 4x + ($________$) = -4$. Completing the squares,

---

63.

x	-3	-2	-1	0	1	2	3	4
y	4	3	2	-1	0	3	8	15

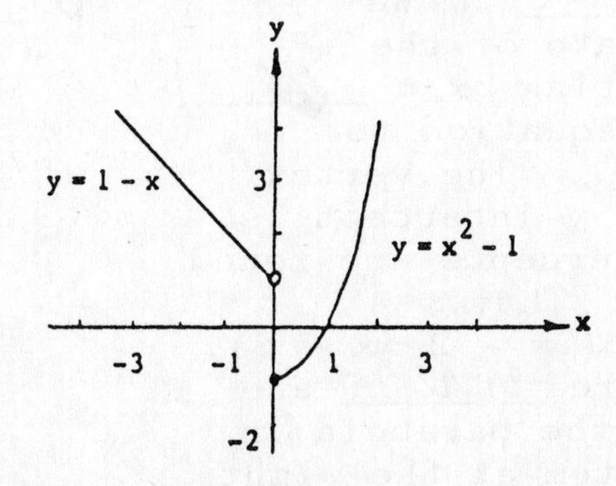

64. down 65. right 66. up 67. left

68. $(x - h)^2 + (y - k)^2 = a^2$ 69. $(x + 1)^2 + (y - 3)^2 = 11, -1$

70. 2, origin, interior

$(x - 2)^2 + (\_\_\_\_)^2 = -4 + \frac{25}{16} + \_\_\_\_$. Therefore, the center is ____ and the radius is ____.

OBJECTIVE C: Given an equation representing a line, circle or parabola, write an equation for a shifted graph when the number of units and directions of the shift are specified.

72. To shift the graph of the line $y = -2x + 1$ horizontally 3 units to the right, we rewrite its equation as

____.

Simplifying algebraically, $y = -2x +$ ____.

73. To shift the graph of the parabola $y = -x^2$ up 2 units and to the left 1 unit, we rewrite its equation as

____.

Simplifying algebraically, $y =$ ____.

OBJECTIVE D: Graph any parabola of the form $y = ax^2 + bx + c$, $a \neq 0$.

74. The direction of opening is ____ if $a < 0$.

75. The y-intercept is the point ____.

76. The axis of symmetry is the line ____.

77. Graph the parabola $y = 2x^2 + 4x - 6$.

Solution. First we identify $a =$ ____, $b =$ ____, and $c =$ ____. The direction of opening is ____ because $a > 0$. The axis of symmetry is the line $x =$ ____. We find the y-coordinate of the vertex by substituting $x =$ ____ in the parabola's equation to obtain $y =$ ____. The vertex is ____. The y-intercept is $(0, -6)$. The x-intercepts are found by setting $y = 0$. Thus, $0 = 2x^2 + 4x - 6 = (2x - 2)(x + 3)$. The x-intercepts are ____ and ____. Graph the parabola in the coordinate system at the right.

---

71. $y^2 + \frac{5y}{2}$, $\left(y + \frac{5}{4}\right)^2$, 4, $\left(2, -\frac{5}{4}\right)$, $\frac{5}{4}$

72. $y - 1 = -2(x - 3)$, 7

73. $y - 2 = -(x + 1)^2$, $-x^2 - 2x + 1$

74. down

75. $(0, c)$

76. $x = -\frac{b}{2a}$

P.5 TRIGONOMETRIC FUNCTIONS

OBJECTIVE A: Convert radian measure to degree measure, and vice versa.

78. The radian measure of 180° is ______ units. Here the symbol ______ represents a real number. This real number corresponds to the length that is subtended by an ______ of a circle of radius 1 with central angle ______. Therefore, 1° corresponds to ______ radians, and 1 radian corresponds to ______ degrees.

79. Converting from degree to radian measure, $60^\circ =$ ______ radians, $-45^\circ =$ ______ radians, and $72^\circ =$ ______ radians.

80. Converting from radian to degree measure, $\frac{\pi}{6}$ radians $=$ ______ degrees, $-\frac{3\pi}{2}$ radians $=$ ______ degrees, and 2 radians $=$ ______ degrees.

Remark. Whenever you encounter $\sin 2$, for instance, you must think the sine of 2 radians not the sine of 2 degrees. The latter is written $\sin 2^\circ$.

OBJECTIVE B: Given an angle in radians, calculate the values of its sine, cosine, tangent, cotangent, secant, and cosecant.

81. $\sin \frac{\pi}{2} =$ ______ .

82. $\tan \frac{\pi}{4} =$ ______ .

83. $\sec \frac{\pi}{3} =$ ______ .

84. $\cos(-\frac{\pi}{6}) =$ ______ .

85. $\cot(-\frac{\pi}{6}) =$ ______ .

86. $\csc(-\frac{\pi}{3}) =$ ______ .

---

77. 2, 4, -6, up, $-\frac{4}{4} = -1$, -1, -8, (-1,-8), (-3,0), (1,0)

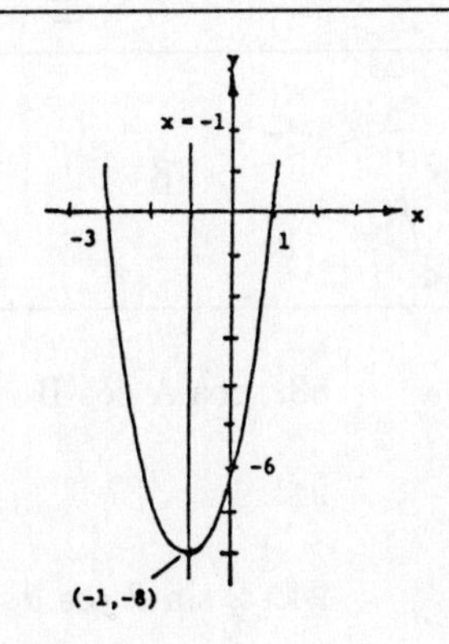

78. π, π, arc, 180°, $\frac{\pi}{180}$, $\frac{180}{\pi}$

79. $\frac{\pi}{3}$, $-\frac{\pi}{4}$, $\frac{2\pi}{5}$

80. 30, -270, $\left(\frac{360}{\pi}\right)$

81. 1

82. 1

83. 2

84. $\frac{\sqrt{3}}{2}$

85. $-\sqrt{3}$

86. $-\frac{2}{\sqrt{3}}$

OBJECTIVE C: Know from memory the most important trigonometric formulas.

Problms 87-96 give the most important trigonometric formulas to remember.

87. $\sin(A + B) =$ ~~sin A cos B + cos A sin B~~ .

88. $\cos(A + B) =$ ~~cos A cos B − sin A sin B~~ .

89. $\sin(-x) =$ ~~−sin x~~ .

90. $\cos(-x) =$ ~~cos x~~ .

91. $\sin^2\theta + \cos^2\theta =$ ~~1~~ .

92. Dividing both sides of the equation in Problem 91 by $\cos^2\theta$ gives $\tan^2\theta + 1 =$ ~~sec²θ~~ .

93. $\cos 2\theta =$ ~~cos²θ − sin²θ~~ .

94. $\sin 2\theta =$ ~~2 sin θ cos θ~~ .

95. $\cos^2\theta =$ ~~(1 + cos 2θ)/2~~ .

96. $\sin^2\theta =$ ~~(1 − cos 2θ)/2~~ .

97. You can use the above results to calculate new formulas. For example,

$$\begin{aligned}\sin(x - \tfrac{\pi}{2}) &= \sin[x + (-\tfrac{\pi}{2})]\\ &= \sin x \cos(-\tfrac{\pi}{2}) + \underline{\cos x \sin(-\pi/2)}\\ &= \sin x \cos(\tfrac{\pi}{2}) - \underline{\cos x \sin \pi/2}\\ &= (\sin x)\cdot 0 - \underline{\cos x}\\ &= \underline{\cos x}\ .\end{aligned}$$

98. If a, b, and c are the sides of a triangle ABC and if θ is the opposite c, then the law of cosines is the equation $\underline{c^2 = a^2 + b^2 - 2ab\cos\theta}$.

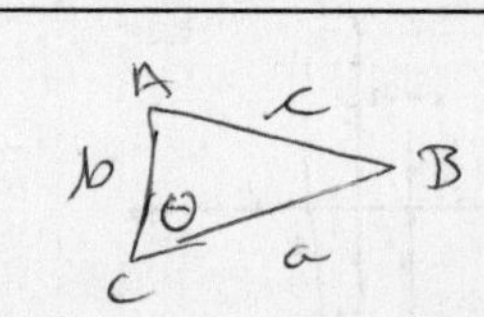

---

87. $\sin A \cos B + \cos A \sin B$

88. $\cos A \cos B - \sin A \sin B$

89. $-\sin x$

90. $\cos x$

91. 1

92. $\sec^2\theta$

93. $\cos^2\theta - \sin^2\theta$

94. $2\sin\theta\cos\theta$

95. $\dfrac{1 + \cos 2\theta}{2}$

96. $\dfrac{1 - \cos 2\theta}{2}$

97. $\cos x \sin(-\frac{\pi}{2})$, $\cos x \sin\frac{\pi}{2}$, $(\cos x)\cdot 1$, $-\cos x$

98. $c^2 = a^2 + b^2 - 2ab\cos\theta$

PRELIMINARIES CHAPTER SELF-TEST

1. Solve the following inequalities:

   (a) $\frac{2x + 3}{7} < \frac{3 - x}{2}$ (b) $-\frac{3 + x}{2} \geq \frac{11 - 3x}{5}$

2. Solve the following inequalities:

   (a) $|2x - 3| \leq 5$ (b) $|x - 2| > 4$ (c) $|2 - 3x| < -1$

3. For each of the following, draw a pair of coordinate axes, plot the given point, and plot the point meeting the specified requirement giving the coordinates of the second point:

   (a) $P(-3,1)$, and $Q(x,y)$ so that PQ is parallel to the x-axis and bisected by the y-axis;

   (b) $P(2,-2)$, and $R(x,y)$ so that PR is perpendicular to the x-axis and bisected by it;

   (c) $P(-1.3,-0.5)$, and $S(x,y)$ so that PS is bisected by the origin.

4. A particle moves in the plane along a straight line from $P(-3,-1)$ to $Q(7,-3)$. Find the net changes Δx and Δy and distance from P to Q.

5. A particle moves from the point $A(2,-3)$ to the x-axis in such a way that $\Delta y = -6\Delta x$. What are its new coordinates?

6. Determine if the points $A(1,-3)$, $B(-2,9)$, and $C(5,-19)$ lie exactly along a single straight line.

7. Find the slope of the line through the points $(1,4)$ and $(-3,2)$, and write an equation of the line.

8. Determine the slope, the x-intercept, and the y-intercept for each of the following equations:

   (a) $3x + 4y = -1$ (b) $x = 2$ (c) $y = -1$ (d) $x^2 = 2y - 1$

9. Find an equation of the line through the point $(5,-7)$ and perpendicular to the line $2y - x = 8$.

10. Let f be defined by the equation $f(x) = x^2 + 3x - 2$. Find the domain and range of f. Also find the values $f(-2)$, $f(-1)$, $f(0)$, $f(2)$, $f(2b)$ and $f(a + b)$, and sketch the graph of f.

11. Find the domain and range of the function $f(x) = \frac{x^2 - x - 6}{x + 2}$, and sketch the graph.

12. Let $f(x) = x^2 + 1$ and $g(x) = (x + 1)^2$. Find

(a) $f(x) - g(x)$ (b) $\frac{f(x)}{g(x)}$ (c) $f(g(x))$

(d) $g(f(x))$ (e) $g(g(x))$ (f) $g(x^2)$

13. Find two functions, f and g, that will produce the given composite $h(x) = f(g(x))$. There is no unique answer.

(a) $h(x) = \sqrt{x^2 - 1}$ (b) $h(t) = \left(t + \frac{1}{t}\right)^5$

14. Investigate any symmetries of the equation $x^{2/3} + y^{2/3} = 1$.

15. Say whether the functions are even, odd, or neither.

(a) $y = 1 + 2 \sin x$ (b) $y = |x| \cos 2x$

16. The following tell how many units and in what directions the graphs of the given equations are to be shifted. Give an equation for the shifted graph.

(a) $y = x^2$, up 2, left 3

(b) $y = \sin x$, down 3, right $\pi/4$

17. Write an equation of the circle with center $(1, -3)$ and of radius 4.

18. Find the coordinates of the center and the radius of the circle

$$x^2 + y^2 + 2x - 4y - 40 = 0 .$$

19. Graph the parabola and label the vertex, axis, and intercepts:

$$y = x^2 - 4x + 5 .$$

20. Graph the function $y = \frac{1}{2} \sin(3x - 2) + \frac{1}{2}$.

SOLUTIONS TO PRELIMINARIES CHAPTER SELF-TEST

1. (a) $\frac{2x + 3}{7} < \frac{3 - x}{2}$

$4x + 6 < 21 - 7x$

$11x < 15$

$x < \frac{15}{11}$

(b) $-\frac{3 + x}{2} \geq \frac{11 - 3x}{5}$

$\frac{3 + x}{2} \leq \frac{3x - 11}{5}$

$15 + 5x \leq 6x - 22$

$37 \leq x$

2. (a) $-5 \leq 2x - 3 \leq 5$ so $-1 \leq x \leq 4$

(b) $x - 2 > 4$ or $-(x - 2) > 4$

so $x > 6$ or $-x > 2$;

that is, x lies within the union of the intervals $(-\infty, -2) \cup (6, \infty)$.

(c) Since for any real number its absolute value is nonnegative, it is impossible for $|2 - 3x|$ to be less than -1; thus there is no solution.

3.

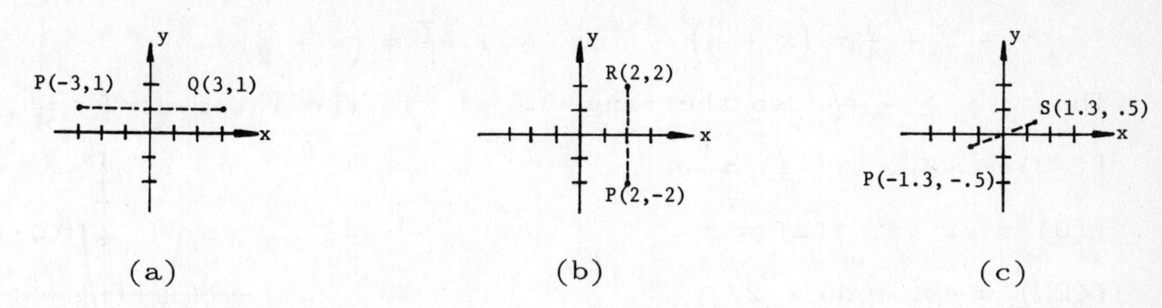

4. $\Delta x = 7 - (-3) = 10$, $\Delta y = -3 - (-1) = -2$,

$d = \sqrt{(\Delta x)^2 + (\Delta y)^2} = \sqrt{104} \approx 10.198$

5. The new coordinates can be written as $(x, 0)$ since the point lies on the x-axis. From $\Delta y = -6\Delta x$, we have $0 - (-3) = -6(x - 2)$ or, solving, $x = 3/2$. Thus $(\frac{3}{2}, 0)$ gives the coordinates of the new position of the particle.

6. The slope of AB is $m_1 = \frac{9 - (-3)}{-2 - 1} = -4$ and the slope of BC is $m_2 = \frac{-19 - 9}{5 - (-2)} = -4$. Since these slopes are equal, the three points do lie along a single straight line.

7. $m = \frac{2 - 4}{-3 - 1} = \frac{1}{2}$ is the slope, and $y = 4 + \frac{1}{2}(x - 1)$ or $2y - x = 7$ is an equation of the line.

8. (a) Solving algebraically for y, $y = -\frac{3}{4}x - \frac{1}{4}$. Thus, the slope is $m = -\frac{3}{4}$, the y-intercept is $b = -\frac{1}{4}$; and when $y = 0$, $x = -\frac{1}{3}$ is the x-intercept.

(b) This is a vertical line so it has no slope and no y-intercept. The x-intercept is 2.

(c) This is a horizontal line. It has slope 0 and y-intercept -1. It has no x-intercept.

(d) Since the variable x is squared, this equation does not represent a straight line. When $x = 0$, $y = \frac{1}{2}$ so the y-intercept is $\frac{1}{2}$. If $y = 0$, $x^2 = -1$ which is impossible, so it has no x-intercept.

9. The slope of $2y - x = 8$ or $y = \frac{1}{2}x + 4$ is $m = \frac{1}{2}$. Therefore, the slope of the perpendicular line is $m_\perp = -2$ and an equation is given by $y + 7 = -2(x - 5)$ or $y = -2x + 3$.

10. The function $f(x) = x^2 + 3x - 2$ is defined for all values of x, so the domain is $-\infty < x < \infty$ (all real numbers). Setting $y = x^2 + 3x - 2$ or $y + 2 = x^2 + 3x$ and completing the square on the righthand side gives

$$y + 2 + \frac{9}{4} = \left(x + \frac{3}{2}\right)^2, \quad \text{or} \quad y + \frac{17}{4} = \left(x + \frac{3}{2}\right)^2 .$$

Thus, $y \geq -\frac{17}{4}$ so the range of f is the interval $[-\frac{17}{4}, \infty)$.

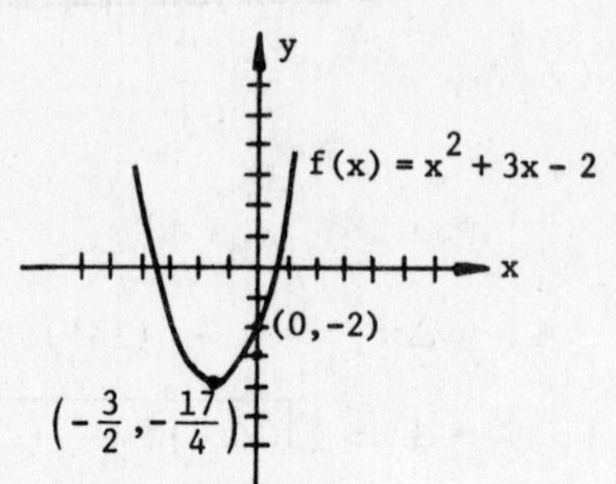

$f(-2) = -4$, $f(-1) = -4$,

$f(0) = -2$, $f(2) = 8$,

$f(2b) = 4b^2 + 6b - 2$,

$f(a + b) = a^2 + 2ab + b^2 + 3(a + b) - 2$

The graph of f is shown at the right.

11. $\dfrac{x^2 - x - 6}{x + 2} = \dfrac{(x - 3)(x + 2)}{x + 2}$.

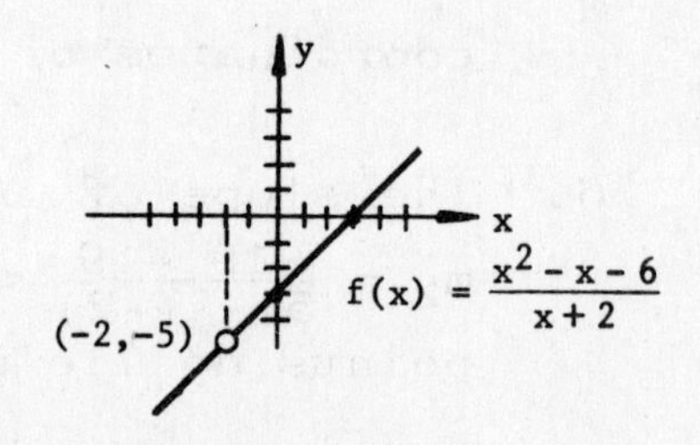

Thus, the domain of f is all real numbers except $x = -2$. Also, for $x \neq -2$, $f(x) = x - 3$. This is a straight line with the point $(-2,-5)$ deleted so the range of f is all real numbers except $y = -5$.

The graph of f is shown at the right.

12. (a) $f(x) - g(x) = -2x$ (b) $\dfrac{f(x)}{g(x)} = \dfrac{x^2 + 1}{(x + 1)^2}$

(c) $f(g(x)) = (x + 1)^4 + 1$ (d) $g(f(x)) = (x^2 + 2)^2$

(e) $g(g(x)) = (x^2 + 2x + 2)^2$ (f) $g(x^2) = (x^2 + 1^2)$

13. (a) $f(x) = \sqrt{x}$ and $g(x) = x^2 - 1$

(b) $f(t) = t^5$ and $g(t) = t + \frac{1}{t}$

14. Since $x^{2/3} + y^{2/3} = 1$ can be written $\left(x^{1/3}\right)^2 + \left(y^{1/3}\right)^2 = 1$, the curve is symmetric with respect to both axes and the origin. Also, $1 - x^{2/3}$ and $1 - y^{2/3}$ must both be nonnegative so $-1 \leq x \leq 1$ and $-1 \leq y \leq 1$ defines the extent of the curve in the x and y directions. The points $(-1,0)$, $(1,0)$, $(0,-1)$, and $(0,1)$ are the intercepts.

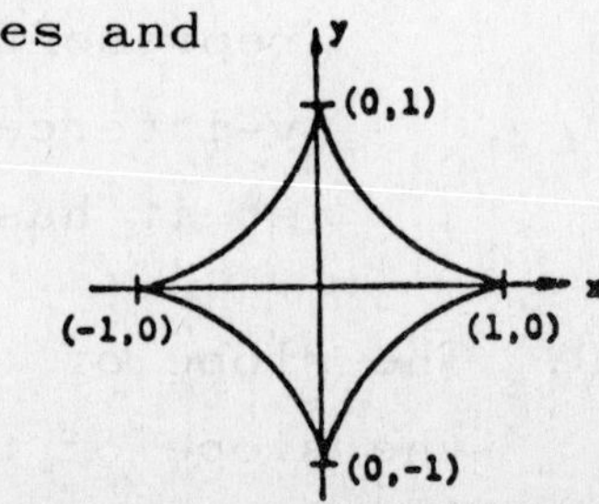

15. (a) neither (b) even

16. (a) $y - 2 = (x + 3)^2$ (b) $y + 3 = \sin(x - \frac{\pi}{4})$

17. $(x - 1)^2 + (y + 3)^2 = 16$ or $x^2 + y^2 - 2x + 6y - 6 = 0$.

18. Completing the square,

$$(x + 1)^2 + (y - 2)^2 = 45 = (3\sqrt{5})^2.$$

The center is $(-1,2)$ and the radius is $3\sqrt{5}$.

19. $y = x^2 - 4x + 5 = (x - 2)^2 + 1$ or $y - 1 = (x - 2)^2$

Vertex: $(2,1)$

Axis of symmetry: $x = -\frac{(-4)}{2(1)} = 2$

Intercepts: $(0,5)$, no x-intercepts

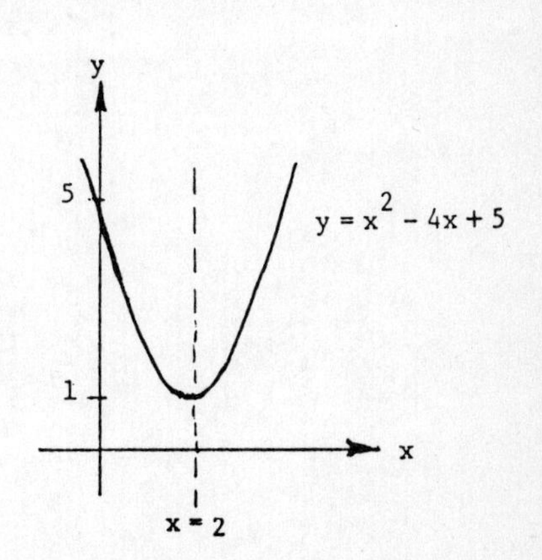

20.

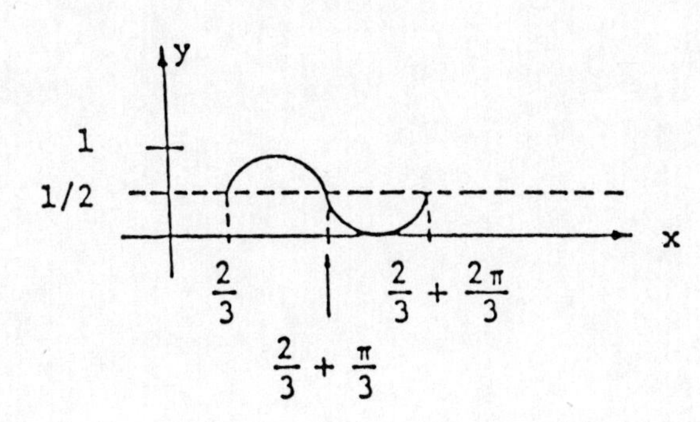

NOTES.

CHAPTER 1 LIMITS AND CONTINUITY

1.1 RATES OF CHANGE AND LIMITS

OBJECTIVE A: Find the average rate of change of a function $y = f(x)$ over an interval $[x_1, x_2]$.

1. The average rate of change of $y = f(x)$ over $[x_1, x_2]$ is the change in y, Δy = ______________ divided by Δx = ______________, the length of the interval over which the change occurred.

2. Geometrically, an average rate of change is a ______________ .

3. The average rate of change of $f(x) = x^2 - x + 1$ over $[0,2]$ is $\frac{\Delta y}{\Delta x} = \frac{f(2) - \_\_\_\_\_\_}{2 - 0} = \frac{(4 - 2 + 1) - \_\_\_\_\_\_\_\_\_\_}{2} = \_\_\_\_\_\_\_\_$.

OBJECTIVE B: Know the informal definition of the limit $\lim_{x \to x_0} f(x) = L$.

4. According to the informal definition, we write

$$\lim_{x \to x_0} f(x) = L$$

if the values of ________ approach the value L as x approaches ________ .

5. If f is the identity function $f(x) = x$, then for any value of x_0, $\lim_{x \to x_0} f(x)$ = ________ .

6. If f is the constant function $f(x) = k$, then for any value of x_0, $\lim_{x \to x_0} f(x)$ = ________ .

OBJECTIVE C: Find limits of elementary functions by substitution, if possible.

7. $\lim_{x \to \frac{1}{4}} (8x - 3)$ = _____ $- 3$ = ______ .

---

1. $f(x_2) - f(x_1)$, $x_2 - x_1$
2. secant slope
3. $f(0)$, $(0^2 - 0 + 1)$, $\frac{2}{2} = 1$
4. $f(x)$, x_0
5. x_0
6. k
7. 2, -1

8. $\lim_{x \to \frac{1}{2}} \frac{6x^2 + \frac{1}{2}}{4x - 1} = \frac{6(\frac{1}{4}) + \frac{1}{2}}{(\_\_\_\_)} = \_\_\_\_\_\_$.

9. $\lim_{x \to \frac{\pi}{4}} (\cos x)(2 \sin x) = \_\_\_\_\_\_ \left(\frac{2}{\sqrt{2}}\right) = \_\_\_\_\_\_$.

10. A function may fail to have a limit as x approaches x_0 because it
   a) ______________________ ,
   b) ______________________ ,
   c) ______________________ .

1.2 RULES FOR FINDING LIMITS

OBJECTIVE A: Specify the five important limit rules related to the arithmetic operations, as stated in Theorem 1 of the text.

11. Sum Rule: ______________________

12. Difference Rule: ______________________

13. Product Rule: ______________________

14. Constant Multiple Rule: ______________________

15. Quotient Rule: ______________________

____

OBJECTIVE B: Evaluate limits $\lim_{x \to c} f(x)$ when $f(x)$ is a sum, difference, product, or quotient of polynomials.

16. To find the limit as x approaches c of any polynomial function

$$f(x) = a_n x^n + a_{n-1} x^{n-1} + \ldots + a_o$$

---

8. 2 - 1, 2

9. $\frac{1}{\sqrt{2}}$, 1

10. a) jumps b) grows too large c) oscillates too much

11. $\lim_{x \to c} [f_1(x) + f_2(x)] = \lim_{x \to c} f_1(x) + \lim_{x \to c} f_2(x)$

12. $\lim_{x \to c} [f_1(x) - f_2(x)] = \lim_{x \to c} f_1(x) - \lim_{x \to c} f_2(x)$

13. $\lim_{x \to c} f_1(x) \cdot f_2(x) = \lim_{x \to c} f_1(x) \cdot \lim_{x \to c} f_2(x)$

14. $\lim_{x \to c} k \cdot f(x) = k \lim_{x \to c} f(x)$

15. $\lim_{x \to c} \frac{f_1(x)}{f_2(x)} = \lim_{x \to c} f_1(x) / \lim_{x \to c} f_2(x)$ provided that $\lim_{x \to c} f_2(x)$ is not zero

you simply ________ the number c for ________ thus evaluating f(c). Hence,

$$\lim_{x\to -1} (2x^3 + x^2 - 4x - 3) = \_\_\_\_\_\_\_\_ = \_\_\_\_\_\_ .$$

17. $\lim_{x\to 1} \dfrac{x^3 - 2x}{x^2 + 3} = \dfrac{\_\_\_\_\_\_}{1 + 3} = \_\_\_\_\_\_ .$

18. $\lim_{x\to -2} (4 + 3x)(x^2 - x + 1) = \lim_{x\to -2} (4 + 3x)$ ________

$= \_\_\_\_\_\_\_\_ (4 + 2 + 1) = -14 .$

19. $\lim_{x\to 4} \dfrac{x^2 + x - 2}{x^2 - 1} = \dfrac{\_\_\_\_\_\_\_\_}{\lim_{x\to 4} (x^2 - 1)} = \dfrac{18}{\_\_\_\_} .$

OBJECTIVE C: Evaluate limits of functions when the denominator is zero at the limit point c by cancelling a common factor, or by creating and cancelling a common factor.

20. $\lim_{x\to 1} \dfrac{x^3 + x^2 - 3x + 1}{x - 1} \neq \dfrac{\lim_{x\to 1} (x^3 + x^2 - 3x + 1)}{\_\_\_\_\_\_\_\_}$

because the limit of the denominator is ________ .

However, $\dfrac{x^3 + x^2 - 3x + 1}{x - 1} = \dfrac{(x - 1)(\_\_\_\_\_\_)}{x - 1}$,

so that $\lim_{x\to 1} \dfrac{x^3 + x^2 - 3x + 1}{x - 1} = \lim_{x\to 1} \_\_\_\_\_\_\_\_ = \_\_\_\_\_\_ .$

21. $\lim_{h\to 0} \dfrac{(1 + h)^3 - 1}{h} = \lim_{h\to 0} \dfrac{(\_\_\_\_\_\_\_\_) - 1}{h}$

$= \lim_{h\to 0} \dfrac{\_\_\_\_\_\_\_\_}{h}$

$= \lim_{h\to 0} \_\_\_\_\_\_\_\_ = \_\_\_\_\_\_ .$

22. $\lim_{x\to 9} \dfrac{x^2 - 81}{3 - \sqrt{x}} = \lim_{x\to 9} \dfrac{x^2 - 81}{3 - \sqrt{x}} \cdot \dfrac{\_\_\_\_\_\_}{3 + \sqrt{x}}$

$= \lim_{x\to 9} \dfrac{(x^2 - 81)(3 + \sqrt{x})}{(\_\_\_\_\_\_\_\_)}$

---

16. substitute, x, $2(-1)^3 + (-1)2 - 4(-1) - 3$, 0

17. 1 - 2, $-\frac{1}{4}$

18. $\lim_{x\to -2} (x^2 - x + 1)$, -2

19. $\lim_{x\to 4} (x^2 + x - 2)$, 15

20. $\lim_{x\to 1} (x - 1)$, 0, $x^2 + 2x - 1$, $x^2 + 2x - 1$, 2

21. $1 + 3h + 3h^2 + h^3$, $3h + 3h^2 + h^3$, $3 + 3h + h^2$, 3

$$= \lim_{x \to 9} \frac{(x - 9)(\_\_\_\_\_\_)(3 + \sqrt{x})}{9 - x}$$

$$= \lim_{x \to 9} -(\_\_\_\_\_\_)(\_\_\_\_\_\_) = \_\_\_\_\_\_\_\_ .$$

OBJECTIVE D: State and use the Sandwich Theorem for limits.

23. If $g(x) \le f(x) \le h(x)$ for $x \ne c$ over some interval containing c, and if $\lim_{x \to c} g(x) = \lim_{x \to c} h(x) = L$, then

____________________ .

24. For $-\frac{\pi}{2} < x < \frac{\pi}{2}$ it is known that

$$1 \le \frac{\tan x}{x} \le \frac{1}{\cos x} .$$

Therefore, since $\lim_{x \to 0} \cos x =$ ______ we have,

$$\lim_{x \to 0} \frac{\tan x}{x} = \_\_\_\_\_\_\_\_ .$$

1.3 TARGET VALUES AND FORMAL LIMITS

OBJECTIVE A: Given a function $y = f(x)$, a positive number ϵ, a point x_o, and a target value L, determine an interval about x_o in which we must hold x to be sure that $y = f(x)$ lies within ϵ units of L.

25. Suppose $y = -3x + 1$, $\epsilon = 1$, $x_o = 2$ and $L = -5$. We must know in what interval of values to hold x to make y satisfy the inequality

$$|y - (-5)| < 1 .$$

Substituting for y, we find $|(-3x + 1) - (-5)| < 1$ or __________ < 1. Thus, $|-3(x - 2)| < 1$ or $|x - 2| <$ ______ . Thus, $-\frac{1}{3} < x - 2 < \frac{1}{3}$, or x satisfies the inequality ______ $< x <$ ______ . In summary, the interval $|x - 2| < \frac{1}{3}$ contains the values near $x_o = 2$ to which x must be held to ensure $y = -3x + 1$ is within $\epsilon = 1$ of $L = -5$.

OBJECTIVE B: Write the formal definition of the limit of a function $f(x)$ as x approaches a number x_o.

---

22. $3 + \sqrt{x}$, $9 - x$, $x + 9$, $(x + 9)(3 + \sqrt{x})$, $-(18)(6) = -108$

23. $\lim_{x \to c} f(x) = L$ 24. 1, 1 25. $|-3x + 6|$, $\frac{1}{3}$, $\frac{5}{3} < x < \frac{7}{3}$

26. Let f be a function defined on an open interval containing the point x_o, except possibly at x_o itself. Then the limit of f as x approaches x_o is L, written

________________ ,

if, given any positive number ϵ, there is a corresponding positive number δ such that ________________ holds whenever ________________ .

27. As an application of the definition, consider the limit of the function $f(x) = 3 - 2x$ as x approaches 5. The limit is $L = -7$. To show this it is required to establish that: For any positive number ϵ, there is a positive number δ such that

________________ when $0 <$ ________________ $< \delta$.

Now, $|(3 - 2x) - (-7)| = 2 \cdot$ ________ . Thus,

$2|x - 5| < \epsilon$ provided $|x - 5| <$ ________ .

Therefore, if $\delta =$ ________, then

$|(3 - 2x) - (-7)| < \epsilon$ whenever ________________ .

That is, $\lim_{x\to 5} (3 - 2x) = -7$.

OBJECTIVE C: Given a function $f(x)$, a point x_0, and a positive number ϵ, find a number $\delta > 0$ such that for all x $0 < |x - x_0| < \delta$ implies $|f(x) - L| < \epsilon$, where $L = \lim_{x\to x_0} f(x)$.

28. For the limit $\lim_{x\to 4} \sqrt{2x + 1} = 3$, find a $\delta > 0$ that works for $\epsilon = 1$.

Solution: STEP 1: We first find an interval about $x_0 = 4$ on which the inequality $|\sqrt{2x + 1} - 3| < 1$ holds for $x \neq 4$.

$|\sqrt{2x + 1} - 2| < 1$

$\Leftrightarrow$ ________ $< \sqrt{2x + 1} - 3 <$ ________

$\Leftrightarrow$ $2 <$ ________ $<$ ________

$\Leftrightarrow$ $4 < \quad < 16$

$\Leftrightarrow$ $2 < 2x < 15$

$\Leftrightarrow$ ________ $< x <$ ________ .

The inequality holds for all x in the open interval $\left(1, \frac{15}{2}\right)$.

---

26. $\lim_{x\to x_o} f(x) = L$, $|f(x) - L| < \epsilon$, $0 < |x - x_o| < \delta$

27. $|(3 - 2x) - (-7)| < \epsilon$, $|x - 5|$, $|x - 5|$, $\frac{\epsilon}{2}$, $\frac{\epsilon}{2}$, $0 < |x - 5| < \delta$

STEP 2: We now find an interval centered at 4. The distance from 4 to the nearest endpoint of $\left(1, \frac{15}{2}\right)$ is ______. If we take δ = ______ or any smaller positive number, the inequality $0 < |x - 4| < \delta$ will automatically place x between 1 and $\frac{15}{2}$ to make the inequality ______________ hold.

1.4 EXTENSIONS OF THE LIMIT CONCEPT

OBJECTIVE A: For elementary functions $y = f(x)$, find the right-hand and left-hand limits as x approaches a, and from these determine if $\lim_{x \to a} f(x)$ exists.

29. Consider the function defined by

$$f(x) = \begin{cases} 4 - x^2 & \text{if } x \le -1 \\ 1 + x^2 & \text{if } x > -1 \end{cases}$$

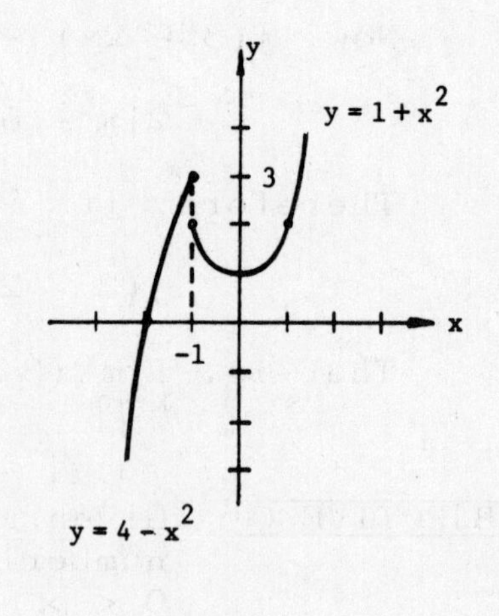

The graph is shown at the right.

$\lim_{x \to -1^+} f(x) = \lim_{x \to -1^+}$ ________ = ________ ,

$\lim_{x \to -1^-} f(x) = \lim_{x \to -1^-}$ ________ = ________ ,

Since $\lim_{x \to -1^+} f(x) \neq \lim_{x \to -1^-} f(x)$,

the limit $\lim_{x \to -1} f(x)$ _____ exist.

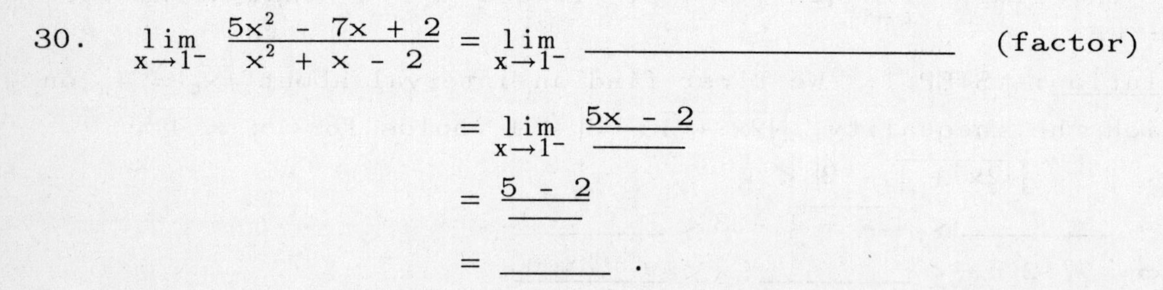

30. $\lim_{x \to 1^-} \frac{5x^2 - 7x + 2}{x^2 + x - 2} = \lim_{x \to 1^-}$ ________________ (factor)

$= \lim_{x \to 1^-} \frac{5x - 2}{\_\_\_\_\_}$

$= \frac{5 - 2}{\_\_\_\_}$

$=$ _____ .

In this problem you <u>must</u> factor the numerator and denominator first and cancel the term $(x-1)$ <u>before</u> calculating the limit because ________________________ is zero in the original expression: $\lim_{x \to 1^-} (x^2 + x - 2) = 1^2 + 1 - 2 = 0$. Note also that the limit as $x \to 1^+$ equals 1.

---

28. $-1 < \sqrt{2x+1} - 3 < 1$, $2 < \sqrt{2x+1} < 4$, $2x + 1$, $1 < x < \frac{15}{2}$, 3, 3, $\left|\sqrt{2x+1} - 3\right| < 1$

29. $1 + x^2$, 2, $4 - x^2$, 3, does not

30. $\frac{(5x-2)(x-1)}{(x+2)(x-1)}$, $x + 2$, $1 + 2$, 1, the limit of the denominator

OBJECTIVE B: Find "infinite limits" such as $\lim_{x\to a} f(x) = \infty$, $\lim_{x\to a^-} f(x) = \infty$, $\lim_{x\to a^+} f(x) = -\infty$, and so forth.

31. $\lim_{x\to 1^+} \frac{|x| + 1}{x^2 - 1} = \lim_{x\to 1^+} \frac{|x| + 1}{(x + 1)(\_\_\_\_\_\_)} = \lim_{x\to 1^+}$ ________ ,
since $|x| + 1 =$ ________ for x near 1. Therefore,

$$\lim_{x\to 1^+} \frac{|x| + 1}{x^2 - 1} = \_\_\_\_\_\_\_\_ .$$

1.5 CONTINUITY

OBJECTIVE A: Specify the test for a function f to be continuous at an interior point $x = c$ of its domain.

32. The three conditions that must be satisfied if the function f is to be continuous at the point $x = c$ are that ________ exists, ________ exists, and ________ .

33. A function is continuous over an interval if it is continuous at ________ within that interval.

34. If a function f is not continuous at the point $x = c$, it is said to be ________ at c.

OBJECTIVE B: Given an elementary function $y = f(x)$, determine its points of continuity and discontinuity. Be able to justify your conclusions.

35. Consider $f(x) = \begin{cases} x + 4 & \text{if } x < -1 \\ -x & \text{if } x \geq -1 \end{cases}$.

Observe that $c = -1$ belongs to the domain of f: $f(-1) = 1$. Does f have a limit as $x \to -1$? To answer that question, we calculate the right and lefthand limits:

$$\lim_{x\to -1^-} f(x) = \lim_{x\to -1^-} (\_\_\_\_\_\_) = \_\_\_\_\_\_ ,$$

$$\lim_{x\to -1^+} f(x) = \lim_{x\to -1^+} (\_\_\_\_\_\_) = \_\_\_\_\_\_ .$$

Since $\lim_{x\to -1^-} f(x) \neq \lim_{x\to -1^+} f(x)$, then $\lim_{x\to -1} f(x)$ ________ .

---

31. $x - 1$, $\frac{1}{x - 1}$, $x + 1$, $+\infty$

32. $f(c)$, $\lim_{x\to c} f(x)$, $\lim_{x\to c} f(x) = f(c)$

33. all points

34. discontinuous

We conclude that f is ________________ at $x = -1$. Sketch a graph of f.

36. Let $f(x) = \frac{x}{x - 1}$. Since $x =$ ________ does not belong to the domain of f we conclude that f is ____________ at 1. Also, $\lim_{x\to 1^-} f(x) =$ ________ and $\lim_{x\to 1^+} f(x) =$ ________ so f does not have a finite limit as $x \to 1$. However, as $x \to +\infty$ or $x \to -\infty$, $f(x) \to$ ________. Sketch a graph of f. Observe that f is continuous at all points except $x = 1$.

OBJECTIVE C: Specify the main algebraic facts related to continuous functions.

37. If f and g are continuous at c, then $f + g$, $f - g$, and $f \cdot g$ are ________________ at c.

38. If f and g are continuous at c, then $\frac{f}{g}$ is ____________ at c provided that ____________.

39. If f is continuous at c, and k is any constant, then kf is ________________ at c.

40. Every constant function is continuous ________________.

41. Every polynomial function is continuous ________________.

42. Every rational function is continuous ________________________________.

43. If f is continuous at c and g is continuous at f(c), then the composite ________ is continuous at ________.

---

35. $x + 4$, 3, $-x$, 1, does not exist, discontinuous

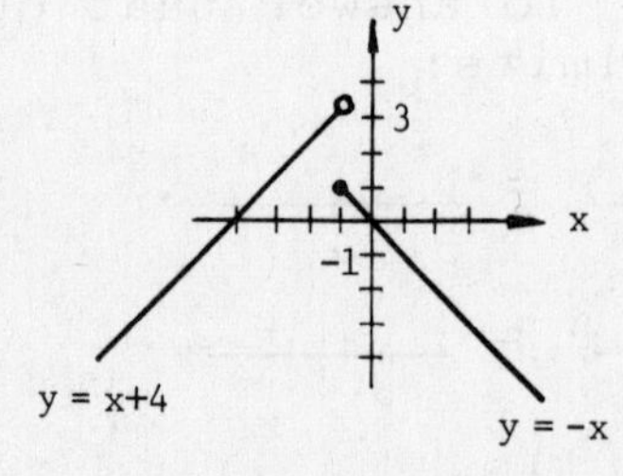

36. 1, discontinuous, $-\infty$, $+\infty$, 1

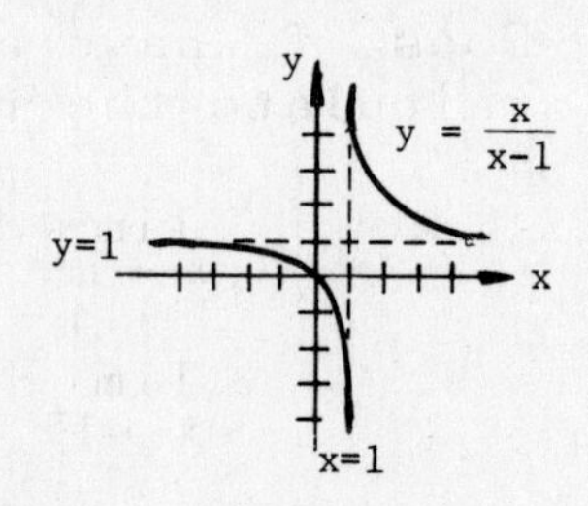

37. continuous

38. continuous, $g(c) \neq 0$

39. continuous

40. at every number

41. at every number

42. at every number at which the denominator is not zero

43. $g(f(x))$, $x = c$

44. Suppose that $f(x)$ is continuous for all x in the closed interval $[a, b]$, and that N is any number between $f(a)$ and $f(b)$. What is your conclusion? ______________________________ .

45. What value must be assigned to $f(x) = \dfrac{(x + 3)(x - 2)}{x - 2}$ at $x = 2$ so the function is continuous there?

1.6 TANGENT LINES

OBJECTIVE A: Find the slope of the function $y = f(x)$ at a given point $P(x_0, f(x_0))$ using the definition

$$m = \lim_{h \to 0} \frac{f(x_0 + h) - f(x_0)}{h}.$$

46. Find the slope of the curve $y = (x + 1)^2$ at $P(0, 1)$.

Solution. Here $f(x) = (x + 1)^2$ and $x_0 =$ ______ .

STEP 1. Calculate $f(x_0)$ and $f(x_0 + h)$:

$f(x_0) = f(0) =$ ____ and $f(x_0 + h) = f(h) =$ ________ .

STEP 2. $\displaystyle\lim_{h \to 0} \frac{f(x_0 + h) - f(x_0)}{h} = \lim_{h \to 0} \frac{\_\_\_\_\_\_\_\_}{h}$

$\displaystyle = \lim_{h \to 0} \frac{h^2 + \_\_\_\_\_\_}{h}$

$\displaystyle = \lim_{h \to 0} (h + \_\_\_\_\_) =$ ______ .

OBJECTIVE B: Find an equation for the tangent to a curve $y = f(x)$ at a given point $P(x_0, f(x_0))$.

47. Find an equation for the tangent to $y = (x + 1)^2$ at $P(0, 1)$.
Solution. From Problem 46, the slope of the tangent line is $m = 2$. An equation of the tangent in point-slope form is

______________________ .

48. Find an equation for the tangent to $y = \dfrac{1}{\sqrt{x - 1}}$ when $x = 2$.

Solution. Here $f(x) = \dfrac{1}{\sqrt{x - 1}}$.

---

44. There is at least one number c between a and b such that $f(c) = N$. 45. $(2 + 3) = 5$

46. 0, 1, $(h + 1)^2$, $(h + 1)^2 - 1$, $2h$, 2, 2 47. $y = 1 + 2(x - 0)$

STEP 1. Calculate $f(2)$ and $f(2 + h)$:

$f(2) =$ ______ and $f(2 + h) =$ ______ .

STEP 2. Calculate the slope m:

$$m = \lim_{h\to 0} \frac{f(2 + h) - f(2)}{h} = \lim_{h\to 0} \frac{\frac{1}{\sqrt{1 + h}} - 1}{h}$$

$$= \lim_{h\to 0} \frac{1 - \underline{\qquad}}{h\sqrt{1 + h}} = \lim_{h\to 0} \frac{(1 - \sqrt{1 + h})(1 + \sqrt{1 + h})}{h\sqrt{1 + h}(1 + \sqrt{1 + h})}$$

$$= \lim_{h\to 0} \frac{\underline{\qquad}}{h\sqrt{1 + h}(1 + \sqrt{1 + h})} = \lim_{h\to 0} \frac{\underline{\qquad}}{\sqrt{1 + h}(1 + \sqrt{1 + h})} = \underline{\qquad} .$$

STEP 3. Find the tangent line using the point-slope equation:

________________ ,

or $y = -\frac{1}{2}x + 2$.

OBJECTIVE C: Find the rate of change of a given function $f(x)$ with respect to x at a specified point $x = x_0$.

49. The rate of change of $f(x)$ with respect to x at $x = x_0$ is the same as the ______ of $y = f(x)$ at $x = x_0$ or the __________ of f at $x = x_0$.

50. Find the rate of change of the area of a square $(A = x^2)$ with respect to its side length when the side length is $x = 5$.

Solution. We first calculate $A(5)$ and $A(5 + h)$:

$A(5) =$ ______ and $A(5 + h) =$ ____________ .

Then the rate of change is

$$\lim_{h\to 0} \frac{A(5 + h) - A(5)}{h} = \lim_{h\to 0} \frac{\underline{\qquad}}{h}$$

$$= \lim_{h\to 0} (10 + h) = \underline{\qquad} .$$

That is, the area changes at the rate 10 sq. units with respect to side length when the side length is 5 units.

---

48. 1, $\frac{1}{\sqrt{1 + h}}$, $\sqrt{1 + h}$, $1 - (1 + h)$, -1, $-\frac{1}{2}$, $y = 1 + \left(-\frac{1}{2}\right)(x - 2)$

49. slope, derivative

50. 25, $(5 + h)^2 = 25 + 10h + h^2$, $10h + h^2$, 10

CHAPTER 1 SELF-TEST

Find the limits in Problems 1-4.

1. $\lim_{t\to 3} \frac{t^2 - 1}{t - 1}$

2. $\lim_{x\to 2} \frac{2x^2 - 3x - 2}{x - 2}$

3. $\lim_{x\to 1^+} \frac{3x - 1}{5x^3 - 2x + 1}$

4. $\lim_{t\to 0^-} \frac{2t^2 + 3t - 1}{t^3 - 2t}$

5. Let f be defined by $f(x) = \begin{cases} 2x - 3, & \text{if } x \geq 0 \\ -1, & \text{if } x < 0 \end{cases}$

   (a) Find $\lim_{x\to 0^+} f(x)$ and $\lim_{x\to 0^-} f(x)$.

   (b) Is f continuous at $x = 0$? Justify your answer.

6. Consider the function $f(x) = \frac{x - 1}{x^2 - x}$.

   (a) For what values of x is f continuous? Justify your conclusion.

   (b) Is f continuous at $x = 1$? If not, what value can be assigned to $f(1)$ so that the resultant function is continuous there?

7. It can be shown that for all values of x

$$\frac{x^2}{2} - \frac{x^4}{24} \leq 1 - \cos x \leq \frac{x^2}{2}.$$

   Use this result to find $\lim_{x\to 0} \frac{1 - \cos x}{x^2}$.

8. In what interval about $x_0 = -1$ must we hold x to be sure that $y = -\frac{x}{3} + \frac{2}{3}$ lies within $\epsilon = 0.5$ units of $y_0 = 1$?

9. Justify that the function $f(x) = x^5 + 1$ has at least one real zero; that is, there is some real number c so that $f(c) = 0$.

10. Given $f(x) = 2x + 7$, $x_0 = -2$, and $\epsilon = 0.01$, find $L = \lim_{x\to x_0} f(x)$. Then find $\delta > 0$ such that $0 < |x - x_0| < \delta \Rightarrow |f(x) - L| < \epsilon$.

11. Find the slope of $f(x) = \frac{2x}{x + 1}$ at $(1,1)$. Then find an equation for the line tangent to the graph there.

12. At t seconds after liftoff, the height of a rocket is $4t^2$ ft. How fast is the rocket climbing after one minute?

SOLUTIONS TO CHAPTER 1 SELF-TEST

1. $\lim_{t\to 3} \frac{t^2 - 1}{t - 1} = \frac{9 - 1}{3 - 1} = 4$

2. $\lim_{x\to 2} \frac{2x^2 - 3x - 2}{x - 2} = \lim_{x\to 2} \frac{(2x + 1)(x - 2)}{x - 2} = \lim_{x\to 2} (2x + 1) = 5$

3. $\lim_{x\to 1^+} \frac{3x - 1}{5x^3 - 2x + 1} = \frac{3 - 1}{5 - 2 + 1} = \frac{1}{2}$

4. $\lim_{t\to 0^-} \frac{2t^2 + 3t - 1}{t^3 - 2t} = \lim_{t\to 0^-} \frac{2t^2 - 3t - 1}{t(t - \sqrt{2})(t + \sqrt{2})} = -\infty$

   because the values are negative for $t < 0$ and t near 0 (approximately $-1/(\text{neg})(\text{neg})(\text{pos})$).

5. (a) From the graph of f shown at the right,

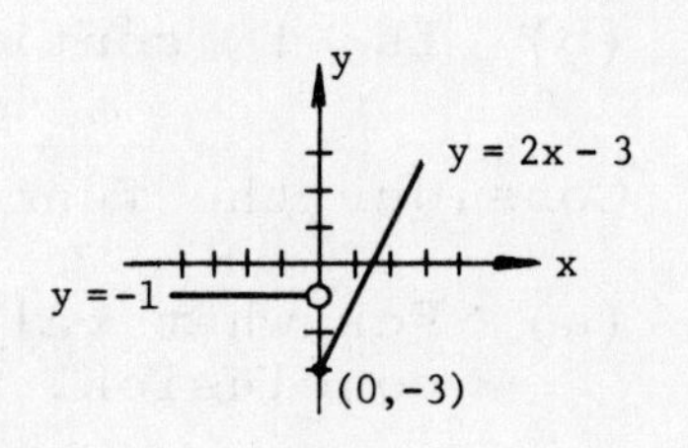

   $\lim_{x\to 0^+} f(x) = \lim_{x\to 0^+} (2x - 3) = -3$

   $\lim_{x\to 0^-} f(x) = \lim_{x\to 0^-} (-1) = -1$

   (b) No, $\lim_{x\to 0} f(x)$ does not exist because the lefthand and righthand limits differ as x tends to zero, so f is not continuous at $x = 0$.

6. (a) Since division by zero is never permitted, the points $x = 0$ and $x = 1$ do not belong to the domain of f, and therefore f is discontinuous at those two values; it is continuous for all other values of x.

   (b) $f(x) = \frac{x - 1}{x^2 - x} = \frac{x - 1}{x(x - 1)} = \frac{1}{x}$ if $x \neq 1$ and $x \neq 0$. Since $\lim_{x\to 1} f(x) = \lim_{x\to 1} \frac{1}{x} = 1$, if we specify $f(1) = 1$ the new function so defined is continuous at $x = 1$.

7. From the assumed inequality, we divide through by the positive number x^2 obtaining,

$$\frac{1}{2} - \frac{x^2}{24} \le \frac{1 - \cos x}{x^2} \le \frac{1}{2}$$

   Applying the Sandwich Theorem,

$$\lim_{x\to 0} \left(\frac{1}{2} - \frac{x^2}{24}\right) = \frac{1}{2} \quad \text{and} \quad \lim_{x\to 0} \frac{1}{2} = \frac{1}{2}$$

   so that

$$\lim_{x\to 0} \frac{1 - \cos x}{x^2} = \frac{1}{2} .$$

8. 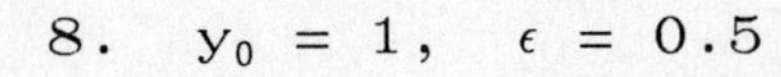
$y_0 = 1, \quad \epsilon = 0.5$

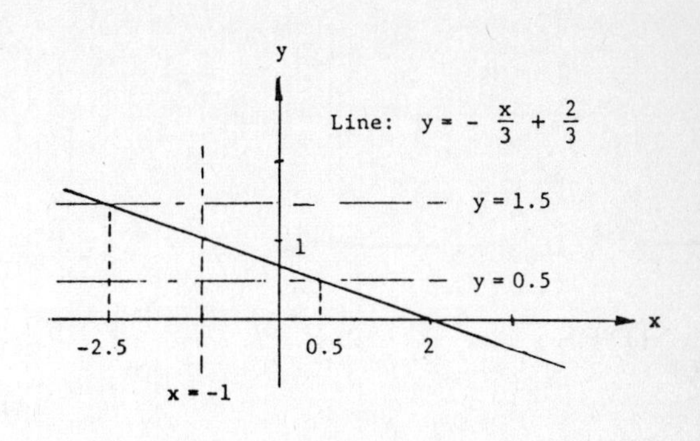

$|y - y_0| = |y - 1| = \left|\left(-\frac{x}{3} + \frac{2}{3}\right) - 1\right| < \epsilon$

$\Leftrightarrow \quad \left|-\frac{x}{3} - \frac{1}{3}\right| = \frac{1}{3}|x + 1| < 0.5$

$\Leftrightarrow \quad |x + 1| < 1.5$

$\Leftrightarrow \quad -1.5 < x + 1 < 1.5$

$\Leftrightarrow \quad -2.5 < x < 0.5$

9. Notice that $f(-2) = (-2)^5 + 1 = -31$ is negative and $f(1) = 1^5 + 1 = 2$ is positive. Since f is a continuous function, the Intermediate Value Theorem guarantees the existence of a real number c satisfying $-2 < c < 1$ and $f(c) = 0$.

10. $\lim_{x \to -2} (2x + 7) = 2 \lim_{x \to -2} x + \lim_{x \to -2} 7 = 2(-2) + 7 = 3.$

$|(2x + 7) - 3| = |2x + 4| = 2|x + 2| = 2|x - (-2)|.$

Thus

$|(2x + 7) - 3| < 0.01 \quad \text{whenever} \quad |x - (-2)| < 0.005.$

That is, $\delta = 0.005$ for $\epsilon = 0.01$ in the formal definition of the limit of $f(x) = 2x + 7$ as x approaches $x_0 = -2$.

11. The slope is

$$m = \lim_{h \to 0} \frac{f(1 + h) - f(1)}{h} = \lim_{h \to 0} \frac{\frac{2(1 + h)}{2 + h} - \frac{2}{2}}{h}$$

$$= \lim_{h \to 0} \frac{(2 + 2h) - (2 + h)}{h(2 + h)} = \lim_{h \to 0} \frac{h}{h(2 + h)}$$

$$= \lim_{h \to 0} \frac{1}{2 + h} = \frac{1}{2}\ .$$

An equation for the tangent line is

$$y = 1 + \frac{1}{2}(x - 1) \quad \text{or} \quad y = \frac{1}{2}(x + 1)\ .$$

12. The rate of change of $H(t) = 4t^2$ at $t = 60$ seconds is

$$\lim_{h \to 0} \frac{H(60 + h) - H(60)}{h} = \lim_{h \to 0} \frac{4(60 + h)^2 - 4(3600)}{h}$$

$$= \lim_{h \to 0} \frac{4(3600 + 120h + h^2) - 4(3600)}{h}$$

$$= \lim_{h \to 0} \frac{4(120h + h^2)}{h} = \lim_{h \to 0} (480 + 4h) = 480\ .$$

The rocket is climbing at the rate 480 ft/sec after 1 minute.

NOTES.

CHAPTER 2 DERIVATIVES

2.1 THE DERIVATIVE OF A FUNCTION

OBJECTIVE A: Use the three algebraic steps to find the derivative dy/dx of a function $y = f(x)$.

1. The derivative of a function f at the point x is given by the limit

$$f'(x) = \_\_\_\_\_\_\_\_\_\_\_\_\_\_\_\_$$

whenever this limit exists.

2. The fraction $(f(x+h) - f(x))/h$ is called the __________ __________ for f at x.

3. When the number $f'(x)$ exists it is called the __________ of the curve $y = f(x)$ at x. The line through the point $(x, f(x))$ with slope $f'(x)$ is the __________ to the curve at x.

4. For the function $f(x) = \sqrt{3 - x}$,

STEP 1. Form $f(x + h) = \_\_\_\_\_\_\_\_\_\_\_\_\_\_\_\_$

and $f(x) = \sqrt{3 - x}$.

STEP 2. Expand and simplify the difference quotient:

$$\frac{f(x + h) - f(x)}{h} = \_\_\_\_\_\_\_\_\_\_\_\_\_\_\_\_$$

$$= \frac{\sqrt{3 - x - h} - \sqrt{3 - x}}{h} \cdot \frac{\sqrt{3 - x - h} + \sqrt{3 - x}}{\sqrt{3 - x - h} + \sqrt{3 - x}}$$

$$= \frac{\_\_\_\_\_\_\_\_\_\_}{h(\sqrt{3 - x - h} + \sqrt{3 - x})}$$

$$= \frac{-1}{\_\_\_\_\_\_\_\_\_\_}$$

STEP 3. Take the limit as $h \to 0$:

$$f'(x) = \lim_{h \to 0} \frac{f(x + h) - f(x)}{h} = \_\_\_\_\_\_\_\_\_\_\_\_\_\_\_\_ .$$

---

1. $\lim_{h \to 0} \frac{f(x + h) - f(x)}{h}$ 2. difference quotient 3. slope, tangent

4. $\sqrt{3 - (x + h)}$, $\frac{\sqrt{3 - (x + h)} - \sqrt{3 - x}}{h}$, $(3 - x - h) - (3 - x)$, $\sqrt{3 - x - h} + \sqrt{3 - x}$, $\frac{-1}{2\sqrt{3 - x}}$

5. For the function $f(x) = \frac{x - 1}{x + 1}$,

STEP 1. Form $f(x + h) =$ ______________

and $f(x) = \frac{x - 1}{x + 1}$.

STEP 2. Expand and simplify the difference quotient:

$$\frac{f(x + h) - f(x)}{h} = \frac{\frac{(x + h) - 1}{(x + h) + 1} - \frac{(x - 1)}{(x + 1)}}{h}$$

$$= \frac{(x + 1)(x + h - 1) - (x - 1)(x + h + 1)}{h(\underline{\qquad\qquad\qquad})}$$

$$= \frac{(x^2 + xh - x + x + h - 1) - (\underline{\qquad\qquad})}{h(x^2 + hx + x + x + h + 1)}$$

$$= \frac{2h}{\underline{\qquad\qquad\qquad}}$$

$$= \frac{2}{x^2 + hx + 2x + h + 1}$$

STEP 3. Take the limit as $h \to 0$:

$$f'(x) = \lim_{h \to 0} \frac{f(x + h) - f(x)}{h} = \underline{\qquad\qquad\qquad} = \frac{2}{(x + 1)^2}.$$

OBJECTIVE B: Write an equation of the tangent line to the curve $y = f(x)$ at a specified value $x = a$.

6. To find an equation of the tangent line to the curve $f(x) = \frac{x - 1}{x + 1}$ when $x = 2$, we first calculate the slope m. By definition the slope is the limit of a secant through $P\left(2, \frac{1}{3}\right)$ and a point Q nearby on the curve. The symbolic notation for this slope is ________. From our calculation in the previous Problem 5, that slope has the value ________ . The point on the curve corresponding to $x = 2$ has coordinates ________ . Therefore, the point-slope form gives an equation of the tangent line as ______________ .

7. If $y = mx + b$ is a straight line, then the derivative dy/dx always has the value ________ . That is, the derivative equals the ________ of the straight line.

---

5. $\frac{(x + h) - 1}{(x + h) + 1}$, $(x + h + 1)(x + 1)$, $(x^2 + xh + x - x - h - 1)$, $h(x^2 + hx + 2x + h + 1)$, $\frac{2}{(x^2 + 2x + 1)}$

6. $f'(2)$, $\frac{2}{9}$, $\left(2, \frac{1}{3}\right)$, $y = \frac{1}{3} + \frac{2}{9}(x - 2)$ or $y = \frac{2}{9}x - \frac{1}{9}$

7. m, slope

OBJECTIVE C: Know the basic elementary facts about the derivative.

8. Differentiable functions are continuous. That is, if f has a ________ at $x = c$, then f is ________ at $x = c$.

9. Can a continuous function fail to have a derivative at a point? ________

10. What are four conditions under which a function whose graph is otherwise smooth fails to have a derivative at a point?
(1) ________,
(2) ________,
(3) ________,
(4) ________.

11. Is every function the derivative of some function?
____
____

2.2 DIFFERENTIATION RULES

OBJECTIVE A: Know from memory the seven derivative rules: derivative of a constant, integer power, constant multiple, sum, difference, product, and quotient.

12. If c is a constant and $y = c$, then $\frac{dy}{dx} =$ ________ .

13. If n is any positive integer and $y = x^n$, then $\frac{dy}{dx} =$ ________ .

14. If u is a differentiable function of x, and if $y = cu$ where c is a constant, then $\frac{dy}{dx} =$ ________ .

15. If u and v are differentiable functions of x, then $y = u + v$ is a ________ function of x, and

---

8. derivative, continuous

9. yes, the absolute value function fails to have a derivative at $x = 0$

10. the graph has a corner, a cusp, a vertical tangent, a discontinuity

11. no, because a function cannot be a derivative on an interval unless it has the intermediate value property there.

12. 0

13. nx^{n-1}

14. $c\frac{du}{dx}$

$\frac{dy}{dx}$ = ______________ .

Likewise, $\frac{d}{dx}(u - v)$ = ______________ .

16. If u and v are differentiable functions of x, then the derivative $\frac{d}{dx}(uv)$ = ______________ .

17. If u and v are differentiable functions of x, then the derivative $\frac{d}{dx}\left(\frac{u}{v}\right)$ = ______________ when $v \neq 0$.

18. If n is any negative integer and $y = x^n$, then

$\frac{dy}{dx}$ = ______________ .

OBJECTIVE B: Calculate the derivative of any polynomial function.

19. $\frac{d}{dx}\left(3x^2 - 12x + 1\right)$ = ______________ .

20. $\frac{d}{dx}\left(\sqrt{3}x^4 - \frac{2}{5}x^3 + \frac{1}{3}x^2 - 15x + 109\right)$ = ______________ .

OBJECTIVE C: Calculate second and higher-order derivatives.

21. If $y = f(x)$, the second derivative of y with respect to x is the derivative of ______________ . The second derivative is denoted by ______ or ______ or ______ .

22. In general, the nth derivative of $y = f(x)$ with respect to x is the derivative of ________ , and is denoted by ________ or ________ or ________ .

23. $\frac{d^2}{dx^2}(4x^5 - 3x^2 + 2x - 20) = \frac{d}{dx}\left(\_\_\_\_\_\_\_\_\_\_\right)$ = ______________ .

24. $\frac{d^2}{dx^2}(2x^2 - 1)(x - 3) = \frac{d^2}{dx^2}\left(\_\_\_\_\_\_\_\_\_\_\right) = \frac{d}{dx}\left(\_\_\_\_\_\_\_\_\_\_\right)$

= ______________ .

---

15. differentiable, $\frac{du}{dx} + \frac{dv}{dx}$, $\frac{du}{dx} - \frac{dv}{dx}$

16. $u\frac{dv}{dx} + v\frac{du}{dx}$

17. $\dfrac{v\frac{du}{dx} - u\frac{dv}{dx}}{v^2}$

18. nx^{n-1}

19. $6x - 12$

20. $4\sqrt{3}\,x^3 - \frac{6}{5}x^2 + \frac{2}{3}x - 15$

21. $\frac{dy}{dx} = f'(x)$, $\frac{d^2y}{dx^2}$, y'', or $f''(x)$

22. $\frac{d^{n-1}y}{dx^{n-1}}$, $\frac{d^ny}{dx^n}$, $y^{(n)}$, or $f^{(n)}(x)$

23. $20x^4 - 6x + 2$, $80x^3 - 6$

24. $2x^3 - 6x^2 - x + 3$, $6x^2 - 12x - 1$, $12x - 12$

OBJECTIVE D: Find the derivative of a product of polynomial or power functions.

25. If $y = (x^2 - 2)(2x^3 - 5)$, then

$y' = (x^2 - 2)\frac{d}{dx}(2x^3 - 5) + (2x^3 - 5)$ __________

$= (x^2 - 2)($________$) + (2x^3 - 5)(2x)$

$=$ __________ $+ 4x^4 - 10x =$ __________ .

26. $\frac{d}{dx}[(3x^2 - 2x + 1)(5x - 4)]$

$= (3x^2 - 2x + 1)($________$) + (5x - 4)($________$)$

$= (15x^2 - 10x + 5) + ($__________$) =$ __________ .

OBJECTIVE E: Find the derivative of a quotient of polynomial or power functions.

27. If $y = \frac{3x}{5x^2 - 1}$, then

$y' = \frac{(5x^2 - 1)(\_\_\_\_\_) - (3x)(\_\_\_\_\_)}{(5x^2 - 1)^2} = \frac{\_\_\_\_\_}{(5x^2 - 1)^2}$.

28. $\frac{d}{dx}\left(\frac{1}{x}\right) = \frac{x(\_\_\_\_\_) - 1(\_\_\_\_\_)}{x^2} =$ __________ .

29. $\frac{d}{dx}\left(\frac{x^2 - 2x + 5}{x^3 + 1}\right) = \frac{(x^3 + 1)(\_\_\_\_\_) - (\_\_\_\_\_)(3x^2)}{(x^3 + 1)^2}$

$= \frac{2(\_\_\_\_\_) - 3x^4 + 6x^3 - 15x^2}{(x^3 + 1)^2}$

$= \frac{(\_\_\_\_\_\_\_\_\_\_)}{(x^3 + 1)^2}$

30. If $y = x^n$, where $x \neq 0$ and n is any nonzero integer, then

$\frac{dy}{dx} =$ __________ .

31. If $y = 2x^3 + 3x^{-2}$, then

$y' =$ __________ and $y'' =$ __________ .

---

25. $\frac{d}{dx}(x^2 - 2)$, $6x^2$, $6x^4 - 12x^2$, $10x^4 - 12x^2 - 10x$

26. 5, $6x - 2$, $30x^2 - 34x + 8$, $45x^2 - 44x + 13$

27. 3, $10x$, $-15x^2 - 3$

28. 0, 1, $-\frac{1}{x^2}$

29. $2x - 2$, $x^2 - 2x + 5$, $x^4 - x^3 + x - 1$, $-x^4 + 4x^3 - 15x^2 + 2x - 2$

30. nx^{n-1}

31. $6x^2 - 6x^{-3}$, $12x + 18x^{-4}$

32. If $y = 3x^2\left(x - \frac{1}{x}\right)$, then by the product rule

$$y' = 6x\left(x - \frac{1}{x}\right) + 3x^2 \cdot \frac{d}{dx}(\_\_\_\_\_\_\_\_)$$
$$= 6x^2 - 6 + 3x^2(\_\_\_\_\_\_\_\_)$$
$$= 6x^2 - 6 + 3x^2 + \_\_\_\_\_\_$$
$$= 9x^2 - 3.$$

OBJECTIVE F: Find an equation of the tangent line to a curve $y = f(x)$ meeting some specified requirement (such as a condition on the slope).

33. Consider the curve $y = x^3 - 9x^2 + 15x - 5$. The derivative y' gives the value of the ________ of the tangent line at any x. For this particular curve, $y' =$ ____________ $= 3($____$)(x - 5)$. Thus, the tangent line is parallel to the x-axis when $x =$ ________ or $x =$ ________ . When the slope of the tangent line to the above curve equals 15 the value of x is ________ or ________ . The corresponding y values are ________ and ________, respectively. Equations of the two tangent lines are then given by ______________ and ______________ .

2.3 RATES OF CHANGE

34. If $s = f(t)$ gives the position of a body moving along a line from position $s = f(t)$ to position $s = f(t + \Delta t)$, then the <u>average velocity</u> over the time interval Δt is ______________________ . The <u>instantaneous velocity</u> at time t is ______________ .

35. The <u>speed</u> is the ________________________ .

36. <u>Acceleration</u> is the derivative of ____________________ .

OBJECTIVE A: If $s = f(t)$ gives the position of a body moving along a line as a function of time t, find and interpret the velocity and acceleration at a specified instant.

37. Suppose a particle is moving along a straight line, negative to the left and positive to the right, according to the law

$$s = t^3 - 3t^2 - 9t + 5 .$$

---

32. $x - \frac{1}{x}$, $1 + \frac{1}{x^2}$, 3

33. slope, $3x^2 - 18x + 15$, $x - 1$, 1, 5, 0, 6, -5, -23, $y + 5 = 15x$ and $y + 23 = 15(x - 6)$

34. $\frac{\Delta s}{\Delta t} = \frac{f(t+\Delta t) - f(t)}{\Delta t}$, $\frac{ds}{dt}$ 35. absolute value of velocity   36. velocity with respect to time

Then the velocity is given by $\frac{ds}{dt}$ = ____________ = ____________ . Thus, the velocity is positive when ______ or __________ and the particle is moving to the _________ ; the velocity is __________ when $-1 < t < 3$ so the particle is moving to the __________ .

38. The acceleration of the particle is $\frac{d^2s}{dt^2}$ = ___________ . When the velocity is zero, t = _________ or __________ and the acceleration has the value ___________ or ___________, respectively.

39. Suppose the law of motion of a particle is given by

$$s = t^3 - 6t^2 + 2 .$$

Then the instantaneous velocity is given by

$$v = \frac{ds}{dt} = \_\_\_\_\_\_\_\_\_\_\_\_\_\_\_\_ .$$

When $t = 2.3$ sec, the velocity of the particle is $v(2.3)$ = ____________ . If our coordinate axis of motion is such that the positive direction is to the right (which is conventional), the interpretation of this negative velocity means that the particle is moving to the ______ . When $t = 4$ sec, the velocity of the particle is _______ and the particle is at rest. When $t = 4.5$ sec, the velocity of the particle is ______ and the particle is moving to the ______ .

40. Suppose a ball is thrown directly upward with a speed of 96 ft/sec and moves according to the law $y = 96t - 16t^2$, where y is the height in feet above the starting point, and t is the time in seconds after it is thrown. The velocity of the ball at any time t is $v(t) = \frac{dy}{dt}$ = ____________ . Hence when $t = 2$ sec, the velocity of the ball is ____________ . Since $v(2)$ is positive, the ball is still rising. At its highest point the velocity of the ball is ________, and this occurs when t = ________ seconds. The height corresponding to this time is y = ________ feet, and this is the highest point reached. Notice the acceleration is a constant _____ ft/sec^2.

OBJECTIVE B: Given a functional relationship $y = f(x)$ between two variables x and y, calculate the average rate of change and the instantaneous rate of change of y with respect to x.

41. Every derivative may be interpreted as the instantaneous rate of change of one variable per unit change in the other. If $y = f(x)$, then

---

37. $3t^2 - 6t - 9 = 3(t - 3)(t + 1)$, $t < -1$ or $t > 3$, right, negative, left   38. $6t - 6$, -1 or 3, -12 or 12

39. $3t^2 - 12t$, -11.73 units/sec., left, 0, $\frac{27}{4}$ units/sec., right

40. $96 - 32t$, 32 ft/sec, zero, 3, $y = 144$, -32

$$\frac{\Delta y}{\Delta x} = \underline{\qquad\qquad}$$

is interpreted as the ________ rate of change of y by a change of one unit in ______ . Passage to the limit as $\Delta x \to 0$ gives

$$\lim_{\Delta x \to 0} \frac{\Delta y}{\Delta x} = \underline{\qquad\qquad}$$

as the ________ rate of change of ______ with respect to ______ .

42. Consider the equilateral triangle pictured to the right. By the Pythagorean theorem $s^2 =$ ________ or, solving for h, h = ________ . Then, the area of the triangle is given by

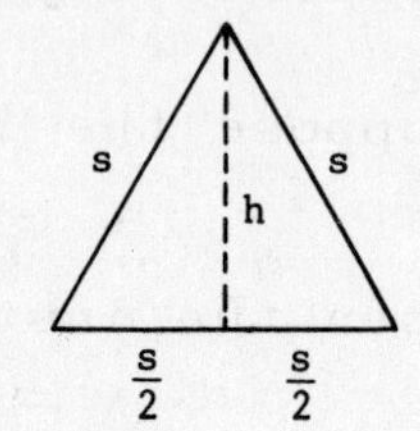

$$A = \frac{1}{2}\text{ base} \cdot \text{height} = \underline{\qquad\qquad} .$$

The average rate of change of area with respect to side length is

$$\frac{\Delta A}{\Delta s} = \frac{\underline{\qquad\qquad}}{\Delta s} = \underline{\qquad\qquad} .$$

Taking the limit as Δs tends to zero gives,

$$\frac{dA}{ds} = \lim_{\Delta s \to 0} \frac{\Delta A}{\Delta s} = \underline{\qquad\qquad} ,$$

the ________ rate of change of area with respect to ______ length for an equilateral triangle.

43. Suppose it costs C(x) thousand dollars per year to produce x thousand gallons of antifreeze, where C(x) is given by the table

x	0.25	0.5	0.75	1.0	1.25	1.50	1.75	2.0	2.25	2.5
C(x)	5.875	8.5	10.875	13.0	14.875	16.5	17.875	19.0	19.875	20.0

The ________ cost at any x is the value of the derivative $C'(x)$. Using the table, we estimate $C'(1.75)$ as follows:

$$C'(1.75) \approx \frac{\Delta C}{\Delta x} = \frac{\underline{\qquad\qquad}}{2.0 - 1.75} = \underline{\qquad\qquad} .$$

Here we have estimated the marginal cost by the ______ cost.

---

41. $\frac{f(x + \Delta x) - f(x)}{\Delta x}$, average, x, $f'(x)$, instantaneous, y, x

42. $s^2 = h^2 + \frac{s^2}{4}$, $h = \frac{\sqrt{3}}{2}s$, $\frac{\sqrt{3}}{4}s^2$, $\frac{\sqrt{3}}{4}(s + \Delta s)^2 - \frac{\sqrt{3}}{4}s^2$, $\frac{\sqrt{3}}{4}(2s + \Delta s)$, $\frac{\sqrt{3}}{2}s$, instantaneous, side

43. marginal, 19.0 - 17.875, 4.5, average

2.4 DERIVATIVES OF TRIGONOMETRIC FUNCTIONS

44. One of the most useful facts in calculus is that

$$\lim_{\theta \to 0} \frac{\sin \theta}{\theta} = \underline{\qquad\qquad} .$$

For this limit the angle θ must be measured in ____________ .

OBJECTIVE A: Evaluate limits of trigonometric functions by making use of appropriate trigonometric identities and the limit theorems.

45. To find $\lim_{x \to 0} \frac{\sin 5x}{3x}$ first substitute $\theta = 5x$, and note that $\theta \to 0$ as $x \to 0$. Then the limit bcomes,

$$\lim_{x \to 0} \frac{\sin 5x}{3x} = \lim_{\theta \to 0} \underline{\qquad\qquad}$$

$$= \underline{\qquad\qquad} \lim_{\theta \to 0} \frac{\sin \theta}{\theta}$$

$$= \frac{5}{3} (\underline{\qquad})$$

$$= \underline{\qquad\qquad} .$$

46. $$\lim_{x \to 0} \frac{1 - \cos x}{1 - \sin x} = \frac{\lim_{x \to 0} (1 - \cos x)}{\underline{\qquad\qquad\qquad}} = \frac{(1 - \underline{\qquad})}{(1 - 0)} = \underline{\qquad\qquad} .$$

47. $$\lim_{h \to 0} \frac{h}{\sin h} = \lim_{h \to 0} \frac{1}{\underline{\qquad\qquad}} = \frac{1}{\lim_{h \to 0} \frac{\sin h}{\underline{\qquad}}} = \frac{1}{\underline{\qquad}} = \underline{\qquad\qquad} .$$

OBJECTIVE B: Calculate the derivatives of functions involving the trigonometric functions, making use of appropriate rules of differentiation and the derivatives of the sine and cosine functions.

48. $\frac{d}{dx}(\sin x) = \underline{\qquad\qquad}$.

49. $\frac{d}{dx}(\cos x) = \underline{\qquad\qquad}$.

---

44. 1, radians

45. $\frac{\sin \theta}{3(\theta/5)}$, $\frac{5}{3}$, 1, $\frac{5}{3}$

46. $\lim_{x \to 0} (1 - \sin x)$, 1, 0

47. $\frac{\sin h}{h}$, h, 1, 1

48. cos x

49. - sin x

50. $\frac{d}{dx}(\tan x) = \frac{d}{dx}\left(\frac{\sin x}{\_\_\_\_}\right) = \frac{\cos x \cdot \frac{d}{dx}(\_\_\_\_) - \sin x \frac{d}{dx}(\cos x)}{\cos^2 x}$

$= \frac{\cos x\,(\_\_\_\_) - \sin x\,(\_\_\_\_)}{\cos^2 x}$

$= \frac{\cos^2 x + \_\_\_\_}{\cos^2 x} = \_\_\_\_\,.$

51. $\frac{d}{dx}(\sec x) = \frac{d}{dx}\left(\frac{1}{\_\_\_\_}\right) = -\frac{1}{\cos^2 x} \cdot \frac{d}{dx}(\_\_\_\_)$

$= \frac{\_\_\_\_}{\cos^2 x} = \_\_\_\_\,.$

Remark: It will be to your advantage in later work to MEMORIZE the derivative formulas in Problems 50 and 51.

52. $\frac{d}{dx}(3x^2 \sin x) = \frac{d}{dx}(\_\_\_\_) \cdot \sin x + 3x^2 \cdot \frac{d}{dx}(\_\_\_\_)$

$= \_\_\_\_ + 3x^2 \cos x$

$= \_\_\_\_\,.$

53. $\frac{d}{dx}(3 \sin x + x \sec x) = 3\frac{d}{dx}(\sin x) + x \cdot \frac{d}{dx}(\sec x) + \_\_\_\_$

$= \_\_\_\_ + x\,\_\_\_\_ + \sec x$

$= 3 \cos x + \sec x(x \tan x + 1)\,.$

54. $\frac{d}{dx}\left(\frac{1 - \sin x}{x}\right) = \frac{x \cdot \frac{d}{dx}(1 - \sin x) - (1 - \sin x) \cdot \frac{d}{dx}(\_\_\_\_)}{x^2}$

$= \frac{x(\_\_\_\_) - (1 - \sin x)}{x^2}$

$= \frac{-x \cos x - 1 + \_\_\_\_}{x^2}$

$= \frac{\sin x - (1 + x \cos x)}{x^2}\,.$

2.5 THE CHAIN RULE

OBJECTIVE: If y is a differentiable function of u, and u is a differentiable function of x, use the chain rule to calculate $\frac{dy}{dx}$.

---

50. $\cos x$, $\sin x$, $\cos x$, $-\sin x$, $\sin^2 x$, $\sec^2 x$

51. $\cos x$, $\cos x$, $\sin x$, $\sec x \tan x$

52. $3x^2$, $\sin x$, $6x \sin x$, $3x(2 \sin x + x \cos x)$

53. $\sec x$, $3 \cos x$, $\sec x \tan x$ (from Problem 51)

54. x, $-\cos x$, $\sin x$

55. If y is a differentiable function of u, and u is a differentiable function of x, then y is a differentiable function of ______, and $\frac{dy}{dx}$ = ________ . This rule is known as the ______ rule for the derivative of a composite function.

56. Consider the chain rule in functional form: let $y = f(u)$ and $u = g(x)$ be differentiable functions. Then the composite $y = (f \circ g)(x) = f(g(x))$ is a differentiable function of ______ . When $x = x_0$, let $u = g(x_0) = u_0$. According to the chain rule, the derivative of $(f \circ g)$ evaluated at $x = x_0$ is given by $(f \circ g)'(x_0)$ = ____________ . In this equation, $(f \circ g)'(x_0)$ corresponds to $\left.\frac{dy}{dx}\right|_{x=x_0}$, $f'(u_0)$ corresponds to ________, and ________ corresponds to $\left.\frac{du}{dx}\right|_{x=x_0}$. It is important that you observe that the derivatives in the chain rule equation $\frac{dy}{dx} = \frac{dy}{du} \cdot \frac{du}{dx}$ are being evaluated at different points: $\frac{dy}{dx}$ and $\frac{du}{dx}$ are evaluated at ________, whereas $\frac{dy}{du}$ is evaluated at $g(x_0)$ = _______ . Failure to understand this fact can lead to serious misuse of the chain rule equation.

57. To find $\frac{dy}{dx}$ if $y = u^2 - 2u + 3$ and $u = \sqrt{x}$, calculate $\frac{dy}{du}$ = ________ and then substitute $u = \sqrt{x}$ to obtain $\left.\frac{dy}{du}\right|_{u=\sqrt{x}}$ = _________ . According to the chain rule,

$$\frac{dy}{dx} = \frac{dy}{du} \cdot \frac{du}{dx} = (2\sqrt{x} - 2) \cdot \underline{\qquad} = \underline{\qquad\qquad} .$$

58. Suppose $y = z^{-2} + 3z^{-1}$ and $z = x^2 + 1$. Then $\frac{dy}{dz}$ = ________ so that when $z = x^2 + 1$, $\left.\frac{dy}{dz}\right|_{z=x^2+1}$ = ____________________ .

Applying the chain rule,

$$\frac{dy}{dx} = \left[-2\left(x^2 + 1\right)^{-3} - 3\left(x^2 + 1\right)^{-2}\right] \cdot \underline{\qquad}$$

$$= \frac{-2x}{\left(x^2 + 1\right)^2}\left[\frac{2}{x^2 + 1} + \underline{\qquad}\right] = \frac{\underline{\qquad}}{\left(x^2 + 1\right)^3} .$$

---

55. x, $\frac{dy}{dx} \cdot \frac{du}{dx}$, chain

56. x, $f'(u_0) \cdot g'(x_0)$, $\left.\frac{dy}{du}\right|_{u=u_0}$, $g'(x_0)$, x_0, u_0

57. $2u - 2$, $2\sqrt{x} - 2$, $\frac{1}{2\sqrt{x}}$, $1 - \frac{1}{\sqrt{x}}$

58. $-2z^{-3} - 3z^{-2}$, $-2(x^2 + 1)^{-3} - 3(x^2 + 1)^{-2}$, $2x$, 3, $-2x(3x^2 + 5)$

59. If u is a differentiable function of x and n is a positive integer, then the derivative

$$\frac{d}{dx}(u^n) = \underline{\qquad\qquad} .$$

60. The above power rule holds when n is a negative integer at all points x where u is $\underline{\qquad\qquad}$.

61. If $y = (5x^3 - x^2 + 7)^4$, then

$$y' = 4(5x^3 - x^2 + 7)^3 \frac{d}{dx}(\underline{\qquad\qquad}) = \underline{\qquad\qquad\qquad} .$$

62. $$\frac{d}{dx}\left[(2x - 1)\right]^{-3} = -3(\underline{\qquad})^{-4} \frac{d}{dx}(\underline{\qquad\qquad})$$

$$= \underline{\qquad\qquad}, \text{ if } x \neq \underline{\qquad\qquad} .$$

63. Let $y = \dfrac{2}{(x - 1)^3} + \dfrac{3}{1 - x^4}$. Then $y = 2(x - 1)^{-3} + 3(1 - x^4)^{-1}$, so that $\dfrac{dy}{dx} = -6(\underline{\qquad\qquad}) \dfrac{d}{dx}(x - 1) - 3(1 - x^4)^{-2} \dfrac{d}{dx}(\underline{\qquad\qquad})$

$= \underline{\qquad\qquad\qquad}$.

64. $$\frac{d}{dx}\, x \tan^2 x = x \frac{d}{dx}(\underline{\qquad}) + \frac{dx}{dx} \tan^2 x = x \cdot 2 \tan \frac{d}{dx}(\underline{\qquad}) + \tan^2 x$$

$$= 2x \underline{\qquad\qquad\qquad} + \tan^2 x$$

$$= \underline{\qquad\qquad\qquad\qquad} .$$

65. $$\frac{d}{dx}\sqrt{\frac{1 - \cos x}{1 + \cos x}} = \frac{1}{2}\, \underline{\qquad\qquad}\, \frac{d}{dx}\left(\frac{1 - \cos x}{1 + \cos x}\right)$$

$$= \frac{1}{2}\left(\frac{1 - \cos x}{1 + \cos x}\right)^{-1/2} \left[\frac{(1 + \cos x)(\underline{\qquad}) - (1 - \cos x)(\underline{\qquad})}{(1 + \cos x)^2}\right]$$

$$= \frac{(1 - \cos x)^{-1/2} \sin x}{\underline{\qquad\qquad\qquad}} .$$

66. $$\frac{d}{dx}\left(\frac{x - 2}{x + 1}\right)^5 = 5(\underline{\qquad\qquad})^4 \frac{d}{dx}(\underline{\qquad\qquad})$$

$$= 5\left(\frac{x - 2}{x + 1}\right)^4 \frac{(x + 1)(1) - (\underline{\qquad\qquad})(1)}{(x + 1)^2} = \frac{\underline{\qquad\qquad}}{(x + 1)^6} .$$

---

59. $nu^{n-1}\dfrac{du}{dx}$

60. not zero

61. $5x^3 - x^2 + 7$, $4(5x^3 - x^2 + 7)^3(15x^2 - 2x)$

62. $2x - 1$, $2x - 1$, $-6(2x - 1)^{-4}$, $\frac{1}{2}$

63. $(x - 1)^{-4}$, $1 - x^4$, $-6(x - 1)^{-4} + 12x^3(1 - x^4)^{-2}$

64. $\tan^2 x$, $\tan x$, $\tan x \sec^2 x$, $\tan x(2x \sec^2 x + \tan x)$

65. $\left(\dfrac{1 - \cos x}{1 + \cos x}\right)^{-1/2}$, $\sin x$, $-\sin x$, $(1 + \cos x)^{3/2}$

66. $\dfrac{x - 2}{x + 1}$, $\dfrac{x - 2}{x + 1}$, $x - 2$, $15(x - 2)^4$

2.6 IMPLICIT DIFFERENTIATION AND RATIONAL EXPONENTS

OBJECTIVE A: Compute first and second derivatives by the technique of implicit differentiation.

67. An equation involving the variables x and y is said to determine y ________ as a function of x, say $y = f(x)$, provided that f satisfies the equation.

68. For instance, consider the equation $x^2 + y^2 = 2$. If we substitute $y = \sqrt{2 - x^2}$ into the equation we obtain

$$x^2 + \left(\sqrt{2 - x^2}\right)^2 = x^2 + (\_\_\_\_\_) = \_\_\_\_\_,$$

so the equation is satisfied. Similarly, if we substitute $y = -\sqrt{2 - x^2}$ into the equation we obtain

$$x^2 + \left(-\sqrt{2 - x^2}\right)^2 = x^2 + (\_\_\_\_\_) = \_\_\_\_,$$

and the equation is again satisfied. Therefore, each of the two functions $y = \sqrt{2 - x^2}$ and $y = -\sqrt{2 - x^2}$ is defined ________ by the equation $x^2 + y^2 = 2$.

69. To calculate the derivative $\frac{dy}{dx}$ for $2xy - y^2 = 3$, differentiate both sides of the equation with respect to x and solve for $\frac{dy}{dx}$:

$$\frac{d}{dx}(2xy - y^2) = \frac{d}{dx}(3) \quad \text{or} \quad \frac{d}{dx}(\_\_\_\_) - \frac{d}{dx}(y^2) = \frac{d}{dx}(3).$$

Thus, $2\left(y + x\frac{dy}{dx}\right) - \_\_\_\_\_ = 0$ and solving for $\frac{dy}{dx}$,

$$\frac{dy}{dx} = \_\_\_\_\_\_\_\_\_\_.$$

This derivative is valid whenever ____________ is not equal to zero.

70. To calculate $\frac{d^2y}{dx^2}$ for $2xy - y^2 = 3$, differentiate both sides of the derivative equation $\frac{dy}{dx} = \frac{y}{x - y}$ with respect to x:

$$\frac{d}{dx}\left(\frac{dy}{dx}\right) = \frac{d}{dx}\left(\frac{y}{x - y}\right) \quad \text{(from Problem 69), or}$$

$$\frac{d^2y}{dx^2} = \frac{(x - y)\frac{dy}{dx} - y\frac{d}{dx}(\_\_\_\_\_)}{(x - y)^2} \quad \text{(quotient rule)}$$

---

67. implicitly

68. $2 - x^2$, 2, $2 - x^2$, 2, implicitly

69. $2xy$, $2y\frac{dy}{dx}$, $\frac{y}{x - y}$, $x - y$

$$= \frac{(x - y)\frac{dy}{dx} - y(\underline{\qquad\qquad})}{(x - y)^2}$$

$$= \frac{x\frac{dy}{dx} - y\frac{dy}{dx} - y + y\frac{dy}{dx}}{(x - y)^2} .$$

Substitution of $\frac{y}{x - y}$ for $\frac{dy}{dx}$ in the last equation gives

$$\frac{d^2y}{dx^2} = \frac{x(\underline{\qquad\qquad}) - y}{(x - y)^2}$$

$$= \frac{xy - y(\underline{\qquad\qquad})}{(x - y)^3}$$

$$= \frac{\underline{\qquad}}{(x - y)^3} .$$

OBJECTIVE B: Find the derivative of $g(x) = x^n$ when n is any rational number $n = p/q$.

71. If $g(x) = x^n$ with $n = 1/m$, where m is a positive odd integer, then $g'(x) =$ ______________ for x satisfying ______________ .

72. If $g(x) = x^n$ with $n = 1/m$, where m is a positive even integer, then $g'(x) =$ ______________ for x satisfying ______________ .

73. $\frac{d}{dx}(x^{1/9}) =$ ____________ provided x satisfies ____________ .

74. $\frac{d}{dx}(x^{1/6}) =$ ____________ provided x satisfies ____________ .

75. $\frac{d}{dx}(x^{3/5}) =$ ____________ provided x satisfies ____________ .

76. $\frac{d}{dx}(x^{-3/4}) =$ ____________ provided x satisfies ____________ .

77. Let $y = \sqrt{\frac{x + 3}{x - 3}}$, so $y = u^{1/2}$ where $u = \frac{x + 3}{x - 3}$. Then

$$\frac{dy}{dx} = \frac{1}{2} u^{-1/2} \frac{du}{dx}$$ whenever $u > 0$. Thus

---

70. $x - y,\ 1 - \frac{dy}{dx},\ \frac{y}{x - y},\ x - y,\ y^2$

71. $nx^{n-1},\ x \neq 0$

72. $nx^{n-1},\ x > 0$

73. $\frac{1}{9}x^{-8/9},\ x \neq 0$

74. $\frac{1}{6}x^{-5/6},\ x > 0$

75. $\frac{3}{5}x^{-2/5},\ x \neq 0$

76. $-\frac{3}{4}x^{-7/4},\ x > 0$

$$\frac{dy}{dx} = \frac{1}{2}\left(\underline{\qquad\qquad}\right)^{-1/2} \frac{(x - 3)(1) - (\underline{\qquad\qquad})}{(x - 3)^2}$$

$$= \frac{1}{2\sqrt{\dfrac{x + 3}{x - 3}}} \underline{\qquad\qquad\qquad} \quad \text{whenever} \quad \frac{x + 3}{x - 3} > 0.$$

OBJECTIVE C: Find the lines that are tangent and normal to a specified curve at a given point.

78. By direct substitution the point $(1, 1)$ lies on the curve $xy = 1$. Differentiating both sides with respect to x yields $y + \underline{\qquad\qquad} = 0$. Solving for dy/dx gives

$$\frac{dy}{dx} = \underline{\qquad\qquad} .$$

Thus, at $(1, 1)$ the slope of the curve is $\left.\frac{dy}{dx}\right|_{(1,1)} = \underline{\qquad\qquad}$.

Therefore, the tangent to the curve at the point $(1, 1)$ is

$$y - 1 = \underline{\qquad\qquad} .$$

79. For the curve in Problem 78, the slope of the normal is $\underline{\qquad\qquad}$ at the point $(1, 1)$. The normal is therefore given by the equation

$$y = 1 + \underline{\qquad\qquad} .$$

2.7 RELATED RATES OF CHANGE

In this section we consider problems that ask us to find the rate at which some variable quantity changes when we know the rate at which another quantity related to it changes. Examples abound for problems of this sort. For instance, the rate of production of a certain commodity may depend upon its rate of sales; the rate of increase or decrease in the water level of a dam or reservoir is essential information to a public utility serving the demands of a growing population; the rate at which oil may be spreading on the sea surface from a stricken tanker depends on the rate at which it may be leaking; and so forth.

---

77. $\frac{x + 3}{x - 3}$, $(x + 3)(1)$, $\frac{-6}{(x - 3)^2}$ 78. $x\frac{dy}{dx}$, $-\frac{y}{x}$, -1, $(-1)(x - 1)$ or $1 - x$

79. 1, $(x - 1)$

OBJECTIVE: Solve related rates problems by using the following strategy, as presented in the text:

1. Draw a picture and name the variables and constants. Use t for time. Assume all variables are differentiable functions of t.
2. Write down the numerical information (in terms of the symbols you have chosen).
3. Write down what you are asked to find (usually a rate, expressed as a derivative).
4. Write an equation that relates the variables. You may have to combine two or more equations to get a single equation that relates the variable whose rate you want to the variable whose rate you know.
5. Differentiate with respect to t to express the rate you want in terms of the rate and variables whose values you know.
6. Evaluate.

80. A plane flying at 1 mile altitude is 2 miles distant from an observer, measured along the ground, and flying directly away from the observer at 400 mph. How fast is the angle of elevation changing?

Solution. We carry out the steps of the basic strategy.

STEP 1: Picture and variables: We picture the plane flying in the coordinate plane using the positive x-axis as the ground pointing in the direction of flight. Let 0 denote the position of the observer at a distance x units (measured along the ground) from the plane P as shown in the figure at the right. Let α denote the angle of elevation.

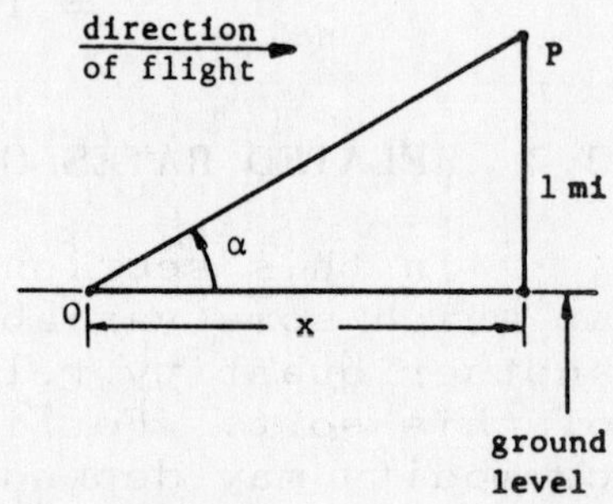

STEP 2: Numerical information: At the time in question,

$$x = \_\_\_\_\_\_ \text{ mi}, \quad y = 1 \text{ mi}, \quad \frac{dx}{dt} = 400 \text{ mi/hr}.$$

STEP 3: To find: the rate __________ .

STEP 4: How the variables are related: The angle α satisfies the equation

$$\tan \alpha = \_\_\_\_\_\_,$$

which holds for all time t.

STEP 5: Differentiate with respect to t:

$$\sec^2 \alpha \cdot (\_\_\_\_\_\_) = -\frac{1}{x^2} \cdot (\_\_\_\_\_\_),$$

or, since $\sec^2\alpha = 1 + \tan^2\alpha =$ ________, we solve to find

$$\frac{d\alpha}{dt} = \left(\underline{\qquad\qquad}\right)\frac{dx}{dt}.$$

STEP 6: Evaluate: When $x = 2$ and $\frac{dx}{dt} = 400$, $\frac{d\alpha}{dt} =$ ______

radians per hour, or ________ rad/sec, or ______ deg/sec. Notice that the angle of elevation is decreasing because $d\alpha/dt$ is __________ .

81. A trough 10 ft long has a cross section that is an isosceles triangle 3 ft deep and 8 ft across. If water flows in at the rate of 2 cu ft/min, how fast is the surface rising when the water is 2 ft deep?

Solution. We carry out the steps of the basic strategy.

STEP 1: Picture and variables: We draw a picture of a partially filled trough. A cross section of the trough is shown in the figure at the right. In the figure h denotes the depth of the water and b its width across the trough at any instant t. Thus, b and h are both functions of ______ . We let V denote the volume of water in the trough at any time t.

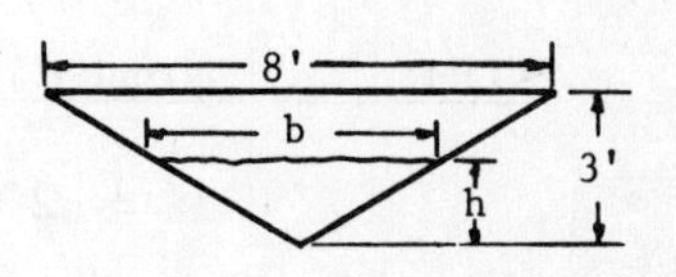

STEP 2: Numerical information: At the time in question,

$$h = 2 \text{ ft}, \quad \frac{dV}{dt} = \underline{\qquad\qquad} \text{ ft}^3/\text{min}.$$

STEP 3: To find: ____________ .

STEP 4: How the variables are related: At any instant of time the volume of water in the trough is given by the formula,

$$V = 10\ (\underline{\qquad\qquad}).$$

Since the formula for V involves both the variables b and h, we need to write down a formula relating these variables. From the geometry of similar triangles in the figure we have

$$\frac{b}{\underline{\quad}} = \frac{8}{3} \quad \text{or,} \quad b = \underline{\qquad\qquad}.$$

Substitution into the formula for V gives V = _______.

STEP 5: Differentiate with respect to t:

$$\frac{dV}{dt} = \underline{\qquad\qquad\qquad}.$$

STEP 6: Evaluate. Solving for dh/dt when dV/dt = 2 and h = 2 gives

dh/dt = ____________ ft/min.

---

80. $2, \frac{d\alpha}{dt}, \frac{1}{x}, \frac{d\alpha}{dt}, \frac{dx}{dt}, 1 + \frac{1}{x^2}, -\frac{1}{x^2}\left(\frac{x^2}{1+x^2}\right), -80, \frac{-80}{3600},$ (approx) -1.3, negative

81. $t, 2, \frac{dh}{dt}, \frac{1}{2}bh, h, \frac{8}{3}h, \frac{40}{3}h^2, \frac{80}{3}h\frac{dh}{dt}, \frac{3}{80}$

82. A walk is perpendicular to a long wall, and a woman strolls along it away from the wall at the rate of 3 ft/sec. There is a light 8 ft from the walk and 24 ft from the wall. How fast is her shadow moving along the wall when she is 20 ft from the wall?

Solution. We carry out the steps of the basic strategy.

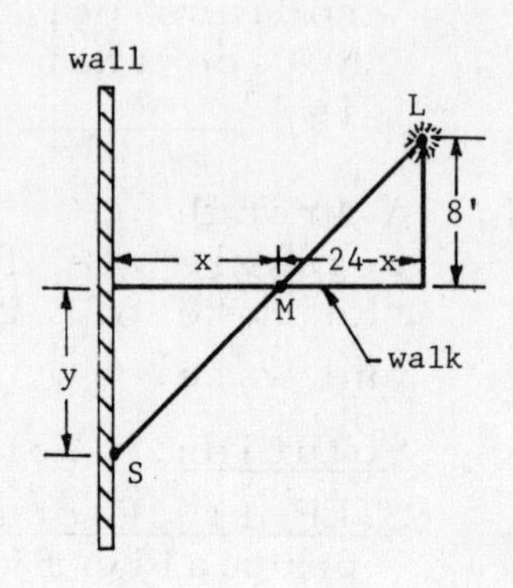

STEP 1: Picture and variables: The situation is pictured in the figure at the right. Here M denotes the position of the woman at a distance x units from the wall, S denotes the position of her shadow on the wall at a distance y units from where the wall and the walk intersect, and L denotes the position of the light.

STEP 2: Numerical information:

$$x = 20 \text{ ft}, \quad \underline{\qquad\qquad} = 3 \text{ ft/sec}.$$

STEP 3: To find: $\underline{\qquad\qquad}$.

STEP 4: How are the variables related: From similar triangles we can establish a relationship between the variables x and y,

$$\frac{x}{y} = \frac{\underline{\qquad}}{8}, \quad \text{or} \quad 8x = \underline{\qquad\qquad} .$$

This equation is valid at any instant of time t.

STEP 5: Differentiate both sides with respect to t:

$$8\frac{dx}{dt} = (\underline{\qquad\qquad})\frac{dy}{dt} - \underline{\qquad\qquad} .$$

STEP 6: Evaluate.

When $x = 20$, $y = \frac{8(20)}{\underline{\qquad}} = \underline{\qquad\qquad}$, so substitution into the previous derivative equation yields

$$8 \cdot 3 = \underline{\qquad\qquad} - 120, \quad \text{or} \quad \frac{dy}{dt} = \underline{\qquad\qquad} \text{ ft/sec} .$$

---

82. $\frac{dx}{dt}$, $\frac{dy}{dt}$, $24 - x$, $(24 - x)y$, $24 - x$, $y\frac{dx}{dt}$, $24 - 20$, 40, $4\frac{dy}{dt}$, 36

CHAPTER 2 SELF-TEST

1. Use the three algebraic steps to find the slope of the curve $y = x^3 - 2x + 5$ at a point (x, y) on the curve.

2. Write an equation of the tangent line to the curve in Problem 1 at the point when $x = -2$.

3. Find $\frac{dy}{dx}$:

   (a) $y = (2x^3 - x^2 + 7x + 3)^6$ (b) $y = (x^2 - 9)(3x^5 + 7x)$

   (c) $y = \frac{3}{x^4 + 1}$ (d) $y = \frac{x^2 - 1}{3x + 1}$

4. Find $\frac{d^2y}{dx^2}$:

   (a) $y = \frac{1}{3}x^3 + \frac{1}{2}x^2 - 6x + 8$ (b) $y = (2x^3 - 11)(x^2 - 3)$

   (c) $y = \cot 3x$ (d) $y = 3 \sin^2 5x - \sec x$

5. Find the limits:

   (a) $\lim_{x \to 0} \frac{\sin x^{1/3}}{x^{1/3}}$ (b) $\lim_{x \to \frac{\pi}{2}} \frac{\cos x}{\pi - 2x}$

6. Find $\frac{dy}{dx}$:

   (a) $y = x^{4/3} - 5x^{4/5}$ (b) $y = \sqrt{x + \frac{1}{x}}$

   (c) $y = \frac{1}{2x - \sqrt{x^2 - 1}}$ (d) $y = x \cos(5x - 2)$

7. Find an equation of the tangent line to the graph of $y = \sqrt{1 - x^2}$ when $x = \frac{1}{2}$.

8. A particle moves along a horizontal line (positive to the right) according to the law

   $$s = t^3 - 6t^2 + 2 \ .$$

   During which intervals of time is the particle moving to the right and during which is it moving to the left? What is the acceleration and the velocity when $t = 2.3$?

9. Calculate the instantaneous rate of change of the volume of a sphere with respect to its radius when the radius is 3 cm.

10. Find $\frac{dy}{dx}$ and $\frac{d^2y}{dx^2}$ when $x + y^2 = xy$.

11. Find an equation of the tangent line to the curve $x^3 + 3xy^3 + xy^2 = xy$ at the point $(1, -1)$.

12. A photographer is televising a 100-yard dash from a position 10 yards from the track in line with the finish line. When the runners are 10 yards from the finish line, the camera is turning at the rate 3/5 rad/sec. How fast are the runners moving then?

13. A swimming pool is 40 ft long, 20 ft wide, 8 ft deep at the deep end, and 3 ft deep at the shallow end, the bottom being rectangular. If the pool is filled by pumping water into it at the rate of 40 cu. ft/min, how fast is the water level rising when it is

 (a) 3 ft deep at the deep end?

 (b) 6 ft deep at the deep end?

14. A guy wire is to pass from the top of a pole 36 ft high to an anchorage on the ground 27 ft from the base of the pole. One end of the wire is made fast to the anchorage, and a man climbs the pole with the wire, keeping it taut. If he climbs 2 ft/sec, how fast is he playing out the wire when he reaches the top of the pole?

SOLUTIONS TO CHAPTER 2 SELF-TEST

1. STEP 1. $f(x + h) = (x + h)^3 - 2(x + h) + 5$

$$= x^3 + 3x^2h^2 + 3xh^2 + h^3 - 2x - 2h + 5$$

STEP 2. Subtracting $f(x + h) - f(x)$:

$$f(x + h) - f(x) = 3x^2h^2 + 3xh^2 + h^3 - 2h$$

Dividing by h yields,

$$\frac{f(x + h) - f(x)}{h} = 3x^2 + 3xh + h^2 - 2$$

STEP 3. As h tends to zero

$$f'(x) = \lim_{h\to 0} \frac{f(x + h) - f(x)}{h} = 3x^2 - 2.$$

2. The derivative is $f'(x) = 3x^2 - 2$. When $x = -2$, $y = (-2)^3 - 2(-2) + 5 = 1$, and the slope $f'(-2)$ is $m = 3(-2)^2 - 2 = 10$. Thus, $y = 1 + 10(x + 2)$ or $y = 10x + 21$ is an equation of the tangent line.

3. (a) $\frac{dy}{dx} = 6(2x^3 - x^2 + 7x + 3)^5(6x^2 - 2x + 7)$

 (b) $\frac{dy}{dx} = 2x(3x^5 + 7x) + (x^2 - 9)(15x^4 + 7)$

 $= 21x^6 - 135x^4 + 21x^2 - 63$

 (c) $\frac{dy}{dx} = \frac{-3(4x^3)}{(x^4 + 1)^2}$

(d) $\frac{dy}{dx} = \frac{(3x + 1)(2x) - (x^2 - 1)(3)}{(3x + 1)^2} = \frac{3x^2 + 2x + 3}{(3x + 1)^2}$

4. (a) $\frac{dy}{dx} = x^2 + x - 6, \quad \frac{d^2y}{dx^2} = 2x + 1$

(b) $\frac{dy}{dx} = 6x^2(x^2 - 3) + (2x^3 - 11)(2x) = 10x^4 - 18x^2 - 22x,$

$\frac{d^2y}{dx^2} = 40x^3 - 36x - 22$

(c) $\frac{dy}{dx} = \frac{d}{dx}\left(\frac{\cos 3x}{\sin 3x}\right) = \frac{(\sin 3x)(-3 \sin 3x) - (\cos 3x)(3 \cos 3x)}{\sin^2 3x}$

$= \frac{-3 \sin^2 3x - 3 \cos^2 3x}{\sin^2 3x} = -3 \csc^2 3x$

$\frac{d^2y}{dx^2} = (-3)(2)(\csc 3x) \cdot \frac{d}{dx}(\csc 3x)$

$= (-6 \csc 3x)(-3 \csc 3x \cot 3x)$

$= 18 \csc^2 3x \cot 3x$

(d) $\frac{dy}{dx} = 30 \sin 5x \cos 5x - \sec x \tan x$

$= 15 \sin 10x - \sec x \tan x$

$\frac{d^2y}{dx^2} = 150 \cos 10x - (\sec x \tan x)(\tan x) - (\sec x)(\sec^2 x)$

$= 150 \cos 10x - (\sec x)(\tan^2 x + \sec^2 x)$

5. (a) Let $u = x^{1/3}$ so that $x \to 0$ is equivalent to $u \to 0$.

Thus $\lim_{x\to 0} \frac{\sin x^{1/3}}{x^{1/3}} = \lim_{u\to 0} \frac{\sin u}{u} = 1$.

(b) $\lim_{x\to \frac{\pi}{2}} \frac{\cos x}{\pi - 2x} = \lim_{x\to \frac{\pi}{2}} \frac{\frac{1}{2} \cos x}{\frac{\pi}{2} - x}$; let $u = \frac{\pi}{2} - x$ so that $x \to \frac{\pi}{2}$ is equivalent to $u \to 0$. Thus,

$\lim_{x\to \frac{\pi}{2}} \frac{\frac{1}{2} \cos x}{\frac{\pi}{2} - x} = \lim_{u\to 0} \frac{\frac{1}{2} \cos\left(\frac{\pi}{2} - u\right)}{u} = \lim_{u\to 0} \frac{1}{2} \frac{\sin u}{u} = \frac{1}{2}$.

6. (a) $\frac{dy}{dx} = \frac{4}{3}x^{1/3} - 4x^{-1/5}$

(b) $\frac{dy}{dx} = \frac{1}{2}\left(x + \frac{1}{x}\right)^{-1/2} \cdot \frac{d}{dx}\left(x + \frac{1}{x}\right) = \frac{1}{2}\left(x + \frac{1}{x}\right)^{-1/2}\left(1 - \frac{1}{x^2}\right)$

(c) $\frac{dy}{dx} = -\left(2x - \sqrt{x^2 - 1}\right)^{-2}\left[2 - \frac{1}{2}(x^2 - 1)^{-1/2} \cdot 2x\right]$

$= -\left(2x - \sqrt{x^2 - 1}\right)^{-2}\left(2 - \frac{x}{\sqrt{x^2 - 1}}\right)$

(d) $\frac{dy}{dx} = \cos(5x - 2) + x[-\sin(5x - 2)\cdot 5]$

$= \cos(5x - 2) - 5x\sin(5x - 2)$

7. $y' = \frac{1}{2}(1 - x^2)^{-1/2}(-2x) = -x(1 - x^2)^{-1/2}$, $y(\frac{1}{2}) = \frac{\sqrt{3}}{2}$ and $y'\left(\frac{1}{2}\right) = -\frac{1}{\sqrt{3}}$ so that $y = \frac{\sqrt{3}}{2} + \left(-\frac{1}{\sqrt{3}}\right)\left(x - \frac{1}{2}\right)$ or $\sqrt{3}y + x - 2 = 0$ is an equation of the tangent line.

8. $\frac{ds}{dt} = 3t^2 - 12t = 3t(t - 4)$; $\frac{d^2s}{dt^2} = 6t - 12$
The particle is moving to the right when $\frac{ds}{dt} > 0$ so $t > 4$ or $t < 0$; it is moving to the left when $0 < t < 4$ and $\frac{ds}{dt} < 0$. At $t = 2.3$,

$$\left.\frac{ds}{dt}\right|_{2.3} = 3(2.3)^2 - 12(2.3) = -11.73, \text{ velocity}$$

$$\left.\frac{d^2s}{dt^2}\right|_{2.3} = 6(2.3) - 12 = 1.8, \text{ acceleration}$$

9. The volume of a sphere is given by $V = \frac{4}{3}\pi r^3$, where r is the radius. We seek the value of $\frac{dV}{dr}$ when $r = 3$. Thus, $\frac{dV}{dr} = V'(r) = 4\pi r^2$, so that $V'(3) = 36\pi$.

10. Differentiating implicitly, $1 + 2yy' = y + xy'$ or $y' = \frac{y - 1}{2y - x}$

$$\frac{d^2y}{dx^2} = \frac{(2y - x)(y') - (y - 1)(2y' - 1)}{(2y - x)^2} = \frac{(2 - x)y' + (y - 1)}{(2y - x)^2}$$

$$= \frac{(2 - x)\left(\frac{y - 1}{2y - x}\right) + (y - 1)}{(2y - x)^2} = \frac{2(y - 1)(y - x + 1)}{(2y - x)^3}.$$

11. Since $1^3 + 3(1)(-1)^3 + 1(-1)^2 = 1(-1)$ is true, the point $(1, -1)$ is on the curve. Differentiating implicitly, $3x^2 + 3y^3 + 9xy^2y' + y^2 + 2xyy' = y + xy'$, so evaluation at $(1, -1)$ yields $3 - 3 + 9y' + 1 - 2y' = -1 + y'$, or $y' = -\frac{1}{3}$. Thus an equation of the tangent line is given by

$$y = 1 + \left(-\frac{1}{3}\right)(x - 1) \text{ or } x + 3y = -2.$$

12. The situation is pictured in the figure at the right. Thus, $\tan\theta = \frac{x}{10}$, or $x = \tan\theta$. $\frac{dx}{dt} = 10\sec^2\theta\,\frac{d\theta}{dt}$. Now, when $x = 10$ yds, $\theta = \frac{\pi}{4}$, and $\frac{d\theta}{dt} = \frac{3}{5}$ rad/sec. Hence,

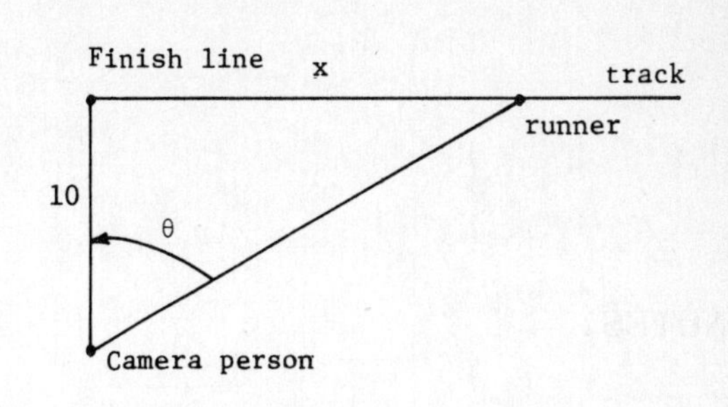

$$\left.\frac{dx}{dt}\right|_{x=10} = 10\left(\sec^2\frac{\pi}{4}\right)\left(\frac{3}{5}\right)$$
$$= 10\left(\sqrt{2}\right)^2\left(\frac{3}{5}\right) = 12 \text{ yd/sec.}$$

13. A vertical cross-section of the pool is pictured in the figure at the right: y denotes the depth of the water at any time t, and x denotes the horizontal length of the water in the bottom of the pool.

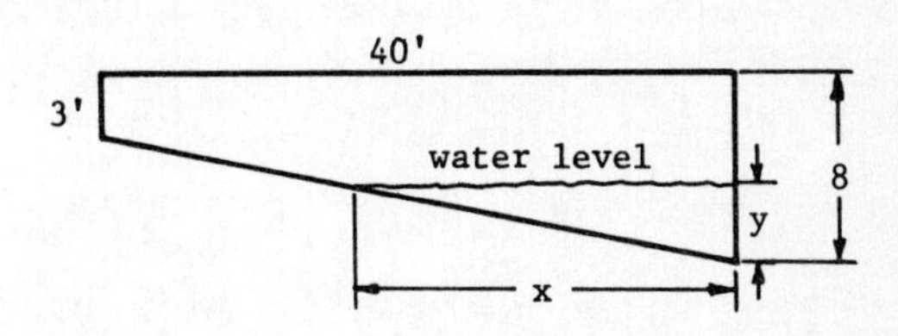

(a) When $y < 5'$, we have from the geometry of similar triangles in the figure that, $\frac{x}{40} = \frac{y}{5}$ or $x = 8y$. The volume of water in the pool is given by $V = \frac{1}{2}x \cdot y \cdot 20 = 80y^2$. Hence $\frac{dV}{dt} = 160y\,\frac{dy}{dt}$, and since $\frac{dV}{dt} = 40$ is given,

solving for dy/dt yields $\left.\frac{dy}{dt}\right|_{y=3} = \frac{40}{160(3)} = \frac{1}{12}$ ft/min.

(b) When $y > 5'$, the total volume of water is given by $V = \frac{1}{2}(40)(5)(20) + (40)(20)(y - 5) = 800y - 2000$. Hence, $\frac{dV}{dt} = 800\,\frac{dy}{dt}$, and since $\frac{dV}{dt} = 40$ is given, solving for dy/dt yields $\left.\frac{dy}{dt}\right|_{y=6} = \frac{40}{800} = \frac{1}{20}$ ft/min.

14. The situation is pictured at the right, where h is the height of the man above the ground and ℓ is the length of guy wire played out at any instant of time t. We want to find $\frac{d\ell}{dt}$ when $h = 36$. Now, $\ell^2 = h^2 + (27)^2$, and differentiation with respect to t gives $2\ell\,\frac{d\ell}{dt} = 2h\,\frac{dh}{dt}$.

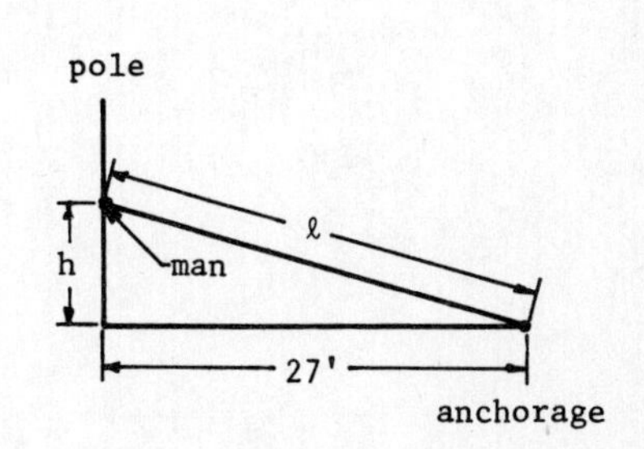

When $h = 36$, we have $\ell = \sqrt{(36)^2 + (27)^2} = 9\sqrt{4^2 + 3^2} = 45$. Thus, for $dh/dt = 2$,

$$\left.\frac{d\ell}{dt}\right|_{h=36} = \frac{h}{\ell}\left.\frac{dh}{dt}\right|_{h=36} = \frac{36}{45}\cdot 2 = \frac{72}{45} = \frac{8}{5} \text{ ft/sec.}$$

NOTES.

CHAPTER 3 APPLICATIONS OF DERIVATIVES

3.1 EXTREME VALUES OF FUNCTIONS

OBJECTIVE A: Define the terms local maximum, local minimum, absolute maximum, and absolute minimum value.

1. A function f is said to have a local maximum at $x = c$ if ____________ for all x in some ______________ I about c.

2. A function f is said to have a _______ maximum over its domain at $x = c$ if $f(c) \geq f(x)$ for all x belonging to the __________ of f.

3. A function f is said to have a _______ minimum over its domain at $x = c$ if $f(c) \leq f(x)$ for all x close to c.

4. If $f(c) \leq f(x)$ for all x in the domain of f, then f is said to have a __________ __________ at $x = c$.

5. Can a local maximum also be an absolute maximum for a function f? Can a local minimum also be an absolute minimum?

OBJECTIVE B: Interpret correctly the First Derivative Theorem relating local extrema at an interior point $x = c$ of the domain $a \leq x \leq b$ of a function f and its derivative $f'(c)$.

Answer questions 6-9 true or false.

6. If $f'(c) = 0$, then f has either a local maximum or a local minimum at the interior point $x = c$. (True or False)

7. If f has a local minimum at the interior point $x = c$, then $f'(c) = 0$. (True or False)

8. If f has a local maximum or local minimum at an endpoint of the interval of definition of the function, then the lefthand (or righthand) tangent must have slope zero there. (True or False)

---

1. $f(c) \geq f(x)$, open interval
2. absolute, domain
3. local
4. absolute minimum
5. yes to both questions
6. False, it could have a point of inflection
7. False, the derivative may fail to exist
8. False

9. If f has an absolute maximum at an interior point $x = c$ and $f'(c)$ exists as a finite number, then $f'(c)$ is necessarily zero. (True or False)

10. If f is continuous over the closed interval $a \le x \le b$, then every point where f has a (local or absolute) maximum or minimum must be an ________ of the interval, a point where f' ________, or an ________ point where f' equals ________.

OBJECTIVE C: Given a function $y = f(x)$ continuous over a closed interval $a \le x \le b$, find the critical points of f and for each critical point, determine whether the function has a local maximum or local minimum there, or neither. If possible, find the absolute maximum and minimum values of the function on the closed interval.

11. Consider $y = x^{3/2}(x - 8)^{-1/2}$ over $10 \le x \le 16$. Now, $y' =$ ________ $= \frac{1}{2} x^{1/2}(x - 8)^{-3/2}$ ________. Thus, $x =$ ________ is the only critical point in the interval $10 \le x \le 16$. Now when $x = 12$, $y \approx 20.78$. Since $y(11) \approx 21.06$ and $y(13) \approx 20.96$, we conclude that the function has a local ________ when $x = 12$. Checking the endpoints of the interval $[10, 16]$ we determine that $y(10) \approx 22.36$ and $y(16) \approx 22.63$. Thus the absolute maximum of y occurs at $x =$ ________ and the absolute minimum of y occurs at $x =$ ________.

3.2 THE MEAN VALUE THEOREM

OBJECTIVE A: Apply Rolle's Theorem to show that a given equation $f(x) = 0$ has exactly one solution in the specified interval $a \le x \le b$.

12. Suppose $y = f(x)$ and its first derivative $f'(x)$ are continuous over $a \le x \le b$. If $f(a)$ and $f(b)$ have opposite sign, then according to the Intermediate Value Theorem there is at least one point c satisfying $a < c < b$ and $f(c) =$ ________.

13. Suppose there is another point d satisfying $a < d < b$ and $f(d) = 0$. Then, according to Rolle's Theorem, there is a point between c and d for which ________ is zero. Thus, if $f'(x)$ is different from zero for all values of x between a and b, there is exactly ________ solution to the equation $f(x) = 0$ in the interval ________.

---

9. True

10. endpoint, does not exist, interior, 0

11. $\frac{3}{2} x^{1/2}(x - 8)^{-1/2} - \frac{1}{2} x^{3/2}(x - 8)^{-3/2}$, $2x - 24$, 12, minimum, 16, 12

12. 0

13. $f'(x)$, one, $a \le x \le b$

14. Consider the equation $x^3 + 2x^2 + 5x - 6 = 0$ for $0 \le x \le 5$. When $x = 0$ the value of the left side is _____; and when $x = 5$, the value is _______ . These values differ in sign. Calculating the derivative, we have

$$\frac{d}{dx}\left(x^3 + 2x^2 + 5x - 6\right) = \underline{\qquad\qquad\qquad},$$

and this is always ______ for $0 < x < 5$. Therefore, we conclude from Problems 12 and 13 that there is exactly one solution to the equation somewhere between $x =$ _____ and $x =$ ______ . We could in fact use Newton's method of Section 3.8 to locate this solution.

OBJECTIVE B: Given a function $y = f(x)$ satisfying the hypotheses of the Mean Value Theorem for $a \le x \le b$, use the theorem to find a number c satisfying the conclusion of the theorem.

15. The hypotheses of the Mean Value Theorem are that f is __________ over the closed interval $a \le x \le b$ and ______________ over the open interval __________ .

16. The conclusion of the Mean Value Theorem is that there is at least one number c in the open interval ________ satisfying __________ . A geometric interpretation of the conclusion is that the slope of the curve $y = f(x)$ when $x = c$ is the same as the slope of the ______ joining the endpoints $(a,f(a))$ and ________ of the curve.

17. Let $f(x) = 3x^2 + 4x - 3$ over $1 \le x \le 3$. Then $f'(x) =$ _____, so that f and f' satisfy the hypotheses of the Mean Value Theorem. To find a value for c, the equation $f(b) - f(a) = f'(c)(b - a)$ becomes $f(3) - f(1) = f'(c)($_____$)$, or $36 - 4 = 2($_______$)$. Solving for c gives $c =$ _____ .

18. Does the Mean Value Theorem apply to the function $f(x) = |x|$ in the interval $[-2,1]$?
No, because the derivative $f'(x)$ is not defined for $x =$ ______ so the function f is not ___________ over the open interval _______ as required by the hypotheses.

OBJECTIVE C: Know the main consequences of the Mean Value Theorem.

19. If $f'(x) = 0$ for all x in an interval I, then ______________________________ .

---

14. -6, 194, $3x^2 + 4x + 5$, positive, 0, 5

15. continuous, differentiable, $a < x < b$

16. (a,b), $f(b) - f(a) = f'(c)(b - a)$, chord, $(b,f(b))$

17. $6x + 4$, 2, $6c + 4$, 2

18. 0, differentiable, $(-2,1)$

19. $f(x)$ is constant for all x in I

20. Functions with the same derivative differ ____________________ .

21. The function f ________________ when $f' < 0$.

22. The function f increases when __________________ .

23. The function f increases over a domain D of real numbers if $x_1 < x_2$ implies ________________ .

OBJECTIVE D: Use the First Derivative Test for Increasing and Decreasing to determine the values of x where the graph of y versus x is increasing and where it is decreasing.

24. Let $y = f(x)$ be a differentiable function of x. When dy/dx has a ________ value, the graph of y versus x is rising (to the right). In this case it is said that the function f is __________ at that point.

25. When $dy/dx < 0$, the graph of y versus x is ________ and the function f is ________________ at that point.

26. Let $y = \frac{1}{3}x^3 - x^2 + 2$. Then $y' =$ ________ $= x($________$)$. The derivative dy/dx is zero when $x =$ _____ or $x =$ _____ . Thus, the curve y is increasing when $x < 0$, it is decreasing when x satisfies __________, and it is increasing again when $x >$ ______ . We construct a table of some values for the curve (complete the table):

x	-2	-1	0	1	2	3	4
y							

Sketch the graph in the coordinate system at the right.

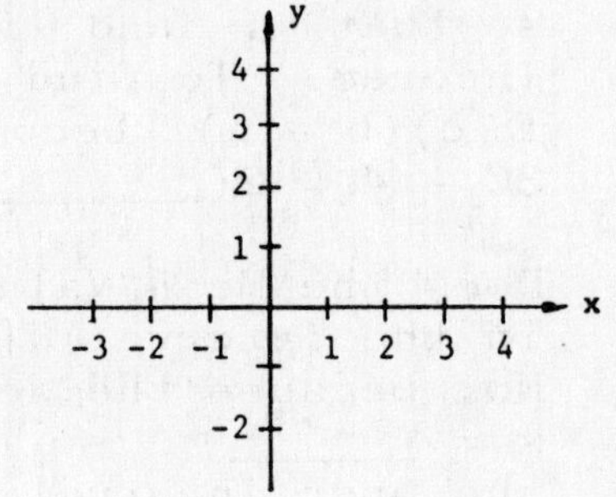

---

20. only by a constant

21. decreases

22. $f' > 0$

23. $f(x_1) < f(x_2)$

24. positive, increasing

25. falling, decreasing

26. $x^2 - 2x$, $x - 2$, 0, 2, $0 < x < 2$, 2

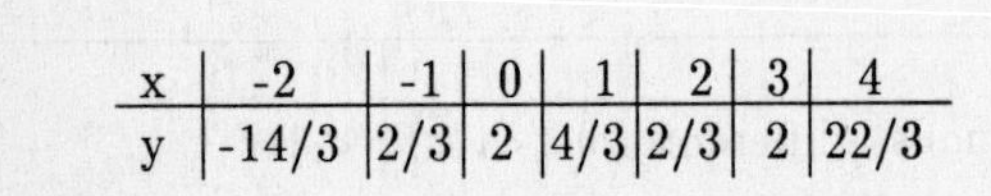

x	-2	-1	0	1	2	3	4
y	-14/3	2/3	2	4/3	2/3	2	22/3

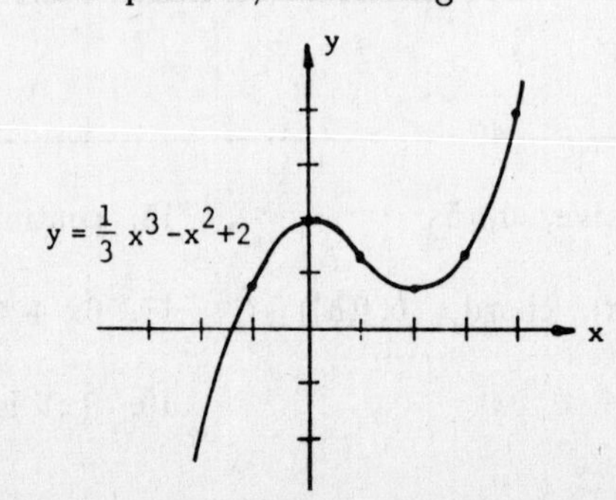

OBJECTIVE E: Find all possible functions with a given derivative. Also find that function whose graph passes through a specified point P.

27. Since $y' = 5x^4 - 2x$ is the derivative of __________, we know that $y =$ __________ for some constant C.

28. If $y = 3$ when $x = -1$ in Problem 27, then the value of C must be __________ .

29. If $y' = \cos 3t$, then $y =$ __________ .

30. If $y' = 3x^4 - x^{-2} + 5$, then $y =$ __________ .

31. If $y' = x^{3/2} + x^{1/2} + \frac{1}{x^2}$, then $y =$ __________ . If the graph of y passes through the point $\left(1, \frac{2}{5}\right)$ then C = __________ .

3.3 THE FIRST DERIVATIVE TEST FOR LOCAL EXTREME VALUES

OBJECTIVE: Use the First Derivative Test for Local Extreme Values to identify the local extreme values of a given function $y = f(x)$.

32. Suppose the function $y = f(x)$ has the derivative $f'(x) = x(x + 1)(x - 2)$. Then the critical points of f are $x = 0$, $x =$ ______, and $x =$ ______ . At $x = 0$, the derivative f' changes from __________ to __________ so f has a local maximum value at $x = 0$. At $x = -1$, the derivative f' changes from negative to positive so f has a local __________ value at $x = -1$. At $x = 2$ the function f has a local __________ value because the derivative f' changes from __________ to __________ .

33. The function f in Problem 32 is increasing on the intervals ________ and ________, and it is decreasing on the intervals ________ and ________ .

34. For the function whose derivative is $f'(x) = x^{-2/3}(2 - x)$ the critical points are $x =$ ______ and $x =$ ______ . The function f is increasing on the interval ________ and decreasing on the interval ________ . Thus the function assumes a local maximum at the critical point $x =$ ______ .

---

27. $x^5 - x^2$, $x^5 - x^2 + C$

28. 5

29. $\frac{1}{3}\sin 3t + C$

30. $\frac{3}{5}x^5 + x^{-1} + 5x + C$

31. $\frac{2}{5}x^{5/2} + \frac{2}{3}x^{3/2} - x^{-1} + C$, $\frac{1}{3}$

32. -1, 2, positive, negative, minimum, minimum, negative, positive

33. $(-1,0)$ and $(2,\infty)$, $(-\infty,-1)$ and $(0,2)$

34. 0, 2, $(-\infty,2)$, $(2,\infty)$, 2

3.4 GRAPHING WITH y' AND y''

OBJECTIVE A: Relate the concavity of a function $y = f(x)$ to the second derivative d^2y/dx^2.

35. If the second derivative d^2y/dx^2 is positive, the y-curve is concave ________ at that point; if d^2y/dx^2 is ________, the curve is concave downward at that point.

36. When a curve is concave upward at a point, locally the curve lies ________ the tangent line; when it is concave ________, locally the curve lies below the tangent line.

37. A point where the curve changes concavity is called a ________ ________________, and is characterized by a change in sign of ____________ .

38. A point of inflection occurs where d^2y/dx^2 is ________ or ________________ .

39. Does the condition $d^2y/dx^2 = 0$ guarantee a point of inflection? ________

OBJECTIVE B: Given a function $y = f(x)$, find the intervals of values of x for which the curve is increasing, decreasing, concave upward, and concave downward. Sketch the curve, showing the points of inflection and the points where the function has local maximum and local minimum values. Use the following strategy to graph $y = f(x)$.

1. Find y' and y''.
2. Find where y' is positive, negative, and zero.
3. Find where y'' is positive, negative, and zero.
4. Make a summary table, and show the curve's general shape.
5. Plot specific points and sketch the graph.

40. Consider the function $f(x) = x^4 - 4x^3 + 10$. We follow the five-step strategy in order to sketch the graph of $y = f(x)$.

STEP 1: $\frac{dy}{dx} =$ ________________ and $\frac{d^2y}{dx^2} =$ ________________ .

STEP 2: In factored terms, $dy/dx = 4x^2(x - 3)$ so that the curve is decreasing when x belongs to the interval ________ and increasing when $x >$ ________ . The slope of the curve is zero when $x =$ ________ or $x =$ ________ .

---

35. upward, negative
36. above, downward
37. point of inflection, d^2y/dx^2
38. zero, fails to exist
39. No, the function $y = x^4$ affords a counterexample at $x = 0$.

STEP 3: In factored form, $d^2y/dx^2 =$ ________ . Thus d^2y/dx^2 is negative when x belongs to the interval ________ and consequently the curve is concave ________ there. The second derivative is positive when x satisfies ________ or ________, and the curve is concave ________. Therefore, the second derivative changes sign when x = ________ or x = ________ so that these are points of inflection of f.

STEP 4: Complete the following table.

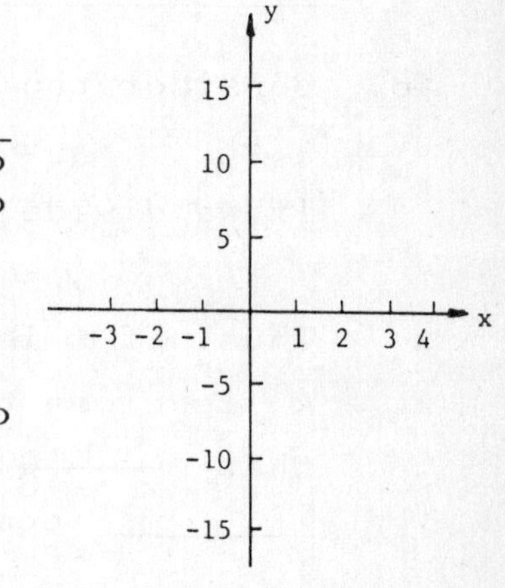

x	y	y′	y″	Conclusions
-2	58	-	+	decreasing; concave up
-1	15	-	+	decreasing; concave up
0				
1				
2				
3				
4	20	+	+	increasing; concave up

STEP 5: Sketch a smooth curve of $y = f(x)$ in the given coordinate system to the right.

3.5 LIMITS AS x → ±∞, ASYMPTOTES, AND DOMINANT TERMS

OBJECTIVE A: Calculate the limit of $f(x)$ as x approaches $+\infty$ or $-\infty$, whenever the limit exists.

41. $\lim_{x\to\infty} \dfrac{5x^3}{1 + 3x - 2x^3} = \lim_{h\to 0} \dfrac{\frac{5}{h^3}}{\rule{2cm}{0.4pt}} = \lim_{h\to 0} \dfrac{5}{\rule{2cm}{0.4pt}} =$ ________ .

42. $\lim_{x\to\infty} \dfrac{1 - 3x^2}{4x^3 + 2x - 5} = \lim_{x\to\infty} \dfrac{\rule{3cm}{0.4pt}}{4 + (2/x^2) - (5/x^3)} = \dfrac{0 - \rule{1cm}{0.4pt}}{4 + 0 - 0} =$ ________ .

43. $\lim_{x\to-\infty} \dfrac{3 - x^2}{6x + 1} = \lim_{x\to-\infty} \dfrac{(3/x) - x}{\rule{2cm}{0.4pt}} =$ ________ .

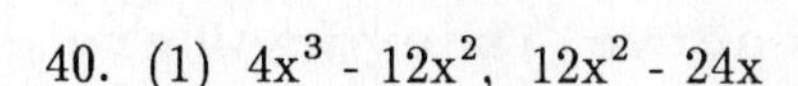

40. (1) $4x^3 - 12x^2$, $12x^2 - 24x$

(2) $(-\infty, 3)$, 3, 0, 3

(3) $12x(x - 2)$, (0,2), downward, $x < 0$, $x > 2$, upward, 0, 2

(4)

x	y	y′	y″	Conclusions
0	10	0	0	point of inflection
1	7	-	-	decreasing; concave down
2	- 6	-	0	point of inflection
3	-17	0	+	"Holds water"; min.

(5)

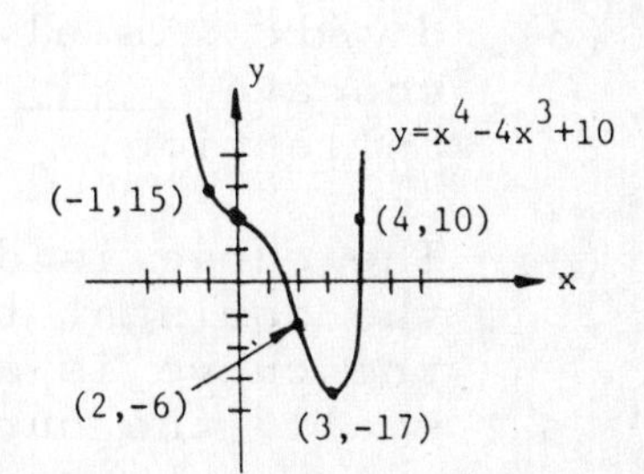

41. $1 + \frac{3}{h} - \frac{2}{h^3}$, $h^3 + 3h^2 - 2$, $-\frac{5}{2}$

42. $\frac{1}{x^3} - \frac{3}{x}$, 0, 0

43. $6 + \frac{1}{x}$, $+\infty$

OBJECTIVE B: Analyze the behavior of the graph of a function $y = f(x)$ as $x \to \pm\infty$.

44. A line $y = b$ is a ________ asymptote of the graph of a function $y = f(x)$ if either $\lim_{x\to\infty} f(x) = b$ or ________.

45. A line $x = a$ is a ________ asymptote of the graph of a function $y = f(x)$ if either $\lim_{x\to a^-} f(x) = \pm\infty$ or ________.

46. Consider the rational function

$$y = \frac{x^2 - 3x - 1}{x - 3}.$$

If we divide $x - 3$ into $x^2 - 3x - 1$ we obtain

$$y = x - \text{________}.$$

Thus, if x is large the curve behaves like $y =$ ________; when x is close to 3 the curve behaves like $y = \frac{1}{x - 3}$. We say that $\frac{1}{x - 3}$ ________ when x is close to 3 and that ________ dominates when x is large.

OBJECTIVE C: Analyze the graph of a given function $y = f(x)$ to investigate the following properties of the curve: (a) symmetry (b) intercepts (c) asymptotes (d) rise and fall (e) concavity, and (f) dominant terms. Using the information you have discovered, sketch the curve.

47. Sketch the graph of $y = x^{-2} + 2x$.
Solution. We follow the five steps as in Problem 40.

(1) $y' =$ ________ and $y'' =$ ________.

(2) In fractional form, $y' = \frac{\text{________}}{x^3}$ so that y' is zero when $x =$ ________. The curve is decreasing when x belongs to the interval ________; it is increasing for x satisfying ________ and ________.

(3) d^2y/dx^2 is always ________ so the curve is everywhere concave ________. Therefore there are no points of inflection.

(4) The curve is discontinuous at $x =$ ________. To identify the dominant terms, note that for large values of $|x|$, the curve is approximately $y \approx$ ________. When x is small, the curve is approximately $y \approx$ ________.

---

44. horizontal, $\lim_{x\to-\infty} f(x) = b$ 45. vertical, $\lim_{x\to a^+} f(x) = \pm\infty$ 46. $\frac{1}{x - 3}$, x, dominates, x

Complete the following table.

x	y	y′	y″	Conclusions
-2	-15/4	+	+	increasing; concave up
-1	-1			
-1/2	3			
1/2	5			
1				
2	17/4	+	+	increasing; concave up

(5) Sketch the graph at the right.

48. Let $y = \frac{3x^2 - 1}{x^3}$. Since $-y = \frac{3(-x)^2 - 1}{(-x)^3}$, the curve is symmetric about the ________ . Moreover, y is undefined when $x =$ ________, but the graph is defined for all other real values of x. Now, $y = 0$ when $3x^2 - 1 = 0$. Thus, the x-intercepts occur at ________ . Since $x \neq 0$, there are no y-intercepts. Now

$$\lim_{x\to 0^-} \frac{3x^2 - 1}{x^3} = \lim_{x\to 0^-} \left(\frac{3}{x} - \_\_\_\_\right) = \_\_\_\_\_\_,$$ and

$$\lim_{x\to 0^+} \frac{3x^2 - 1}{x^3} = \_\_\_\_\_\_ .$$

Therefore, the line ________ is a ________ asymptote. Also,

$$\lim_{x\to \pm\infty} \frac{3x^2 - 1}{x^3} = \_\_\_\_\_\_$$

so the line ________ is a ________ asymptote.

Finally, $y' = \frac{x^3(6x) - (\quad)3x^2}{x^6} = \frac{3(1 - x^2)}{\_\_\_\_}$. Thus, $y' = 0$ when $x =$ ________ . The derivative is ________ for

---

47. (1) $-2x^{-3} + 2,\ 6x^{-4}$

(2) $2(x^3 - 1),\ 1,\ (0,1),\ x < 0,\ x > 1$

(3) positive, upward

(4) $0,\ 2x,\ 1/x^2$

x	y	y′	y″	Conclusions
-1	-1	+	+	increasing; concave up
-1/2	3	+	+	increasing, concave up
1/2	5	-	+	decreasing; concave up
1	3	0	+	min.; concave up

(5)

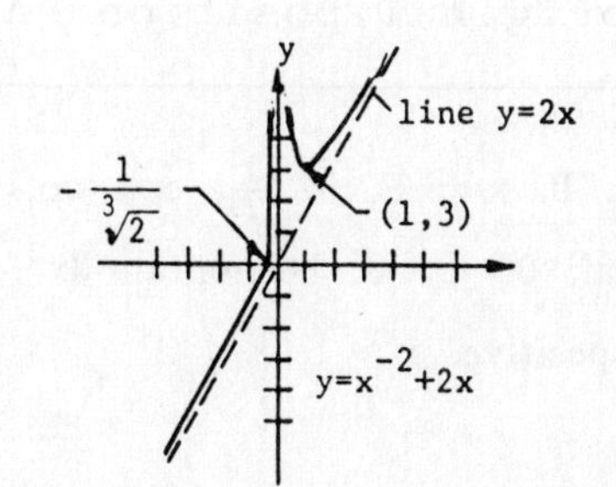

$0 < x < 1$ and negative for __________ and for __________ . A sketch of the graph is shown below.

3.6 OPTIMIZATION

OBJECTIVE: Solve a max-min problem by the following strategy:

1. Reading the problem until you understand it.
2. Drawing a figure, if possible, to illustrate the problem.
3. Introducing variables and listing every relation in your picture and in the problem as an equation or algebraic expression.
4. Identifying and writing an equation for the unknown that is to be a maximum or minimum.
5. Finding and testing the critical and endpoints for a possible maximum or minimum. Use what you know about the shape of the function's graph and the physics of the problem.

49. At 9:00 A.M. ship B was 65 miles due east of another ship A. Ship B was then sailing due west at 10 miles per hour, and ship A was sailing due south at 15 miles per hour. If they continue to follow their respective courses, when will they be nearest one another and how near?

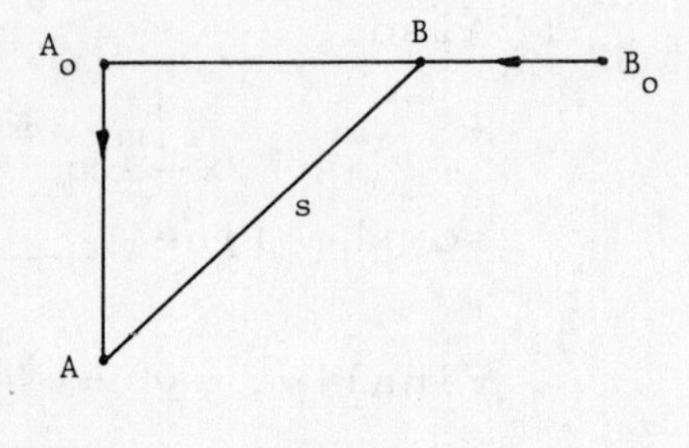

Solution. Let A_0 and B_0 denote the original positions of the ships at 9:00 A.M., and let A and B denote their new positions, respectively, at t hours later. This is pictured in the figure at the right. Let s denote the distance between A and B. The problem is to minimize ________ and to find the time when its minimum occurs. Since rate x time = distance, the distance covered by ship A in t hours is ________ miles, and by ship B ________ miles. The original distance between A and B is given as 65 miles, so the distance between the original position A_0 and ship B after t hours is ________ .

---

48. origin, 0, $x = \pm\sqrt{3}$, $\frac{1}{x^3}$, ∞, $-\infty$, $x = 0$, vertical, 0, $y = 0$, horizontal, $3x^2 - 1$, x^4, ± 1, positive, $x > 1$, $x < -1$.

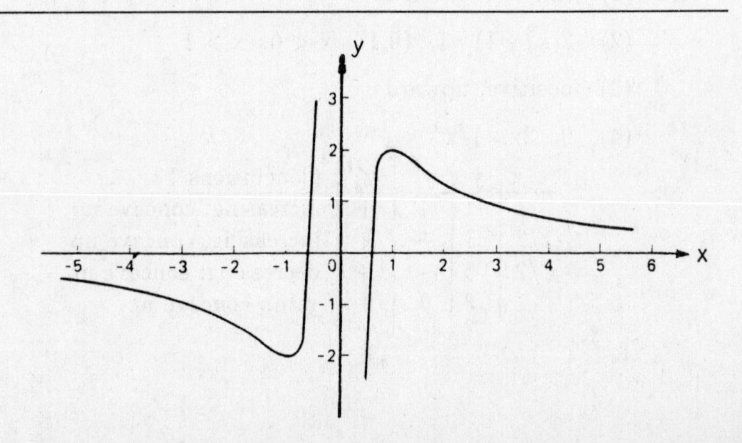

Fill this information into the figure and then calculate the square of the distance: $s^2 =$ ________ . Differentiation of both sides of this equation with respect to t gives $2s\, ds/dt =$ ________ . Thus, $ds/dt = 0$ when $30(15t) - 20(65 - 10t)$ equals 0, or $t =$ ____ hours. Simplifying ds/dt algebraically, we see that $ds/dt =$ ______ . Thus, ds/dt is ________ when $t < 2$ and ________ when $t > 2$. Therefore, a relative ______ distance occurs for s at $t = 2$ hours. Solving for the distance s after two hours, we find

$$s^2 = (30)^2 + (\_\_\_\_\_\_)^2 \quad \text{or} \quad s = \_\_\_\_\_\_ \text{ miles,}$$

the distance the ships are apart at 11:00 A.M. when they are nearest each other.

50. A company's cost function is $C(x) = 10x + 3$, and its revenue function is $R(x) = 50x - 0.5x^2$, both in thousands of dollars per thousand items. Find the company's maximum profit.

<u>Solution</u>. If $P(x)$ denotes the profit function, then $P(x) = R(x) -$ ________ $=$ ________ . The maximum profit occurs when $P'(x) =$ ______, so $P'(x) =$ ______ $= 0$. Thus, $x =$ ______ thousand items. Since $P''(x) =$ ______ is always negative, this yields a maximum profit of $P(40) =$ _____ thousand dollars.

51. It is known that the population P for the fur-bearing snowshoe hare in the Hudson Bay area will grow to $f(P) = -0.025P^2 + 4P$ in one year. If they "harvest" the amount $f(P) - P$ so the initial population is not depleted, then the harvest is said to be "sustained." Find the population at which the maximum sustainable harvest occurs, and find the maximum sustainable harvest for the snowshoe hare. Assume P is measured in thousands.

<u>Solution</u>. The harvest function $H(P) = f(P) - P =$ ________ . The maximum sustainable harvest occurs when $H'(P) =$ ______, so $H'(P) =$ ________ $= 0$, or $P =$ ________ thousand hares. This is the population at which the maximum sustainable harvest occurs, since $H''(P) =$ ______ is always ______ . The maximum harvest is $H(60) =$ ______ thousand animals.

---

49. s, 15t, 10t, 65 - 10t, $(15t)^2 + (65 - 10t)^2$, $30(15t) - 20(65 - 10t)$, 2, $\frac{325t - 650}{s}$, negative, positive, minimum, 45, $15\sqrt{13}$

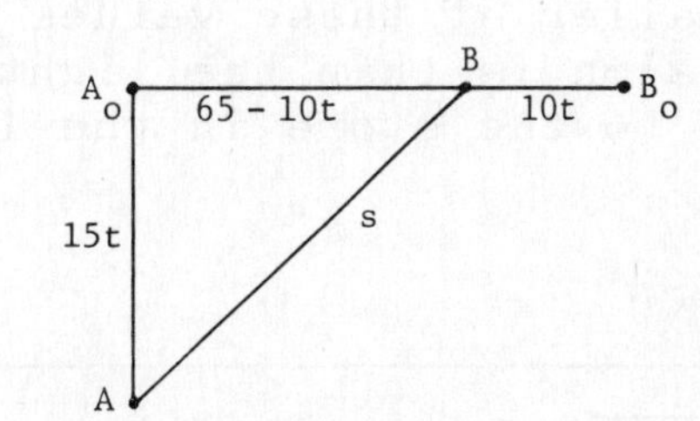

50. C(x), $40x - 0.5x^2 - 3$, 0, $40 - x$, 40, -1, 797

51. $-0.025P^2 + 3P$, 0, $-0.05P + 3$, 60, -0.05, negative, 90

52. Determine the point on the ellipse $4x^2 + 9y^2 = 36$ that is nearest the origin.

Solution. The problem is to ________ the distance s from a point (x,y) on the ellipse to the origin. That is, find the minimum of $s =$ ________ subject to the auxiliary condition that $4x^2 + 9y^2 = 36$. We can just as well minimize $s^2 = S$ since that will also minimize s. Since $S = x^2 + y^2$ is a function of both the variables x and y, we use the equation of the ellipse to eliminate the variable y: $y^2 = 4 - \frac{4}{9}x^2$ so that substitution gives

$$S(x) = x^2 + y^2 = \text{\_\_\_\_\_\_\_\_\_\_}.$$

The minimum distance occurs when $dS/dx =$ ________, so $(10/9)x = 0$ or $x =$ ________ . Since $S''(x) > 0$ this value of x yields a ________ . When $x = 0$, $y^2 = 4$ on the ellipse so that $(0,2)$ and $(0,-2)$ are the points on the ellipse that are nearest to the origin.

53. A lighthouse is at a point A, 4 miles offshore from the nearest point 0 of a straight beach; a store is at point B, 4 miles down the beach from 0. If the lighthouse keeper can row 4 miles/hour and walk 5 miles/hour, find the point C on the beach to which the lighthouse keeper should row to get from the lighthouse to the store in the least possible time.

Solution. The information is sketched in the figure at the right. From the diagram and the Pythagorean theorem, the distance from A to C is ________ . The total time required to get from A to C to B is

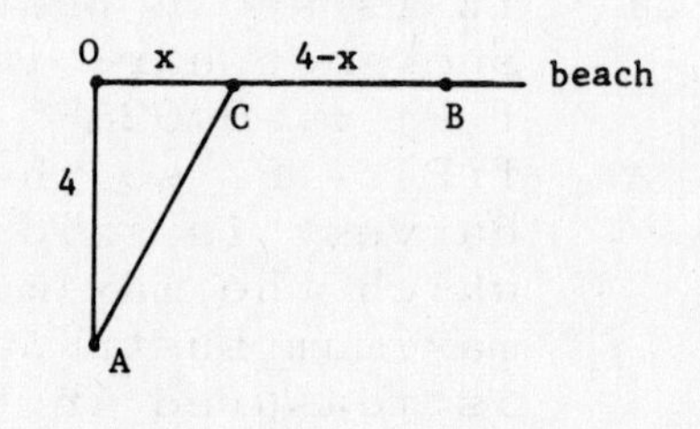

$$T = \frac{\sqrt{16 + x^2}}{4} + \left(\frac{\text{\_\_\_\_}}{5}\right),$$

where $0 \le x \le 4$. The minimum time occurs when $dT/dx = 0$, or

$$0 = T'(x) = \text{\_\_\_\_\_\_\_\_\_\_}.$$

Simplifying algebraically, $5x =$ ________ or $x^2 =$ ________ or $x =$ ________ . However, $x = \frac{16}{3}$ is outside the allowable range of values $0 \le x \le 4$. Therefore, the minimum must be taken on at one of the ________ of the interval. Checking each point, $T(0) =$ ________ hours and $T(4) =$ ________ hours. The smaller of these values occurs when $x =$ ________ so our conclusion is that the lighthouse keeper should row all the way to get to the store in the least possible time.

---

52. minimize, $\sqrt{x^2 + y^2}$, $\frac{5}{9}x^2 + 4$, 0, 0, minimum

53. $\sqrt{16 + x^2}$, $4 - x$, $\frac{1}{8}(16 + x^2)^{-1/2}(2x) - \frac{1}{5}$, $4\sqrt{16 + x^2}$, $\frac{256}{9}$, $\frac{16}{3}$, endpoints, 1.8, $\sqrt{2}$, 4

54. The cost per hour of driving a ship through the water varies approximately as the cube of its speed in the water. Suppose a ship runs into a current of V miles per hour, measured relative to the ocean bottom. Find the total cost for the ship to travel M miles, and find the most economical speed of the ship relative to the ocean bottom.

Solution. Let x denote the speed of the ship relative to the water. Then ________ will be its speed relative to the bottom. The time taken to travel M miles will be ________ . The cost per hour in fuel will be kx^3 for some constant of proportionality k, so the total cost function is given by

$C(x) =$ ________ .

To find the most economical speed, minimize the cost. Now,

$C'(x) =$ ________________ .

The minimum cost occurs when $dC/dx = 0$, or $kMx^2[3(x - V) - x] = 0$. Thus, $x =$ ________ or $x =$ ________ . Since $x = 0$ is ruled out if the ship moves, and since $C(x) \to +\infty$ as $x \to V^+$, we see that $x = 1.5V$ must provide the minimum cost.

3.7 LINEARIZATIONS AND DIFFERENTIALS

OBJECTIVE A: Given a function $y = f(x)$ and a point $x = a$, find the linearization of $f(x)$ at a. Use your linearization to estimate a specified function value.

55. If $y = f(x)$ is differentiable at $x = a$, then the linearization of f at a is given by

$L(x) =$ ________________ .

56. To find the linearization to $f(x) = \frac{1}{2}x^2 - 7x + 9$ at $x = 4$, first calculate the derivative $f'(x) =$ ________ . The value $f'(4) =$ ________ is the slope of the linearization at the point $(4,$ ______$)$ on the graph of f. Thus, an equation of the linearization is $L(x) = -11 +$ ________ $(x - 4)$, or $L(x) =$ ________ .

57. From the result in Problem 56, an estimate to $\frac{1}{2}\left(\frac{1}{6}\right)^2 - \frac{7}{6} + 9$ is ____________ .

OBJECTIVE B: Given $y = f(x)$, find the differential dy.

58. If $y = x^2 + \sin 3x$, then $\frac{dy}{dx} =$ ____________ . Thus, $dy =$ ____________ .

---

54. $x - V$, $M/(x - V)$, $\frac{kMx^3}{x - V}$, $\frac{(x - V)3kMx^2 - kMx^3}{(x - V)^2}$, 0, 1.5V

55. $f(a) + f'(a)(x - a)$

56. $x - 7$, -3, -11, -3, $-3x + 1$

57. $-3\left(\frac{1}{6}\right) + 1 = \frac{1}{2}$

58. $2x + 3\cos 3x$, $(2x + 3\cos 3x)\,dx$

59. If $y = f(x)$ is differentiable at $x = a$, and x changes from a to $a + \Delta x$, the error $|\Delta y - dy|$ in the approximation $f'(a)\Delta x$ is given by ________ , where ________ as $\Delta x \to 0$.

OBJECTIVE C: Estimate the change Δf produced in a function $y = f(x)$ when $x = x_0$ changes by a small amount dx.

60. An estimate of $\Delta f = f(x_0 + \Delta x) - f(x_0)$ is given by the <u>differential</u> $df =$ ________ . Thus, df denotes the change in the linearization of f that results from the change dx in x.

61. The equation $\frac{df}{dx} = f'(x)$ says we may regard the derivative as a ________ of differentials.

62. Suppose we wish to estimate the change in $y = x^3$ when x changes by $dx = 0.1$ at $x = 2$. Now $\Delta y \approx dy = \frac{dy}{dx}\,dx$ $=$ ________ dx. When $x = 2$ and $dx = 0.1$, $dy = 3(\_\_\_\_)^2(\_\_\_\_) = 1.2$. Therefore, since $f(x + dx) = y + \Delta y \approx y + dy$, $(2 + 0.1)^3 \approx 2^3 +$ ________ $=$ ________ . The actual value of $(2.1)^3$ is ________ giving an error in our estimate of $\epsilon \cdot dx = \Delta y - dy =$ ________ . The positive sign of $\epsilon \cdot dx$ indicates that our estimate 9.2 is too small.

63. To estimate the value of $\sqrt{16.56}$, let $y = \sqrt{x}$, $x = 16$, and $dx = 0.56$. Then $dy = (\_\_\_\_\_\_)dx$, so when $x = 16$ and $dx = 0.56$, $dy = (\_\_\_\_\_\_)(0.56) = .07$. Thus, $\sqrt{16.56} = \sqrt{\_\_\_\_}$ $+ .07 =$ ________ . (The actual value of $\sqrt{16.56}$ is 4.0694 correct to 5 decimal places, so our estimate is fairly accurate.)

3.8 NEWTON'S METHOD

OBJECTIVE: Use Newton's method to estimate the root of an equation $f(x) = 0$ within specified $a \le x \le b$.

64. In using Newton's method, to go from the nth approximation x_n of the root to the next approximation x_{n+1}, use the formula $x_{n+1} =$ ________ . This formula fails if the derivative $f'(x_n)$ equals ________ .

---

59. $|\epsilon\Delta x|$, $\epsilon \to 0$

60. $f'(x_0)\Delta x$ or $f'(x_0)\,dx$

61. quotient

62. $3x^2$, 2, 0.1, 1.2, 9.2, 9.261, 0.061

63. $\frac{1}{2\sqrt{x}}$, $\frac{1}{8}$, 16, 4.07

64. $x_n - \frac{f(x_n)}{f'(x_n)}$, 0

65. Suppose it is required to find a real root to the equation $f(x) = x^3 + x - 1$. Since $f(0) =$ ______ and $f(1) =$ ______ differ in sign, an unknown root lies somewhere in the interval $0 < x < 1$. As a first guess, choose $x_1 = 0.5$. To apply Newton's formula, we calculate the derivative $f'(x) =$ ______________ . Then,

$$x_2 = x_1 - \frac{x_1^3 + x_1 - 1}{3x_1^2 + 1} = \frac{1}{2} - \frac{(1/2)^3 + (1/2) - 1}{\rule{3cm}{0.4pt}} = \frac{1}{2} + \frac{\rule{1cm}{0.4pt}}{14}$$

$$= \frac{\rule{1cm}{0.4pt}}{7} \approx 0.71429.$$

$$x_3 = x_2 - \frac{\rule{2cm}{0.4pt}}{3x_2^2 + 1} = \frac{5}{7} - \frac{\rule{2cm}{0.4pt}}{3(5/7)^2 + 1}$$

$$= \frac{5}{7} - \frac{\rule{1cm}{0.4pt}}{7 \cdot 124} = \frac{\rule{1cm}{0.4pt}}{7 \cdot 124} \approx 0.68318 .$$

With the aid of a calculator, we have computed the following iterations in the same way: $x_4 = 0.68233$ and $x_5 = 0.68233$. Thus, a root to $f(x) = x^3 + x - 1$ is $r = 0.68233$ correct to 5 decimal places. The method is easy, but the arithmetic can be cumbersome without the aid of a calculator.

66. The speed with which Newton's method converges to a root r is expressed by the formula

$$|r - x_{n+1}| \le \rule{5cm}{0.4pt}$$

in an interval surrounding r. For the function $f(x) = x^3 + x - 1$ in Problem 65, on $0 < x < 1$,

$$\min f'(x) = \min(3x^2 + 1) = \rule{2cm}{0.4pt} \quad \text{and}$$

$$\max f''(x) = \max 6x = \rule{2cm}{0.4pt} .$$

Thus,

$$|r - x_{n+1}| \le \frac{1}{2}\left|\frac{6}{1}\right|(r - x_n)^2 .$$

---

65. -1, 1, $3x^2 + 1$, $3\left(\frac{1}{2}\right)^2 + 1$, 3, 5, $x_2^3 + x_2 - 1$, $\left(\frac{5}{7}\right)^3 + \left(\frac{5}{7}\right) - 1$, 27, 593

66. $\frac{1}{2}\frac{\max|f''|}{\min|f'|}(r - x_n)^2$, 1, 6

CHAPTER 3 SELF-TEST

1. Find the absolute maximum and minimum values (if they exist) of $f(x) = x^3 - x^2 - x + 2$ over the interval $0 \le x < 2$.

In Problems 2-4, sketch the curves. Find the intervals of values of x for which the curve is increasing, decreasing, concave upward, and concave downward. Locate all asymptotes.

2. $y = \dfrac{x}{\sqrt{1 + x^2}}$

3. $y = \dfrac{4x}{x^2 + 1}$

4. $y = 1 - (x + 1)^{1/3}$

5. Apply Rolle's Theorem to show that the equation $\cos x = \sqrt{x}$, $x \ge 0$, has exactly one real solution.

6. Find all numbers c which satisfy the conclusion of the Mean Value Theorem for $f(x) = 1 + 2x^2$ over $-1 \le x \le 1$.

7. Let $f(x) = \frac{1}{x}$. Show that there is no c in the interval $-1 < x < 2$ such that $f'(c) = \dfrac{f(2) - f(-1)}{2 - (-1)}$. Explain why this does <u>not</u> contradict the Mean Value Theorem.

8. Find a reasonable estimate to $\sqrt[3]{25}$.

9. Evaluate the limits.

   (a) $\lim_{t \to \infty} \dfrac{t^2}{4 - t^2}$

   (b) $\lim_{x \to \infty} \frac{1}{x} \cos x^2$

10. Suppose a company can sell x items per week at a price $P = 200 - 0.01x$ cents, and that it costs $C = 50x + 20{,}000$ cents to produce the x items. How much should the company charge per item in order to maximize its profits?

11. The weight W (lbs/sec) of flue gas passing up a chimney at different temperatures T is represented by

$$W = A(T - T_0)(1 + \alpha T)^{-2},$$

where A is a positive constant, T the absolute temperature of the hot gases passing up the chimney, T_0 the temperature of the outside air (all in °C), and $\alpha = 1/273$ is the coefficient of expansion of the gas. For a given $T_0 = 15$°C, find the temperature T at which the greatest amount of gas will pass up the chimney.

12. Find the linearization of $f(x) = \sqrt{x + \frac{1}{x}}$ at $x = 4$.

13. Use the linearization $L(x)$ to estimate the value of $\sin 29°$.

14. Beginning with the estimate $x_1 = \frac{\pi}{2}$, apply Newton's method once to calculate a positive solution to the equation $\sin x = \frac{2}{3}x$.

SOLUTIONS TO CHAPTER 3 SELF-TEST

1. $f(x) = x^3 - x^2 - x + 2$ for $0 \leq x < 2$, and $f'(x) = 3x^2 - 2x - 1 = (3x + 1)(x - 1)$. Thus, $f'(x) = 0$ implies $x = -\frac{1}{3}$ or $x = 1$. Then $x = 1$ is the only critical point in the interval $[0,2)$. Next note that $f'(x) < 0$ in $[0,1)$ so f is decreasing to the left of $x = 1$, and $f'(x) > 0$ in $(1,2)$ so f is increasing to the right of $x = 1$. Also, $f(0) = 2$, $f(1) = 1$ and $f(2) = 4$. Since $x = 2$ is not in the interval, there is no absolute maximum. The absolute minimum value is $f(1) = 1$ (which is also a relative minimum).

2. $y = \dfrac{x}{(1 + x^2)^{1/2}}$

$$y' = \frac{(1 + x^2)^{1/2} - x \cdot \frac{1}{2}(1 + x^2)^{-1/2}2x}{1 + x^2} = \frac{1}{(1 + x^2)^{3/2}}, \text{ and}$$

$$y'' = \frac{-3x}{(1 + x^2)^{5/2}}.$$

Note that $y = \dfrac{1}{(1/x^2 + 1)^{1/2}}$ for $x \geq 0$ $\left(\text{since } \sqrt{x^2} = |x|\right)$

and that $y = \dfrac{-1}{(1/x^2 + 1)^{1/2}}$ for $x < 0$. Thus $\lim_{x\to\infty} y = 1$ and $\lim_{x\to-\infty} y = -1$. Hence the lines $y = 1$ and $y = -1$ are horizontal asymptotes. Since y' exists for all x and is never zero, there are no critical points. At $x = 0$, $y'' = 0$ so that $x = 0$ is a point of inflection where the graph has slope 1. On $(-\infty, 0)$ $y'' > 0$ and the function is concave upward; on $(0, \infty)$ it is concave downward. Since $y' > 0$ for all x, the graph is everywhere an increasing function of x. This information yields the graph sketched at the right.

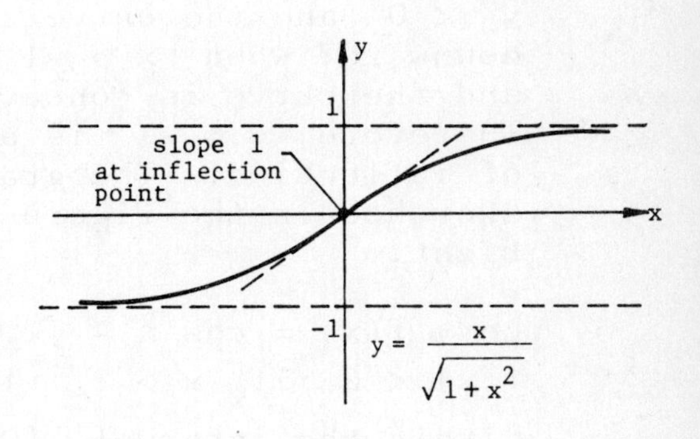

3. Since $\lim_{x\to\pm\infty} y = \lim_{x\to\pm\infty} \dfrac{4/x}{1 + 1/x^2} = 0$, the x-axis is a horizontal asymptote. Next, $y' = \dfrac{4(1 - x^2)}{(x^2 + 1)^2}$ and $y'' = \dfrac{8x(x^2 - 3)}{(x^2 + 1)^3}$.

Hence, $y' = 0$ implies $x = \pm 1$. Since $y'' > 0$ at $x = -1$ and $y'' < 0$ at $x = 1$, it follows from the second derivative test that $y(-1) = -2$ is a relative minimum and $y(1) = 2$ is a relative maximum.

Next, $y'' = 0$ when $x = 0$, $-\sqrt{3}$, and $\sqrt{3}$ so that these values for x are points of inflection. Moreover, for

$x < -\sqrt{3}$, $y'' < 0$ and the graph of y is concave downward;

$-\sqrt{3} < x < 0$, $y'' > 0$ and the graph of y is concave upward;

$0 < x < \sqrt{3}$, $y'' < 0$ and the graph of y is concave downward;
$x > \sqrt{3}$, $y'' > 0$ and the graph of y is concave upward.

Note that at $x = 0$, $y' = 4$. The graph of y is sketched at the right. Note the symmetry about the origin.

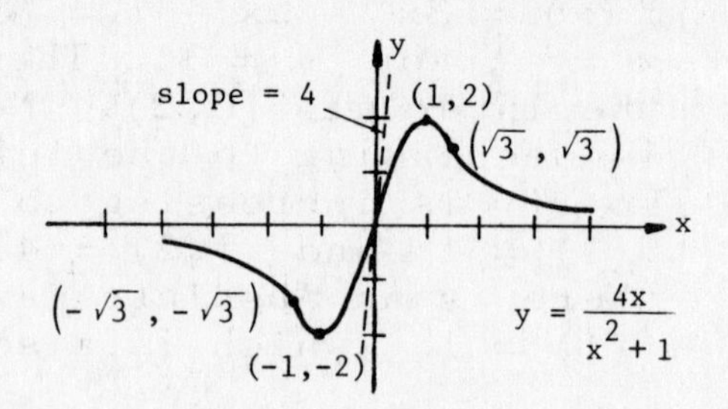

4. $y = 1 - (x + 1)^{1/3}$, $y' = -\frac{1}{3}(x + 1)^{-2/3}$, and $y'' = \frac{2}{9}(x + 1)^{-5/3}$.

The derivative y' does not exist when $x = -1$, although the curve y is continuous at $x = -1$. Since

$$\lim_{x\to-1} \frac{dx}{dy} = \lim_{x\to-1} -3(x + 1)^{2/3} = 0,$$

the graph has a vertical tangent at $x = -1$. Since $y' < 0$ for all $x \neq -1$, the curve is everywhere decreasing.

We note that y'' is never zero. However, y'' fails to exist at $x = -1$. When $x < -1$, $y'' < 0$ and the curve is concave downward; when $x > -1$, $y'' > 0$ and the curve is concave upward. Therefore $x = -1$ is a point of inflection. The graph is sketched in the figure at the right.

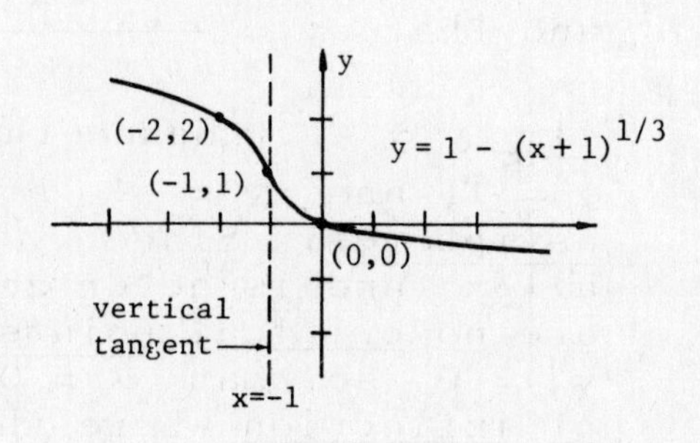

5. Let $f(x) = \cos x - \sqrt{x}$. Since $|\cos x| \leq 1$, we see that $f(x) < 0$ if $x > 1$. Thus, the only possible root must lie within the interval $[0, 1]$. Now, $f(0) = 1$ and $f(\frac{\pi}{2}) = -\sqrt{\frac{\pi}{2}}$ so the Intermediate Value Theorem guarantees a root in the interval $\left[0, \frac{\pi}{2}\right]$: we know in fact that the root must lie in $[0, 1]$. Calculating the derivative, $f'(x) = -\sin x - \frac{1}{2\sqrt{x}}$, we see that f' is negative in the interval $(0, 1)$. Since f' is different from zero for all values of x between 0 and 1, we conclude there is exactly one real root to the equation $f(x) = 0$ for $x \geq 0$.

6. $f(-1) = 3$ and $f(1) = 3$; $f'(x) = 4x$. Hence $\frac{f(1) - f(-1)}{1 - (-1)} = f'(c)$ translates into $0 = 4c$, or $c = 0$.

7. $\frac{f(2) - f(-1)}{2 - (-1)} = \frac{\frac{1}{2} - (-1)}{3} = \frac{1}{2}$ and $f'(c) = -\frac{1}{c^2}$. Since $-\frac{1}{c^2} = \frac{1}{2}$ is impossible to solve for real values of c, there is no number c in the interval $(-1, 2)$ satisfying the conclusion of the Mean Value Theorem. However, this does not contradict the Theorem because $f(x) = 1/x$ is not continuous over the closed interval $[-1, 2]$: it fails to be continuous at $x = 0$. Thus the hypotheses of the theorem are not satisfied.

8. Let $f(x) = x^{1/3}$. Then by the Mean Value Theorem, $f(25) \approx f(27) + (25 - 27)f'(27)$ or,
$\sqrt[3]{25} \approx \sqrt[3]{27} + (-2)\frac{1}{3(27)^{2/3}} = 3 - \frac{2}{3 \cdot 9} = \frac{81 - 2}{27} = \frac{79}{27} \approx 2.926.$

(A calculator gives $\sqrt[3]{25} \approx 2.924017738.$)

9. (a) $\lim_{t\to\infty} \frac{t^2}{4 - t^2} = \lim_{h\to 0} \frac{\frac{1}{h^2}}{4 - \frac{1}{h^2}} = \lim_{h\to 0} \frac{1}{4h^2 - 1} = \frac{1}{0 - 1} = -1$

(b) $0 \le \left|\frac{1}{x} \cos x^2\right| \le \frac{1}{|x|}$ because $|\cos x^2| \le 1$ for all values of x. Since $\lim_{x\to\infty} \frac{1}{|x|} = 0$, we have $\lim_{x\to\infty} \frac{1}{x} \cos x^2 = 0$.

10. Let Q denote the profit function. Then, $Q(x) = xP - C = 150x - 0.01x^2 - 20{,}000$. The maximum occurs when $dQ/dx = 0$, or $150 - .02x = 0$; thus $x = 7500$ items. Since $d^2Q/dx^2 = -.02 < 0$, this provides a maximum profit. The price per item is then given by $P(7500) = 200 - (.01)(7500) = 125$ cents, the price required to obtain the maximum profit $Q(7500) = \$5{,}425.00$.

11. We want to maximize the weight function W. Now,

$$W = A(T - T_0)(1 + \alpha T)^{-2}$$

$$\frac{dW}{dT} = A(1 + \alpha T)^{-2} - 2A\alpha(T - T_0)(1 + \alpha T)^{-3}.$$

Setting $dW/dt = 0$, and simplifying algebraically, gives $(1 + \alpha T) - 2\alpha(T - T_0) = 0$, or $T = (1 + 2\alpha T_0)/\alpha$. Thus, for $T_0 = 15^\circ C$ and $\alpha = 1/273$ as given,

$$T = \frac{1}{\alpha} + 2T_0 = 273 + 30 = 303^\circ C .$$

Since $\frac{dW}{dT} > 0$ if $T < 303$, and $\frac{dW}{dT} < 0$ if $T > 303$, it is clear that $T = 303^\circ$ provides an absolute maximum for W.

12. $f'(x) = \frac{1}{2}\left(x + \frac{1}{x}\right)^{-1/2} \cdot \frac{d}{dx}\left(x + \frac{1}{x}\right) = \frac{1}{2}\left(x + \frac{1}{x}\right)^{-1/2}\left(1 - \frac{1}{x^2}\right)$

Thus, $f'(4) = \frac{1}{2}\left(4 + \frac{1}{4}\right)^{-1/2}\left(1 - \frac{1}{16}\right) = \frac{15}{16\sqrt{17}} \approx 0.227$

and $f(4) = \sqrt{\frac{17}{4}} \approx 2.062$. Therefore, $L(x) = \frac{\sqrt{17}}{2} + \frac{15}{16\sqrt{17}}(x - 4)$
$\approx 2.062 + 0.227(x - 4)$.

13. The calculation must be done when $y = \sin x$ for x measured in <u>radians</u>. Thus,

$$\sin 29^\circ \approx \sin \frac{\pi}{6} + dy ,$$

where $dy = \frac{dy}{dx}\,dx$ when $x = \frac{\pi}{6}$ and $dx = -\frac{\pi}{180}$ radians.

Now, $\left.\frac{dy}{dx}\right|_{\pi/6} = \cos \frac{\pi}{6} = \frac{\sqrt{3}}{2}$ so that

$$\sin 29^\circ \approx \frac{1}{2} + \left(\frac{\sqrt{3}}{2}\right)\left(-\frac{\pi}{180}\right) \approx .48489 .$$

14. Let $f(x) = \sin x - \frac{2}{3}x = 0$, $f'(x) = \cos x - \frac{2}{3}$. By Newton's method, $x_2 = x_1 - \frac{\sin x_1 - (2/3)x_1}{\cos x_1 - (2/3)} = \frac{\pi}{2} - \frac{1 - \pi/3}{0 - 2/3}$

$$= \frac{\pi}{2} + \frac{3}{2}\left(1 - \frac{\pi}{3}\right) = 1.5.$$

CHAPTER 4 INTEGRATION

4.1 INDEFINITE INTEGRALS

OBJECTIVE A: Find an antiderivative for a given function based on your knowledge of derivatives of elementary power and trigonometric functions.

1. If $F'(x) = f(x)$ at every point of an interval I, then the function $F(x)$ is known as an ______________ of the function $f(x)$ over the interval I.

2. An antiderivative of the function x^n, $n \neq -1$, is ______________.

3. An antiderivative of the function $\cos kx$ is ______________.

4. An antiderivative of $3x^4 - x^{-2} + 5$ is ______________.

5. An antiderivative of $x^{3/2} + x^{1/2} + \frac{1}{x^2}$ is ______________.

6. An antiderivative of $\cos \frac{x}{2} + \sec^2 3x$ is ______________.

OBJECTIVE B: Evaluate indefinite integrals using Table 4.1.

7. If the function f is a derivative, then the indefinite integral of f is ______________. It is denoted by the symbol ______________.

8. If $F(x)$ is an antiderivative of f, then

$$\int f(x)\,dx = \_\_\_\_\_\_\_\_\_\_\_\_ .$$

---

1. antiderivative
2. $\frac{x^{n+1}}{n+1}$
3. $\frac{1}{k} \sin kx$
4. $\frac{3}{5}x^5 + x^{-1} + 5x$
5. $\frac{2}{5}x^{5/2} + \frac{2}{3}x^{3/2} - x^{-1}$
6. $2 \sin \frac{x}{2} + \frac{1}{3} \tan 3x$
7. the set of all antiderivatives of f, $\int f(x)\,dx$
8. $F(x) + C$

9. $\int (3x^5 - x^2 + \sin \pi x)\, dx$

$$= 3 \int x^5 dx - \underline{\hspace{3cm}} + \int \sin \pi x\, dx$$

$$= 3(\underline{\hspace{2.5cm}}) - \frac{1}{3}x^3 + (\underline{\hspace{2.5cm}}) + C .$$

10. $\int \left(x^{3/2} - 2x^{2/3} + 5\right) \sqrt{x}\, dx = \int \left(x^2 - \underline{\hspace{1.5cm}} + 5x^{1/2}\right) dx$

$$= (\underline{\hspace{1.2cm}}) - 2(\underline{\hspace{1.5cm}}) + 5(\underline{\hspace{1.5cm}}) + C$$

$$= \underline{\hspace{4cm}} .$$

11. $\int \left(\frac{1}{\sqrt{x}} + x\right) dx = \int x^{-1/2} dx + \underline{\hspace{2.5cm}}$

$$= \underline{\hspace{2.5cm}} + \frac{1}{2}x^2 + C$$

$$= 2\sqrt{x} + \frac{1}{2}x^2 + C .$$

12. $\int (\sin^2 x - \sec^2 x)\, dx = \int \sin^2 x\, dx - \int \sec^2 x\, dx$

$$= \underline{\hspace{6cm}} .$$

4.2 DIFFERENTIAL EQUATIONS, INITIAL VALUE PROBLEMS AND MATHEMATICAL MODELING

OBJECTIVE A: Solve elementary initial value problems.

13. Suppose that $\frac{dy}{dx} = \frac{1}{x^3} + 2x$, $x > 0$ and that $y = \frac{5}{2}$ when $x = 1$. To solve for y we first solve the differential equation by finding an antiderivative of $\frac{1}{x^3} + 2x$. The indefinite integral is

$$y(x) = \underline{\hspace{4cm}} .$$

Then we substitute $x = 1$ and $y = \frac{5}{2}$ to find the constant C:

$$\frac{5}{2} = -\frac{1}{2}(1)^{-2} + \underline{\hspace{2cm}} + C$$

or, solving for C,

---

9. $\int x^2 dx$, $\frac{x^6}{6}$, $-\frac{1}{\pi} \cos \pi x$

10. $2x^{7/6}$, $\frac{x^3}{3}$, $\frac{x^{13/6}}{13/6}$, $\frac{x^{3/2}}{3/2}$, $\frac{1}{3}x^3 - \frac{12}{13}x^{13/6} + \frac{10}{3}x^{3/2} + C$

11. $\int x\, dx$, $\frac{x^{1/2}}{1/2}$

12. $\left(\frac{x}{2} - \frac{\sin 2x}{4}\right) - \tan x + C$

$C =$ ____________ .

The function we want is then $y = -\frac{1}{2}x^{-2} + x^2 + 2$.

14. Solve the following initial value problem for y as a function of x:

Differential equation: $\frac{d^2y}{dx^2} = 2x - 1$

Initial conditions: $\frac{dy}{dx} = 1$ and $y = 0$ when $x = -1$

Solution. We integrate the differential equation with respect to x to find dy/dx:

$$\int \frac{d^2y}{dx^2}\,dx = \int (2x - 1)\,dx$$

$$\frac{dy}{dx} = \underline{\qquad\qquad} + C_1$$

We apply the initial condition $\frac{dy}{dx} = 1$ when $x = -1$ to find C_1:

$$1 = \underline{\qquad\qquad} + C_1 \quad \text{or} \quad C_1 = -1.$$

Thus, $\frac{dy}{dx} =$ ____________ .

We integrate dy/dx with respect to x to find y:

$$\int \frac{dy}{dx}\,dx = \int (x^2 - x - 1)\,dx$$

$$\underline{\qquad\quad} = \frac{1}{3}x^3 - \frac{1}{2}x^2 - x + C_2 \,.$$

We apply the initial condition $y = 0$ when $x = -1$ to find C_2:

$$0 = \frac{1}{3}(-1)^3 - \frac{1}{2}(-1)^2 - (-1) + C_2 \,.$$

Thus, $C_2 =$ ________ . The formula for y as a function of x is:

$$y = \underline{\qquad\qquad\qquad\qquad} \,.$$

OBJECTIVE B: Given (a) the velocity and initial position, or (b) the acceleration, initial velocity, and initial position, find the position at any time t of a body moving on a coordinate (straight) line.

---

13. $-\frac{1}{2}x^{-2} + x^2 + C$, 1^2, 2

14. $x^2 - x$, $(-1)^2 - (-1)$, $x^2 - x - 1$, y, $-\frac{1}{6}$, $\frac{1}{3}x^3 - \frac{1}{2}x^2 - x - \frac{1}{6}$

15. The velocity of a body moving along a coordinate line is known to be $v = t + \cos t$. The initial position is $s = -1$ when $t = 0$. To find the body's position at any time t, we first solve the differential equation $\frac{ds}{dt} = t + \cos t$. Thus,

$$s = \_\_\_\_\_\_\_\_\_\_\_\_ .$$

To evaluate the constant C, we substitute $s = -1$ and $t = 0$:

$$-1 = \tfrac{1}{2}(0)^2 + \_\_\_\_\_\_\_\_\_\_\_\_ + C$$

or, solving for C,

$$C = \_\_\_\_\_\_\_\_ .$$

The body's position at time t is $s = \frac{1}{2}t^2 + \sin t - 1$.

16. Suppose the acceleration of a moving body is given by the equation $a = \sqrt{t}$, and that when $t = 0$ it is known that the body has initial velocity $v = v_0$ and initial position $s = s_0$. Let us find its position at any time. Since $a = dv/dt$, substitution gives $dv/dt = \sqrt{t}$. The indefinite integral is $v = \_\_\_\_\_\_ + C_1$. Imposing the initial condition $v = v_0$ when $t = 0$ yields $v_0 = \_\_\_\_\_\_$; so $C_1 = \_\_\_\_$. Hence the velocity equation becomes $v = \_\_\_\_\_\_\_$. Now, $v = ds/dt$ so the last equation can be written

$$\frac{ds}{dt} = \frac{2}{3}t^{3/2} + v_0 .$$

We next solve this differential equation to find s:

$$s = \_\_\_\_\_\_\_\_\_\_\_\_ + C_2 .$$

From the initial condition $s = s_0$ when $t = 0$ it is readily seen that $C_2 = \_\_\_\_$. Therefore, the body's position is $s = \_\_\_\_\_\_\_$ valid for all $t \geq 0$.

4.3 INTEGRATION BY SUBSTITUTION -- RUNNING THE CHAIN RULE BACKWARD

OBJECTIVE: Evaluate indefinite integrals using the substitution method of integration to reduce the integrals to standard form.

17. To evaluate $\int (3x - 1)^2\,dx$ substitute $u = \_\_\_\_\_\_$ and $du = 3\,dx$. Then,

---

15. $\frac{1}{2}t^2 + \sin t + C$, $\sin 0$, -1

16. $\frac{2}{3}t^{3/2}$, $\frac{2}{3}(0)^{3/2} + C_1$, v_0, $\frac{2}{3}t^{3/2} + v_0$, $\frac{4}{15}t^{5/2} + v_0 t$, s_0, $\frac{4}{15}t^{5/2} + v_0 t + s_0$

$$\int (3x - 1)^2\,dx = \frac{1}{3}\int (3x - 1)^2 \cdot 3\,dx$$

$$= \frac{1}{3}\int \underline{\hspace{3cm}} = \underline{\hspace{3cm}}$$

$$= \frac{1}{9}(3x - 1)^3 + C \quad \text{(Replace u by } \underline{\hspace{2cm}}\text{.)}$$

18. To evaluate $\int 5x\cos(x^2 + 1)\,dx$ substitute $u =$ ________ and $du =$ ________ . Then,

$$\int 5x\cos(x^2 + 1)\,dx = \frac{5}{2}\int \cos(x^2 + 1) \cdot 2x\,dx$$

$$= \frac{5}{2}\int \underline{\hspace{3cm}}$$

$$= \underline{\hspace{3cm}}$$

$$= \frac{5}{2}\sin(x^2 + 1) + C \quad \text{(Replace u by } \underline{\hspace{1cm}}\text{.)}$$

19. $\int \sin(3 - 2x)\,dx$. Let $u = 3 - 2x$. Then $du =$ ________ so $dx = -\frac{1}{2}du$. Thus the integral becomes

$$\int \sin(3 - 2x)\,dx = \int \underline{\hspace{3cm}} = \frac{1}{2}\int \underline{\hspace{3cm}}$$

$$= \frac{1}{2}\underline{\hspace{3cm}} + C = \underline{\hspace{3cm}}.$$

20. $\int x^2\cos(4x^3)\,dx$. Let $u = 4x^3$. Then $du =$ ________ so $x^2\,dx =$ ________ . Thus the integral becomes

$$\int x^2\cos(4x^3)\,dx = \int \cos(4x^3) \cdot x^2\,dx = \int \underline{\hspace{2cm}}\,du$$

$$= \frac{1}{12}\underline{\hspace{2cm}} + C = \underline{\hspace{3cm}}.$$

21. $\int (3 - \sin 2t)^{1/3}\cos 2t\,dt$. Let $u = 3 - \sin 2t$. Then $du =$ ________ . Substitution into the integral gives

$$\int (3 - \sin 2t)^{1/3}\cos 2t\,dt = \int \underline{\hspace{2cm}}\,du = \underline{\hspace{2cm}} + C$$

$$= \underline{\hspace{4cm}}.$$

---

17. $3x - 1$, $u^2\,du$, $\frac{1}{9}u^3 + C$, $3x - 1$

18. $x^2 + 1$, $2x\,dx$, $\cos u\,du$, $\frac{5}{2}\sin u + C$, $x^2 + 1$

19. $-2dx$, $\sin u \cdot \left(-\frac{1}{2}\right)du$, $-\sin u\,du$, $\cos u$, $\frac{1}{2}\cos(3 - 2x) + C$

20. $12x^2\,dx$, $\frac{1}{12}du$, $\frac{1}{12}\cos u$, $\sin u$, $\frac{1}{12}\sin(4x^3) + C$

21. $-2\cos 2t\,dt$, $-\frac{1}{2}u^{1/3}\,du$, $-\frac{3}{8}u^{4/3}$, $-\frac{3}{8}(3 - \sin 2t)^{4/3} + C$

22. $\int \sec^{5/2} x \tan x\,dx$. Let $u = \sec x$. Then $du =$ ________ .

Now, $\sec^{5/2} x \tan x = \sec^{3/2} x \cdot$ ________ . Hence substitution into the integral gives,

$$\int \sec^{5/2} x \tan x\,dx = \int \text{\_\_\_\_\_\_}\,du = \text{\_\_\_\_\_\_\_\_} + C$$

$$= \text{\_\_\_\_\_\_\_\_\_\_\_\_\_\_} .$$

23. $\int x\sqrt{3x^2 + 1}\,dx$. Let $u = 3x^2 + 1$. Then $du =$ ________ so $x\,dx = \frac{1}{6}\,du$. Thus the integral becomes

$$\int x\sqrt{3x^2 + 1}\,dx = \int \text{\_\_\_\_\_\_\_\_} = \frac{1}{6}\,\text{\_\_\_\_\_\_\_\_} + C$$

$$= \text{\_\_\_\_\_\_\_\_} .$$

24. $\int \frac{1}{z^2 - 6z + 9}\,dz$. Let $u = z - 3$ so $u^2 =$ ________ and $du = dx$. Thus the integral becomes

$$\int \frac{1}{z^2 - 6z + 9}\,dz = \int \text{\_\_\_\_\_\_\_\_}\,du = \text{\_\_\_\_\_\_\_} + C$$

$$= \frac{1}{3 - z} + C .$$

4.4 ESTIMATING WITH FINITE SUMS

OBJECTIVE: Estimate a final result (e.g., distance traveled, area under a curve, volume of a solid, average value of a function) by summing a finite number of close estimates made with a standard formula.

25. Use the following table to estimate the distance traveled by a car moving down a highway for one hour. Use the left-end beginning velocity values for each subinterval.

Time (min)	Velocity (mi/min)	Time (min)	Velocity (mi/min)
0	0.2	35	0.8
5	1.0	40	0.7
10	1.1	45	0.9
15	1.2	50	1.0
20	1.0	55	0.6
25	1.1	60	0.5
30	0.9		

---

22. $\sec x \tan x\,dx$, $\sec x \tan x$, $u^{3/2}$, $\frac{2}{5}u^{5/2}$, $\frac{2}{5}\sec^{5/2} x + C$

23. $6x\,dx$, $\frac{1}{6}\sqrt{u}\,du$, $\frac{2}{3}u^{3/2}$, $\frac{1}{9}(3x^2 + 1)^{3/2} + C$

24. $z^2 - 6z + 9$, u^{-2}, $-u^{-1}$

Solution. We use the standard formula

Distance = rate × time .

The total distance S traveled from $t = 0$ to $t = 60$ minutes is given by

$$S \approx (0.2)(5) + (1.0)(5) + (1.1)(5) + (1.2)(5) + (1.0)(5)$$
$$+ \underline{\qquad\qquad\qquad\qquad}$$
$$= 52.5 \text{ miles.}$$

26. Estimate the average value of the function $f(x) = 3x + 2$ over the interval $1 \le x \le 4$ by partitioning the interval into 6 subintervals of equal length and evaluating f at the subinterval midpoints.

Solution. The partition consists of the subintervals

1 1.5 2 2.5 3 3.5 4

The midpoints of the subintervals are 1.25, 1.75, ______, 2.75, 3.25, ______ . The subintervals are each of length $\Delta x = 1/2$ and the length of the entire interval is $4 - 1 = 3$. Thus,

$$\text{av}(f) \approx \frac{f(1.25) + f(1.75) + f(2.25) + f(2.75) + f(3.25) + f(3.75)}{2 \cdot 3}$$

$$= \frac{5.75 + \underline{\qquad} + \underline{\qquad} + 10.25 + 11.75 + \underline{\qquad}}{6}$$

$$= 9.5.$$

We find where f assumes this value by solving the equation $f(x) = 9.5$ for x:

$$3x + 2 = 9.5 \quad \text{implies} \quad x \approx \underline{\qquad\qquad} .$$

4.5 RIEMANN SUMS AND DEFINITE INTEGRALS

OBJECTIVE A: Interpret and utilize the sigma notation to express or write out sums, and determine (if possible) the value of a sum expressed in sigma notation.

27. To write out the sum $\sum_{k=3}^{7} 2^{k-2}$,

Replace the k in 2^{k-2} by 3 and obtain ______ .

Replace the k in 2^{k-2} by 4 and obtain ______ .

Replace the k in 2^{k-2} by 5 and obtain ______ .

Replace the k in 2^{k-2} by 6 and obtain ______ .

Replace the k in 2^{k-2} by 7 and obtain ______ .

---

25. (1.1)(5) + (0.9)(5) + (0.8)(5) + (0.7)(5) + (0.9)(5) + (1.0)(5) + (0.6)(5)

26. 2.25, 3.75, 7.25, 8.75, 13.25, 2.5

The expanded form is $\sum_{k=3}^{7} 2^{k-2} =$ ______________________ .

This finite sum is equal to ______ .

28. To express the finite sum $1 + \frac{1}{2} + \frac{1}{4} + \frac{1}{8} + \frac{1}{16}$ in sigma notation, we may observe that the sum can be written as

$$\left(\tfrac{1}{2}\right)^0 + \left(\tfrac{1}{2}\right)^1 + \left(\tfrac{1}{2}\right)^2 + (\_\_\_\_)^3 + (\_\_\_\_)^4 .$$

The kth term in this expression is $\left(\frac{1}{2}\right)^k$, and we see that k starts at ______ and ends at ______ . Therefore, the required sigma notation is ______________ .

29. To write out the sum $\sum_{k=0}^{3} \frac{(-1)^k}{k!} x^k$, first recall that $k!$ means $1\cdot2\cdot3\cdots k$. Also, $0! = 1$. Then,

Replace the k in $\frac{(-1)^k}{k!}$ by 0 and obtain ______ = ______ .

Replace the k in $\frac{(-1)^k}{k!}$ by 1 and obtain ______ = ______ .

Replace the k in $\frac{(-1)^k}{k!}$ by 2 and obtain ______ = ______ .

Replace the k in $\frac{(-1)^k}{k!}$ by 3 and obtain ______ = ______ .

The expanded form is $\sum_{k=0}^{3} \frac{(-1)^k}{k!} x^k =$ ______________________ .

Substitution of $x = 1$ gives the value ______ .

OBJECTIVE B: Apply the three important formulas for summing consecutive integers, their squares, or their cubes.

30. Three important formulas for sums are given by

$$\sum_{k=1}^{n} k = \_\_\_\_\_\_\_\_\_\_\_\_\_\_\_\_$$

$$\sum_{k=1}^{n} k^2 = \_\_\_\_\_\_\_\_\_\_\_\_\_\_\_\_$$

---

27. $2^1,\ 2^2,\ 2^3,\ 2^4,\ 2^5,\ 2^1 + 2^2 + 2^3 + 2^4 + 2^5,\ 62$

28. $\frac{1}{2},\ \frac{1}{2},\ 0,\ 4,\ \sum_{k=0}^{4} \left(\frac{1}{2}\right)^k$

29. $\frac{(-1)^0}{0!} x^0 = 1,\ \frac{(-1)^1}{1!} x^1 = -x,\ \frac{(-1)^2}{2^1} x^2 = \frac{1}{2} x^2,\ \frac{(-1)^3}{3!} x^3 = -\frac{1}{6} x^3,\ 1 - x + \frac{1}{2} x^2 - \frac{1}{6} x^3,\ \frac{1}{3}$

$\sum_{k=1}^{n} k^3 =$ ______________________ .

31\. To find $\sum_{k=1}^{6} (k + 3)$ we first use the sum rule and write

$$\sum_{k=1}^{6} (k + 3) = \sum_{k=1}^{6} k + \text{\_\_\_\_\_\_\_\_\_\_} .$$

Then, using the rule for the sum of the first 6 integers we substitute and find

$$\sum_{k=1}^{6} (k + 3) = \text{\_\_\_\_\_\_\_\_} + (3 + 3 + 3 + 3 + 3 + 3)$$

$$= 21 + \text{\_\_\_\_\_\_} = 39.$$

OBJECTIVE C: Express a limit of Riemann sums over an interval as a definite integral.

32\. Let $f(x)$ be a function defined on a closed interval $[a, b]$. The expression $\sum_{k=1}^{n} f(c_k)\Delta x_k$ is called a ______________ on the interval $[a, b]$. In this expression the interval has been partitioned

$$a = x_0 < x_1 < x_2 < \ldots < x_{n-1} < x_n = b$$

into n subintervals. The typical closed subinterval $[x_{k-1}, x_k]$ is called the ________________ . Its length is ________________ . The number c_k is <u>any</u> point lying within the ______________ .

33\. The ________ of a partition of $[a, b]$ is the length of the ________________ subinterval.

34\. The limit (provided it exists) of any Riemann sum for f as the norm of the partition tends to zero is called the ________________ over $[a, b]$. It is denoted by the symbol

________________________ .

35\. The numbers a and b in the symbol $\int_a^b f(x)\,dx$ are called the ____________________ . The function f is the ______________ of the definite integral.

30\. $\frac{n(n + 1)}{2}$, $\frac{n(n + 1)(2n + 1)}{6}$, $\left(\frac{n(n + 1)}{2}\right)^2$ 31. $\sum_{k=1}^{6} 3$, $\frac{6(6 + 1)}{2}$, 18

32\. Riemann sum for f, kth subinterval, $\Delta x_k = x_k - x_{k-1}$, kth subinterval 33. norm, longest

34\. definite integral for f, $\int_a^b f(x)\,dx$ 35. limits of integration, integrand

36. The limit

$$\lim_{\|P\|\to 0} \sum_{k=1}^{n} (c_k \sin 2c_k)\Delta x_k ,$$

where P denotes a partition of the interval $[0, 2\pi]$, is the definite integral

_______________________ .

37. The definite integral $\int_a^b f(x)\,dx$ is the area of the region between the graph of f and the x-axis from a to b provided that $y = f(x)$ is integrable and ____________ on the closed interval $[a, b]$. When $f(x) \leq 0$ over $[a, b]$ the area is given by

_______________________ .

38. For any integrable function the interpretation of the definite integral in terms of areas is

$$\int_a^b f(x)\,dx = \_\_\_\_\_\_\_\_\_\_\_\_\_\_\_\_\_\_\_\_\_\_\_\_ .$$

OBJECTIVE D: Find the area bounded by a curve $y = f(x)$ positive over $a \leq x \leq b$ by finding the limit of the sum of inscribed rectangles, if f is of the form $y = mx^k$ for $k = 0, 1, 2, 3$.

39. Let $y = f(x)$ define a positive continuous function of x on the closed interval $a \leq x \leq b$. The area under the curve and above the x-axis from $x = a$ to $x = b$ is defined to be the ________ of the sums of the areas of inscribed __________ as their number __________ without bound.

40. To find the area under the graph $y = x^2$, $-1 \leq x \leq 0$ using inscribed rectangles, divide the interval $-1 \leq x \leq 0$ into n subintervals each of equal length $\Delta x =$ ________ by inserting the points $x_1 = -1 + \Delta x$, $x_2 = -1 + 2\Delta x$, ..., $x_{n-1} =$ ______. Notice that $y = x^2$ is a ________ function of x over the interval $-1 \leq x \leq 0$. Thus the height of the first inscribed rectangle is __________ .

---

36. $\int_0^{2\pi} x \sin 2x\,dx$

37. nonnegative, $-\int_a^b f(x)\,dx$

38. (area above x-axis) - (area below x-axis)

39. limit, rectangles, increases

The inscribed rectangles have areas

$f(x_1)\Delta x = (-1 + \Delta x)^2 \cdot \Delta x$

$f(x_2)\Delta x = (-1 + 2\Delta x)^2 \cdot \Delta x$

$\vdots$

$f(x_{n-1})\Delta x = (-1 + (n-1)\Delta x)^2 \cdot \Delta x$

$f(\_\_\_\_)\Delta x = \_\_\_\_\_\_\_\_\_\_\_\_\_\_\_\_$,

y=x^2
y
x
-1 x_1 x_2 ... x_{n+1} 0

whose sum is

$$S_n = \left[\left(-1 + \frac{1}{n}\right)^2 + \left(-1 + \frac{2}{n}\right)^2 + \ldots + \left(-1 + \frac{n-1}{n}\right)^2 + 0\right] \cdot \_\_\_\_\_\_$$

Expanding each term on the right side and collecting like terms gives,

$$S_n = \left((-1)^2(n - 1) + 2(-1)\left(\frac{1}{n} + \frac{2}{n} + \ldots + \frac{n-1}{n}\right) + \left(\frac{1^2}{n^2} + \frac{2^2}{n^2} + \ldots + \frac{(n-1)^2}{n^2}\right)\right)\frac{1}{n}$$

$$= \frac{1}{n}\left[(n - 1) - \frac{2}{n}\left(\_\_\_\_\_\_\right) + \frac{1}{n^2}\left(\_\_\_\_\_\_\_\_\right)\right]$$

$$= \frac{1}{n}\left[(n - 1) - \frac{2}{n} \cdot \left(\_\_\_\_\_\right) + \frac{1}{n^2} \cdot \left(\_\_\_\_\_\_\_\_\right)\right]$$

The area under the graph is defined to be the limit of ______ as $n \to \infty$. This limit equals ______ .

4.6 PROPERTIES, AREA, AND THE MEAN VALUE THEOREM FOR INTEGRALS

OBJECTIVE A: Know from memory the rules for working with definite integrals.

41. $\int_a^a f(x)\,dx = \_\_\_\_\_\_\_\_$.

42. $\int_a^b [f(x) \pm g(x)]\,dx = \_\_\_\_\_\_\_\_\_\_\_\_\_\_\_\_$.

43. $\int_a^b kf(x)\,dx = \_\_\_\_\_\_\_\_\_\_\_\_$.

---

40. $1/n$, $-1 + (n - 1)\Delta x$, decreasing, $(-1 + \Delta x)^2$ or x_1^2, 0, $0 \cdot \Delta x$ or $(-1 + n\Delta x)^2 \cdot \Delta x$, $1/n$, $1 + 2 + 3 + \ldots + (n - 1)$, $1^2 + 2^2 + \ldots + (n - 1)^2$, $\frac{(n-1)n}{2}$, $\frac{(n-1)n\left(2(n-1)+1\right)}{6}$, $(n - 1)$, $\frac{1}{3} - \frac{1}{2n} + \frac{1}{6n^2}$, S_n, 1/3

41. 0

42. $\int_a^b f(x)\,dx \pm \int_a^b g(x)\,dx$

43. $k\int_a^b f(x)\,dx$

44. $\int_b^a f(x)\,dx =$ ______________ .

45. $f(x) \geq g(x)$ on $[a,b]$ implies $\int_a^b f(x)\,dx$ ______________ .

46. $\int_a^b -f(x)\,dx =$ ______________ .

47. $f(x) \geq 0$ on $[a,b]$ implies $\int_a^b f(x)\,dx$ ______________ .

48. $\int_a^b f(x)\,dx + \int_b^c f(x)\,dx =$ ______________ .

49. $\min f \cdot (b - a) \leq$ ______________ $\leq \max f \cdot (b - a)$.

OBJECTIVE B: Use the rules in Table 4.5 to find values of definite integrals.

50. Suppose f and g are continuous and that

$$\int_1^3 f(x)\,dx = -5, \qquad \int_1^6 f(x)\,dx = 2, \qquad \int_1^3 g(x)\,dx = 4.$$

Then,

$$\int_3^6 f(x)\,dx = \int_3^1 f(x)\,dx + \text{______________}$$
$$= - \text{______________} + 2$$
$$= \text{__________} .$$

51. Evaluate $\int_0^3 \left(3x^2 - \frac{x}{2} + 5\right) dx$.

Solution.

$$\int_0^3 \left(3x^2 - \frac{x}{2} + 5\right) dx = 3\int_0^3 x^2\,dx - \frac{1}{2}\text{__________} + \int_0^3 5\,dx$$
$$= 3(\text{__________}) - \frac{1}{2}\left(3^2 - 0^2\right) + \text{__________}$$
$$= 27 - \frac{9}{2} + \text{__________} = \frac{75}{2} .$$

---

44. $-\int_a^b f(x)\,dx$ 45. $\geq \int_a^b g(x)\,dx$ 46. $-\int_a^b f(x)\,dx$

47. ≥ 0 48. $\int_a^c f(x)\,dx$ 49. $\int_a^b f(x)\,dx$

50. $\int_1^6 f(x)\,dx$, $\int_1^3 f(x)\,dx$, 7 51. $\int_0^3 x\,dx$, $\frac{3^3}{3}$, $5(3 - 0)$, 15

OBJECTIVE C: Find the total area between the x-axis and the graph of a continuous function $y = f(x)$ with both positive and negative values on $[a, b]$ using the following steps:

STEP 1: Partition $[a, b]$ into subintervals with the zeros of f.
STEP 2: Integrate f over each subinterval.
STEP 3: Add the absolute values of the integrals.

52. Find the total area between the curve $y = x^2 + x + 1$, the x-axis, and the lines $x = 2$, $x = 3$.

Solution. The curve $y = x^2 + x + 1$ is everywhere positive over the interval $[2,3]$, so it has no zeros there. The area is

$$\int_2^3 (x^2 + x + 1)\,dx = \underline{\hspace{3cm}} + \int_2^3 x\,dx + \underline{\hspace{3cm}}$$

$$= \left(\frac{3^3}{3} - \underline{\hspace{1.5cm}}\right) + \left(\underline{\hspace{1.5cm}} - \frac{2^2}{2}\right) + \left(\underline{\hspace{2cm}}\right)$$

$$= \frac{19}{3} + \underline{\hspace{1.5cm}} + 1 = \frac{59}{6}.$$

53. Find the area of the region between the curve $y = (x - 1)(x - 3)$, $0 \le x \le 5$, and the x-axis.

Solution. The zeros of the curve are $x =$ _____ and $x =$ _____. Thus we integrate $f(x) = (x - 1)(x - 3) = x^2 - 4x + 3$ over the intervals ________, $[1,3]$, and ________. Then we add the absolute values of the results.

$$\int_0^1 (x^2 - 4x + 3)\,dx = \underline{\hspace{2.5cm}} - 4\int_0^1 x\,dx + \int_0^1 3\,dx$$

$$= \frac{1}{3} - 4\left(\underline{\hspace{1.5cm}}\right) + 3(1 - 0) = \frac{4}{3};$$

$$\int_1^3 (x^2 - 4x + 3)\,dx = \int_1^3 x^2\,dx - \underline{\hspace{2.5cm}} + \int_1^3 3\,dx$$

$$= \left(\underline{\hspace{1.5cm}} - \cdot\frac{1}{3}\right) - 4\left(\frac{9}{2} - \frac{1}{2}\right) + 3\left(\underline{\hspace{1.5cm}}\right)$$

$$= \frac{26}{3} - 16 + 6 = -\frac{4}{3};$$

$$\int_3^5 (x^2 - 4x + 3)\,dx = \int_3^5 x^2\,dx - 4\int_3^5 x\,dx + \int_3^5 3\,dx$$

$$= \left(\underline{\hspace{2cm}}\right) - 4\left(\frac{25}{2} - \frac{9}{2}\right) + 3\left(\underline{\hspace{1.5cm}}\right)$$

$$= \frac{98}{3} - 32 + 6 = \frac{20}{3}.$$

---

52. $\int_2^3 x^2\,dx$, $\int_2^3 dx$, $\frac{2^3}{3}$, $\frac{3^2}{2}$, $3 - 2$, $\frac{5}{2}$

Therefore the region's total area is

$$\text{Area} = \underline{\qquad} + \underline{\qquad} + \underline{\qquad} = \frac{28}{3}.$$

OBJECTIVE D: Calculate the average value of a given continuous function $y = f(x)$ over a specified interval $a \le x \le b$.

54. Find the average value of $f(x) = x^2 - 4x + 3$ over the interval $[0,2]$.

Solution. $\text{av}(f) = \frac{1}{2-0} \underline{\qquad\qquad}$

$$= \frac{1}{2}\left[\int_0^2 x^2\,dx - \underline{\qquad\qquad} + \underline{\qquad\qquad}\right]$$

$$= \frac{1}{2}\left[\left(\frac{2^3}{3}\right) - 4\left(\underline{\qquad}\right) + 3\left(\underline{\qquad}\right)\right]$$

$$= \frac{1}{2}\left(\frac{8}{3} - \underline{\qquad} + 6\right) = \frac{1}{3}.$$

55. If f is continuous on $[a,b]$, then the Mean Value Theorem for Definite Integrals guarantees that there is a point c in $[a,b]$ such that

$\underline{\qquad\qquad\qquad\qquad\qquad}$.

56. To find the point or points in the interval $[0,2]$ for which the function $f(x) = x^2 - 4x + 3$ in Problem 54 assumes its average value, we must solve the equation

$\underline{\qquad\qquad\qquad\qquad\qquad}$.

Using the quadratic formula, the solution is

$$x = \frac{4 \pm \sqrt{16 - 32/3}}{2} = 2 \pm \frac{1}{2}\sqrt{\frac{16}{3}}$$

$$= 2 \pm \frac{2}{\sqrt{3}} = \frac{6 \pm 2\sqrt{3}}{3}.$$

The only solution in the interval $[0,2]$ gives the point

$c = \underline{\qquad\qquad\qquad} \approx 0.85$.

---

53. 1, 3, $[0,1]$, $[3,5]$, $\int_0^1 x^2\,dx$, $\frac{1}{2}$, $4\int_1^3 x\,dx$, $\frac{27}{3}$, 2, $\frac{125}{3} - \frac{27}{3}$, 2, $\frac{4}{3} + \left|-\frac{4}{3}\right| + \frac{20}{3}$

54. $\int_0^2 (x^2 - 4x + 3)\,dx$, $4\int_0^2 x\,dx$, $\int_0^2 3\,dx$, $\frac{2^2}{2}$, 2, 8

55. $\frac{1}{b-a}\int_a^b f(x)\,dx = f(c)$

56. $x^2 - 4x + 3 = \frac{1}{3}$ or, equivalently, $x^2 - 4x + \frac{8}{3} = 0$, $\frac{6 - 2\sqrt{3}}{3}$

4.7 THE FUNDAMENTAL THEOREM

57. The Fundamental Theorem of Calculus, Part 1 concerns the integral

$$F(x) = \int_a^x f(t)\,dt ,$$

where f is continuous on $[a, b]$. This theorem says that F is ____________ at every point x in $[a, b]$ and

$$F'(x) = \_\_\_\_\_\_\_\_\_\_ .$$

OBJECTIVE A: Use the Fundamental Theorem, Part 1 to calculate the derivative of an integral $\int_0^{v(x)} f(t)\,dt$ with respect to x. Assume that the integrand f is continuous and that v is a differentiable function of x.

58. If $F(x) = \int_1^x (t^5 - 2t^3 + 1)^4\,dt$, then we may find $F'(x)$ by replacing t by x in the integrand. Thus, $F'(x) =$ ________ .

59. Let $F(x) = \int_0^{x^2} \sqrt{1 + t}\,dt$. If $u = x^2$, then by the chain rule, $\frac{dF}{dx} = \frac{dF}{du} \cdot$ ____ . Now, $\frac{dF}{du} = \frac{d}{du} \int_0^u \sqrt{1 + t}\,dt =$ ________ . Thus, $F'(x) = \sqrt{1 + u} \cdot$ ____ $=$ ________ .

OBJECTIVE B: Evaluate definite integrals of elementary continuous functions, using the Fundamental Theorem, Part 2.

60. The Fundamental Theorem of Calculus, Part 2 gives a rule for calculating the definite integral $\int_a^b f(x)\,dx$ of a continuous function. The rule states that you must first find an __________ F of f. That is, the relationship between F and f is ________ . Next, calculate the number $F(b) -$ _____ . This computation gives

$$\int_a^b f(x)\,dx = \_\_\_\_\_\_\_\_\_\_ .$$

61. The notation $F(x)\Big]_d^c$ means ____________ .

---

57. differentiable, $f(x)$ 58. $(x^5 - 2x^3 + 1)^4$ 59. $\frac{du}{dx}$, $\sqrt{1 + u}$, $2x$, $2x\sqrt{1 + x^2}$

60. antiderivative, $F'(x) = f(x)$, $F(a)$, $F(b) - F(a)$ 61. $F(c) - F(d)$

62. Find $\int_{-1}^{2} (x^3 - 2x + 5)\,dx$.

$$\int_{-1}^{2} (x^3 - 2x + 5)\,dx = \underline{\qquad\qquad}\,\Big]_{-1}^{2}$$

$$= \left(\tfrac{1}{4}(2^4) - 2^2 + 5\cdot 2\right) - \left(\underline{\qquad\qquad}\right)$$

$$= 10 - (\underline{\qquad}) = \underline{\qquad}\,.$$

63. Find $\int_{0}^{\pi/4} (\sin t + \cos t - \frac{t}{2})\,dt$.

$$\int_{0}^{\pi/4} (\sin t + \cos t - \tfrac{t}{2})\,dt = \underline{\qquad\qquad\qquad}\,\Big]_{0}^{\pi/4}$$

$$= \left(-\frac{\sqrt{2}}{2} + \frac{\sqrt{2}}{2} - \frac{\pi^2}{64}\right) - (\underline{\qquad\qquad})$$

$$= \underline{\qquad\qquad}$$

$$\approx 0.84579.$$

64. Find the area of the region between the x-axis and the curve

$$y = x^{1/3} + x^{2/3}, \quad -1 \le x \le 1\,.$$

<u>Solution</u>. The graph of $y = f(x)$ is shown below. The zeros exist at $x = \underline{\qquad}$ and $x = \underline{\qquad}$.

Thus,

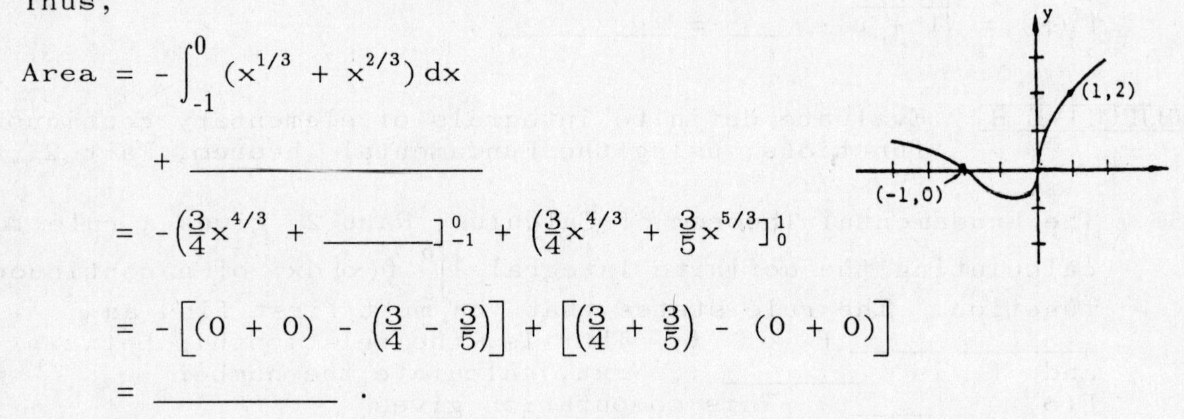

$$\text{Area} = -\int_{-1}^{0} (x^{1/3} + x^{2/3})\,dx$$

$$+ \underline{\qquad\qquad\qquad}$$

$$= -\left(\tfrac{3}{4}x^{4/3} + \underline{\qquad\qquad}\right]_{-1}^{0} + \left(\tfrac{3}{4}x^{4/3} + \tfrac{3}{5}x^{5/3}\right]_{0}^{1}$$

$$= -\left[(0 + 0) - \left(\tfrac{3}{4} - \tfrac{3}{5}\right)\right] + \left[\left(\tfrac{3}{4} + \tfrac{3}{5}\right) - (0 + 0)\right]$$

$$= \underline{\qquad\qquad}\,.$$

65. To find the average value of $y = \sin x - \cos x$ over $0 \le x \le \pi/4$ we have

Av. val. of y

$$\text{on } \left[0, \tfrac{\pi}{4}\right] = \frac{1}{\frac{\pi}{4} - 0}\int_{\underline{\quad}}^{\underline{\quad}} (\underline{\qquad\qquad\qquad})\,dx$$

---

62. $\frac{1}{4}x^4 - x^2 + 5x,\ \frac{1}{4}(-1)^4 - (-1)^2 + 5(-1),\ -\frac{23}{4},\ \frac{63}{4}$

63. $-\cos t + \sin t - \frac{t^2}{4},\ -1 + 0 - 0,\ 1 - \frac{\pi^2}{64}$

64. $-1,\ 0,\ \int_0^1 \left(x^{1/3} + x^{2/3}\right)dx,\ \frac{3}{5}x^{5/3},\ \frac{3}{2}$

$= \frac{4}{\pi}$ (______________)]$\overline{\quad}$

$= \frac{4}{\pi}\left[\left(-\frac{\sqrt{2}}{2} - \frac{\sqrt{2}}{2} - (\_\_\_\_\_\_\_\_)\right)\right] = \frac{4}{\pi}(\_\_\_\_\_\_) \approx -.52739.$

4.8 SUBSTITUTION IN DEFINITE INTEGRALS

OBJECTIVE: Evaluate definite integrals using the substitution method of integration.

66. Find $\int_0^{\pi/4} (3 - \sin 2t)^{1/3} \cos 2t\,dt$.

Solution. From Problem 21 of this chapter, the indefinite integral $\int (3 - \sin 2t)^{1/3} \cos 2t\,dt =$ ______________ . Therefore, the definite integral is given by

$\int_0^{\pi/4} (3 - \sin 2t)^{1/3} \cos 2t\,dt =$ ______________$]_0^{\pi/4}$

$= -\frac{3}{8}(\_\_\_\_)^{4/3} + \frac{3}{8}(\_\_\_\_)^{4/3}$

$=$ ______________ $\approx 0.67759.$

67. If $u = g(x)$ then $du =$ __________ and

$\int_a^b f(g(x)) \cdot g'(x)\,dx =$ ______________________ .

68. $\int_0^{\pi/6} \frac{\cos x\,dx}{\sqrt{1 - \sin x}}$. Let $u = g(x) = \sin x$. Then

$du = g'(x)\,dx =$ __________ . Also, $g(0) = 0$ and $g\left(\frac{\pi}{6}\right) =$ ________ .

Thus,

$\int_0^{\pi/6} \frac{\cos x\,dx}{\sqrt{1 - \sin x}} = \int_{\_\_}^{\_\_} \frac{du}{\sqrt{1 - u}} =$ __________ $= -2(\_\_\_\_\_\_) = 2 - \sqrt{2}.$

69. $\int_0^1 \frac{x\,dx}{\sqrt{4 - x^2}}$. Let $u = 4 - x^2$. Then,

$\int_0^1 \frac{x\,dx}{\sqrt{4 - x^2}} = -\frac{1}{2}\int_{\_\_}^{\_\_} \frac{du}{\sqrt{u}} =$ __________ $= -\sqrt{3} + 2$.

---

65. $\int_0^{\pi/4} (\sin x - \cos x)\,dx$, $(-\cos x - \sin x)]_0^{\pi/4}$, $(-1 - 0)$, $(1 - \sqrt{2})$

66. $-\frac{3}{8}(3 - \sin 2t)^{4/3} + C$, $-\frac{3}{8}(3 - \sin 2t)^{4/3}$, $3 - 1$, $3 - 0$, $\frac{3}{8}(3^{4/3} - 2^{4/3})$

67. $g'(x)\,dx$, $\int_{g(a)}^{g(b)} f(u)\,du$ 68. $\cos x\,dx$, $\frac{1}{2}$, $\int_0^{1/2}$, $-2\sqrt{1 - u}]_0^{1/2}$, $\frac{1}{\sqrt{2}} - 1$ 69. $\int_4^3$, $-\sqrt{u}]_4^3$

4.9 NUMERICAL INTEGRATION

OBJECTIVE A: Approximate a given definite integral by using the trapezoidal rule with a specified number n of subintervals. Estimate the error in this approximation.

70. Let $y = f(x)$ be defined and continuous over the interval $a \le x \le b$. Divide the interval $[a, b]$ into n subintervals, each of length $h = (b - a)/n$, by inserting the points $x_1 = a + h$, $x_2 = a + 2h$, ..., $x_{n-1} = a + (n - 1)h$. Set $x_0 = a$ and $x_n = b$ for convenience in notation. Define $y_k = f(x_k)$ for each $k = 0, 1, 2, \ldots, n$. Then the trapezoidal approximation for the definite integral is

$$\int_a^b f(x)\, dx \approx \underline{\hspace{6em}} .$$

The error estimate is $E_T = \underline{\hspace{4em}}$, where M is any upper bound for the values of $\underline{\hspace{4em}}$ on $[a, b]$.

71. Use the trapezoidal rule to approximate $\int_0^1 \sqrt{1 + x^2}\, dx$, $n = 5$.

Solution. Here, $h = \frac{1 - 0}{5} = \underline{\hspace{3em}}$, $x_0 = \underline{\hspace{3em}}$, $x_5 = \underline{\hspace{3em}}$. The subdivision points are $x_1 = \frac{1}{5}$, $x_2 = \underline{\hspace{3em}}$, $x_3 = \underline{\hspace{3em}}$, $x_4 = \underline{\hspace{3em}}$. The corresponding function values are computed as,

$y_0 = \sqrt{1 + 0^2} = 1$, $y_1 = \sqrt{1 + \left(\frac{1}{5}\right)^2} = \underline{\hspace{3em}} \approx \underline{\hspace{4em}}$

$y_2 = \sqrt{1 + \left(\frac{2}{5}\right)^2} = \underline{\hspace{2em}} \approx \underline{\hspace{5em}}$, $y_1 \approx \underline{\hspace{4em}}$,

$y_4 \approx \underline{\hspace{4em}}$, and $y_5 \approx \underline{\hspace{4em}}$. Therefore, the trapezoidal approximation is

$$T = \frac{1}{2(5)} \cdot (y_0 + 2y_1 + 2y_2 + 2y_3 + 2y_4 + y_5), \quad \text{or}$$

$$T = \frac{1}{10} \cdot (1 + 2.03960 + \underline{\hspace{4em}} + \underline{\hspace{4em}} + \underline{\hspace{4em}} + \underline{\hspace{4em}})$$

$$= \frac{1}{10} \cdot (\underline{\hspace{5em}}) = \underline{\hspace{5em}} .$$

Therefore,

$$\int_0^1 \sqrt{1 + x^2}\, dx \approx \underline{\hspace{5em}} .$$

---

70. $\frac{h}{2}(y_0 + 2y_1 + 2y_2 + \ldots + 2y_{n-1} + y_n)$, $\frac{b - a}{12} h^2 M$, $|f''|$

71. $\frac{1}{5}$, 0, 1, $\frac{2}{5}$, $\frac{3}{5}$, $\frac{4}{5}$, $\frac{\sqrt{26}}{5}$, 1.01980, $\frac{\sqrt{29}}{5}$, 1.07703, 1.16619, 1.28062, 1.41421, 2.15407, 2.33238, 2.56124, 1.41421, 11.50150, 1.15015, 1.15015

72. To estimate the error in the approximation of Problem 71, let $f(x) = \sqrt{1 + x^2}$. Then, $f'(x) =$ ______ and $f''(x) =$ _______ .

Therefore, for $0 \le x \le 1$, we see that

$$|f''(x)| = \frac{1}{\left(1 + x^2\right)^{3/2}} < \_\_\_\_\_\_ .$$

Thus, the error E_T satisfies

$$|E_T| \le \frac{b-a}{12} h^2 M = \_\_\_\_ \cdot \_\_\_\_ \cdot \_\_\_\_ \approx \_\_\_\_\_\_\_ .$$

Then, $\int_0^1 \sqrt{1 + x^2}\, dx = 1.15015 \pm |E_T|$, or

$$\_\_\_\_\_\_ \le \int_0^1 \sqrt{1 + x^2}\, dx \le \_\_\_\_\_\_ .$$

(The value of the integral to five decimal places is 1.14779, so the error is about 2/10 of one percent.)

73. How many subintervals are required to obtain $\int_0^1 \sqrt{1 + x^2}\, dx$ to 5 decimal places of accuracy by the trapezoidal rule?

<u>Solution</u>. From Problem 72, $|E_T| \le \frac{1}{12}h^2 M \le \frac{1}{12}h^2$. To obtain 5-place accuracy, we need $|E_T| < 5 \cdot 10^{-6}$. Thus, $\frac{1}{12}h^2 < 5 \cdot 10^{-6}$ implies $h^2 <$ ______ or, since $h = \frac{b - a}{n}$ $=$ ______, $n^2 >$ ______ . Thus, $n >$ ____ $\approx$ ____, and choosing $n =$ ____ as the number of subintervals ensures 5-place accuracy. (This is only an upper estimate: fewer subintervals may work, but there are no guarantees.)

OBJECTIVE B: Approximate a given definite integral by use of Simpson's rule with a specified even number n of subintervals. Estimate the error in this approximation.

74. Let $y = f(x)$ be defined and continuous over the interval $a \le x \le b$. Divide the interval $[a,b]$ into n subintervals, where n is an <u>even</u> number, each of length $h = (b - a)/n$, using the points $x_0 = a$, $x_1 = a + h$, $x_2 = 1 + 2h$, ..., $x_n = a + nh = b$. Define $y = f(x_k)$ for each $k = 0,1,2, \ldots,n$. Then the Simpson approximation for the definite integral is

$$\int_a^b f(x)\, dx \approx \_\_\_\_\_\_\_\_\_\_\_\_ .$$

---

72. $x(1 + x^2)^{-1/2}$, $(1 + x^2)^{-3/2}$, 1, $\frac{1}{12} \cdot \frac{1}{25} \cdot 1$, 0.00333, 1.14682, 1.15348

73. $60 \cdot 10^{-6}$, $\frac{1}{n}$, $\frac{1}{60} \cdot 10^6$, $\frac{1}{2\sqrt{15}}\, 10^3$, 129, 130

The error estimate is $|E_S| \leq$ ______ where M is any upper bound on the values of ______ on $[a, b]$.

75. Approximate $\int_0^1 \sqrt{1 + x^2}\, dx$ by Simpson's rule with $n = 6$.
Solution. Here $h =$ ___, $x_0 =$ ____, and $x_6 =$ ___ . The subdivision points are,

$x_1 = \frac{1}{6}$, $x_2 =$ ____, $x_3 =$ ____, $x_4 =$ ____, and $x_5 =$ ____ .

The corresponding function values are $y_0 = 1$,

$y_1 = \sqrt{1 + \left(\frac{1}{6}\right)^2} =$ _____ $\approx$ ________, $y_2 \approx$ ________,
$y_3 \approx$ ________, $y_4 \approx$ ________, $y_5 \approx$ ________, and
$y_6 =$ ______ $\approx$ ______ . Therefore, the Simpson approximation is,

$S = \frac{1}{6 \cdot 3}(y_0 + 4y_1 + 2y_2 + 4y_3 + 2y_4 + 4y_5 + y_6)$

$= \frac{1}{18}(1 + 4.05518 +$ ______ $+$ ______ $+$ ______ $+$ ______ $+$ ______$)$

$= \frac{1}{18}($______$) =$ ______ .

Compare the answer with that found in Problem 71 where we employed the trapezoidal rule.

76. To estimate the error in the approximation in Problem 75, let $f(x) = \sqrt{1 + x^2}$. Then, from Problem 72,
$f''(x) = \left(1 + x^2\right)^{-3/2}$, so that $f^{(3)}(x) =$ ____________ and
$f^{(4)}(x) =$ ____________ . Now,

$$\frac{4x^2 - 1}{\left(1 + x^2\right)^{7/2}} < \frac{4x^2}{\left(1 + x^2\right)^{7/2}} < \frac{4x^2}{1 + x^2} < 4 \quad \text{for} \quad 0 \leq x \leq 1.$$

Thus, the error E_S satisfies

$$|E_S| = \frac{b - a}{180} h^4 M = \_\_\_ \cdot \_\_\_ \cdot \_\_\_ \approx \_\_\_\_ .$$

Therefore, we observe that Simpson's rule, with $n = 6$, provides the value of the integral to at least 3 decimal places of accuracy; with $n = 10$ we will obtain at least 4 place accuracy, according to our error estimates.

---

74. $\frac{h}{3}(y_0 + 4y_1 + 2y_2 + 4y_3 + \ldots + 2y_{n-2} + 4y_{n-1} + y_n)$, $\frac{b-a}{180} h^4 M$, $\left|f^{(4)}\right|$

75. $\frac{1}{6}$, 0, 1, $\frac{1}{3}$, $\frac{1}{2}$, $\frac{2}{3}$, $\frac{5}{6}$, $\frac{\sqrt{37}}{6}$, 1.01379, 1.05409, 1.11803, 1.20185, 1.30171, $\sqrt{2}$, 1.41421, 2.10819, 4.47212, 2.40370, 5.20684, 1.41421, 20.66021, 1.14779

76. $-3x\left(1 + x^2\right)^{-5/2}$ $3\left(4x^2 - 1\right)\left(1 + x^2\right)^{-7/2}$, $\frac{1}{180} \cdot \frac{1}{6^4} \cdot 12$, 0.00005

CHAPTER 4 SELF-TEST

1. Find an antiderivative of the following functions.

   (a) $7 - 4x + 5x^2$ (b) $\sin \frac{x}{3} + \frac{1}{\sqrt{x}} + x^{1/3}$

2. Find the following indefinite integrals.

   (a) $\int (x - 1)(2 + x)\, dx$ (b) $\int \sqrt{2x - 1}\, dx$

   (c) $\int x^2(5 - 3x^3)^{-1/2}\, dx$ (d) $\int x^{-1/2} \sin(\sqrt{x} - 3)\, dx$

3. Solve the initial value problems.

   (a) $\frac{dy}{dx} = \sec^2 x, \quad y = 3 \quad \text{when} \quad x = 0$

   (b) $\frac{dy}{dx} = \frac{x^3 + 1}{x^3}, \quad y = \frac{7}{2} \quad \text{when} \quad x = 1$

4. Approximate the area under the curve $y = x^2 - 2x + 4$ between $x = 1$ and $x = 4$ by summing $n = 6$ inscribed rectangles of uniform width.

5. Find the numerical values of each of the following.

   (a) $\sum_{k=1}^{4} \frac{1}{2k}$ (b) $\sum_{n=1}^{5} n(n - 3)$ (c) $\sum_{k=5}^{6} (2k - 1)$

6. Find the area under the curve $y = x\sqrt{x^2 + 1}$, above the x-axis, between $x = 1$ and $x = 4$.

7. Evaluate the definite integrals.

   (a) $\int_1^4 \frac{(x - 2)^2}{\sqrt{x}}\, dx$ (b) $\int_{-2}^{0} x^2(4 - x)\, dx$

8. Find $\frac{d}{dx} \int_0^{1-x^2} \sqrt[3]{t^2 + 1}\, dt$.

9. Suppose g and h are continuous and that $\int_0^4 g(x)\, dx = 3$,

   $\int_1^4 g(x)\, dx = -5$, $\int_1^0 h(x)\, dx = 2$. Find $\int_0^1 [2g(x) - h(x)]\, dx$.

10. Find the area of the planar region bounded by the curves $y = x^3 + 1$ and $y = x^2 + x$.

11. A train leaving a railroad station has an acceleration of $a = 0.5 + 0.02t$ ft/sec^2. How far will the train move in the first 20 sec of motion? What is its velocity after 20 seconds?

12. Find the average value of the function $y = x^2 - x + 1$ over the interval $0 \le x \le 2$.

13. Use the trapezoidal rule with $n = 4$ to approximate $\int_0^1 \sqrt{1 + x^3}\, dx$.

14. Use Simpson's rule with $n = 6$ to approximate $\int_1^2 \frac{dx}{x}$. Estimate the error in your approximation.

15. Compute $\int_0^{\pi} f(x)\, dx$ where $f(x) = \begin{cases} \sin x, & 0 \le x < \frac{\pi}{2} \\ \pi x, & \frac{\pi}{2} \le x \le \pi. \end{cases}$

SOLUTIONS TO CHAPTER 4 SELF-TEST

1. (a) $7x - 2x^2 + \frac{5}{3}x^3$

   (b) $-3 \cos \frac{x}{3} + 2x^{1/2} + \frac{3}{4}x^{4/3}$

2. (a) $\int (x - 1)(2 + x)\, dx = \int (x^2 + x - 2)\, dx = \frac{1}{3}x^3 + \frac{1}{2}x^2 - 2x + C$

   (b) $\int \sqrt{2x - 1}\, dx = \frac{1}{2}\int \sqrt{u}\, du = \frac{1}{3}(2x - 1)^{3/2} + C \quad (u = 2x - 1)$

   (c) $\int x^2(5 - 3x^3)^{-1/2}\, dx = -\frac{1}{9}\int u^{-1/2}\, du \quad (u = 5 - 3x^3)$

   $= -\frac{2}{9}(5 - 3x^3)^{1/2} + C$

   (d) $\int x^{-1/2} \sin(\sqrt{x} - 3)\, dx = 2\int \sin u\, du \quad (u = \sqrt{x} - 3)$

   $= -2 \cos(\sqrt{x} - 3) + C$

3. (a) $\frac{dy}{dx} = \sec^2 x$, so $y = \tan x + C$
   Since $y = 3$ when $x = 0$, $3 = \tan(0) + C$ or $C = 3$.
   Hence, the solution of the initial value problem is
   $y = \tan x + 3$.

   (b) $\frac{dy}{dx} = 1 + \frac{1}{x^3}$ has the general solution $y = x - \frac{1}{2}x^{-2} + C$.
   Substituting $y = \frac{7}{2}$ and $x = 1$ gives $\frac{7}{2} = 1 - \frac{1}{2} + C$ or
   $C = 3$. Hence, $y = x - \frac{1}{2x^2} + 3$.

4. The partition points are $x_0 = 1$, $x_1 = \frac{3}{2}$, $x_2 = 2$, $x_3 = \frac{5}{2}$, $x_4 = 3$, $x_5 = \frac{7}{2}$, and $x_6 = 4$. Since $\frac{dy}{dx} > 0$ on the interval $1 \le x \le 4$, the curve is increasing so the altitude of each rectangle is its left edge. Thus, the areas of the inscribed rectangles for $y = f(x)$ are,

$$f(1)\, \Delta x = 3 \cdot \frac{1}{2} = \frac{3}{2}$$

$$f(\tfrac{3}{2})\, \Delta x = \frac{13}{4} \cdot \frac{1}{2} = \frac{13}{8}$$

$$f(2)\, \Delta x = 4 \cdot \frac{1}{2} = 2$$

$$f(\tfrac{5}{2})\,\Delta x = \frac{21}{4}\cdot\frac{1}{2}\cdot\frac{21}{8}$$

$$f(3)\,\Delta x = 7\cdot\frac{1}{2} = \frac{7}{2}$$

$$f(\tfrac{7}{2})\,\Delta x = \frac{37}{4}\cdot\frac{1}{2} = \frac{37}{8}$$

$$\text{Sum} = \frac{127}{8} = 15.875, \quad \text{approximate area}$$

5. (a) $\frac{1}{2}+\frac{1}{4}+\frac{1}{6}+\frac{1}{8} = \frac{25}{24}$

(b) $1(-2) + 2(-1) + 3(0) + 4(1) + 5(2) = 10$

(c) $(2\cdot 5 - 1) + (2\cdot 6 - 1) = 9 + 11 = 20$

6. $$\int_1^4 x\sqrt{x^2+1}\,dx = \frac{1}{3}(x^2+1)^{3/2}\Big|_1^4 = \frac{1}{3}(17^{3/2} - 2^{3/2}) \approx 22.42146.$$

7. (a) $$\int_1^4 \frac{(x-2)^2}{\sqrt{x}}\,dx = \int_1^4 \frac{x^2-4x+4}{\sqrt{x}}\,dx$$

$$= \int_1^4 (x^{3/2} - 4x^{1/2} + 4x^{-1/2})\,dx$$

$$= \frac{2}{5}x^{5/2} - \frac{8}{3}x^{3/2} + 8x^{1/2}\Big|_1^4$$

$$= (\tfrac{2}{5}(32) - \tfrac{8}{3}(8) + 16) - (\tfrac{2}{5} - \tfrac{8}{3} + 8)$$

$$\approx 1.73333.$$

(b) $$\int_{-2}^0 x^2(4-x)\,dx = \int_{-2}^0 (4x^2 - x^3)\,dx = \frac{4}{3}x^3 - \frac{1}{4}x^4\Big|_{-2}^0$$

$$= -\left[\frac{4}{3}(-2)^3 - \frac{1}{4}(-2)^4\right] = \frac{44}{3} \approx 14.66667.$$

8. $$\frac{d}{dx}\int_0^{1-x^2} \sqrt[3]{t^2+1}\,dt = \sqrt[3]{(1-x^2)^2+1}\cdot\frac{d}{dx}(1-x^2)$$

$$= -2x\sqrt[3]{x^4 - 2x^2 + 2}\,.$$

9. $$\int_0^1 [2g(x) - h(x)]\,dx = 2\int_0^1 g(x)\,dx - \int_0^1 h(x)\,dx$$

$$= 2\left[\int_0^1 g(x)\,dx + \int_1^4 g(x)\,dx - \int_1^4 g(x)\,dx\right] - \left[-\int_1^0 h(x)dx\right]$$

$$= 2\left[\int_0^4 g(x)\,dx - \int_1^4 g(x)\,dx\right] + \int_1^0 h(x)\,dx$$

$$= 2[3 - (-5)] + 2$$

$$= 18.$$

10. The graph of $y = x^3 + 1$ crosses the x-axis at $(-1,0)$ and so does the graph of $y = x^2 + x$. (See the figure at the right.) Solving for the other point of intersection of the two curves, $x^3 + 1 = x^2 + x$ or $x^3 - x^2 - x + 1 = 0$. Since $x = -1$ is a root, by division $x^3 - x^2 - x + 1 = (x + 1)(x^2 - 2x + 1) = 0$ or $(x + 1)(x - 1)^2 = 0$. Thus, the other point of intersection is $(1,2)$. The area between the curves is then given by

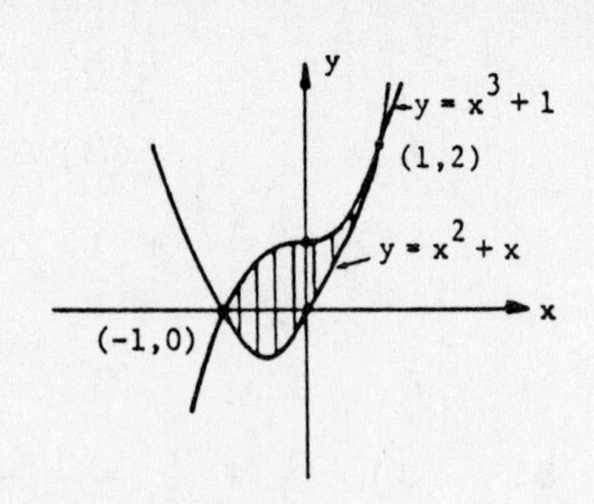

$$A = \int_{-1}^{1} \left[(x^3 + 1) - (x^2 + x)\right] dx = \left(\tfrac{1}{4}x^4 + x - \tfrac{1}{3}x^3 - \tfrac{1}{2}x^2\right)\Big]_{-1}^{1}$$
$$= \frac{4}{3} \text{ square units.}$$

11. Since $v = \int a\,dt$ we have $v = 0.5t + 0.01t^2 + C_1$. At $t = 0$ the train is at rest, so $v = 0$; hence $C_1 = 0$. Next, $s = \int v\,dt$ or $s = 0.25t^2 + \frac{1}{300}t^3 + C_2$. At $t = 0$, $s = 0$ so that $C_2 = 0$. Thus, when $t = 20$ seconds,
$s = \frac{1}{4}(400) + \frac{1}{300}(800) = 126\frac{2}{3}$ feet, the distance traveled by the train in the first 20 seconds. Its velocity at that time is

$$v = (0.5)(20) + (0.01)(400) = 14 \text{ ft/sec.}$$

12. The average value is

$$\frac{1}{2 - 0}\int_0^2 (x^2 - x + 1)\,dx = \frac{1}{2}\left[\frac{1}{3}x^3 - \frac{1}{2}x^2 + x\right]_0^2 = \frac{4}{3}\,.$$

13. Subdivision points are $x_0 = 0$, $x_1 = 1/4$, $x_2 = 1/2$, $x_3 = 3/4$, $x_4 = 1$ and $h = (1 - 0)/4 = 1/4$. Then,
$y_0 = \sqrt{1 + 0} = 1$, $y_1 = \sqrt{1 + (1/64)} \approx 1.00778$,
$y_2 = \sqrt{1 + (1/8)} \approx 1.06066$, $y_3 = \sqrt{1 + (27/64)} \approx 1.19242$,
$y_4 = \sqrt{1 + 1} \approx 1.41421$. Thus,

$$\int_0^1 \sqrt{1 + x^3}\,dx \approx T = \frac{1}{8}(y_0 + 2y_1 + 2y_2 + 2y_3 + y_4)$$
$$\approx \frac{1}{8}(8.93593) \approx 1.11699.$$

14. Subdivision points are $x_0 = 1$, $x_1 = 7/6$, $x_2 = 4/3$, $x_3 = 3/2$, $x_4 = 5/3$, $x_5 = 11/6$, $x_6 = 2$, and $h = 1/6$. Then, $y_0 = 1$, $y_1 = 6/7 \approx .85714$, $y_2 = 3/4 = .75$, $y_3 = 2/3 \approx .66667$, $y_4 = 3/5 = .6$, $y_5 = 6/11 \approx .54545$, $y_6 = 1/2 = .5$. Thus,

$$\int_1^2 \frac{dx}{x} \approx \frac{1}{3\cdot 6}\left[y_0 + 4y_1 + 2y_2 + 4y_3 + 2y_4 + 4y_5 + y_6\right]$$

$$\approx \frac{1}{18}(12.47706) \approx 0.69317.$$

To estimate the error, $f(x) = \frac{1}{x}$, $f'(x) = -\frac{1}{x^2}$, $f^{(3)}(x) = \frac{2}{x^3}$, and $f^{(4)}(x) = -\frac{6}{x^4}$. Since $\left|-\frac{6}{x^4}\right| < 6$ on $1 \le x \le 2$, the error satisfies

$$|E_S| \le \frac{b-a}{180} h^4 M = \frac{1}{180} \cdot \frac{1}{1296} \cdot 6 \approx .00003.$$

Therefore, the approximation is accurate to 4 decimal places.

15. $$\int_0^{\pi} f(x)\,dx = \int_0^{\pi/2} \sin x\,dx + \int_{\pi/2}^{\pi} \pi x\,dx$$

$$= [-\cos x]_0^{\pi/2} + [\tfrac{1}{2}\pi x^2]_{\pi/2}^{\pi} = 1 + \frac{3\pi^3}{8} \approx 12.62735.$$

NOTES.

CHAPTER 5 APPLICATIONS OF INTEGRALS

5.1 AREAS BETWEEN CURVES

OBJECTIVE A: Find the area bounded by two given continuous curves $y = f(x)$ and $y = g(x)$ over an interval $a \le x \le b$. It may be required to calculate the endpoints a and b.

1. Find the area between the curves $y = 1$ and $y = 1 - x^{-2}$ for $1 \le x \le 4$.

Solution.

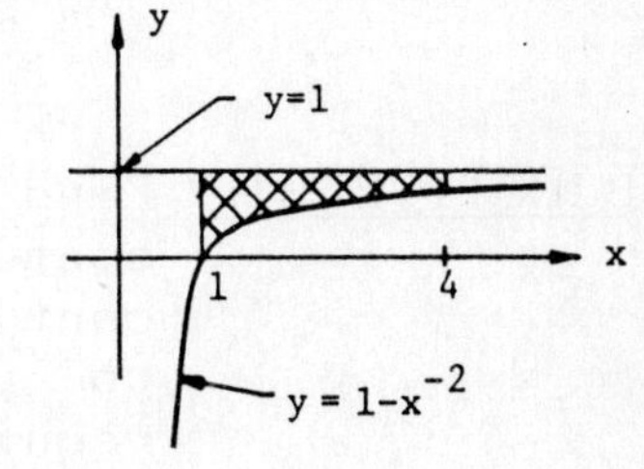

STEP 1: The desired region is shown in the figure at the right.

STEP 2: The limits of integration are already given: a = ______ and b = ______ .

STEP 3: $f(x) - g(x) =$ ____________________

$=$ ____________ (simplified).

STEP 4: The area is given by

$A =$ __________________

$=$ ________$\Big]_1^4 = -\frac{1}{4} - ($________$)$

$=$ __________ .

2. Find the area of the region in the first quadrant bounded above by the curves $y = x$ and $y = 2 - x^2$ and below by the x-axis.

Solution.

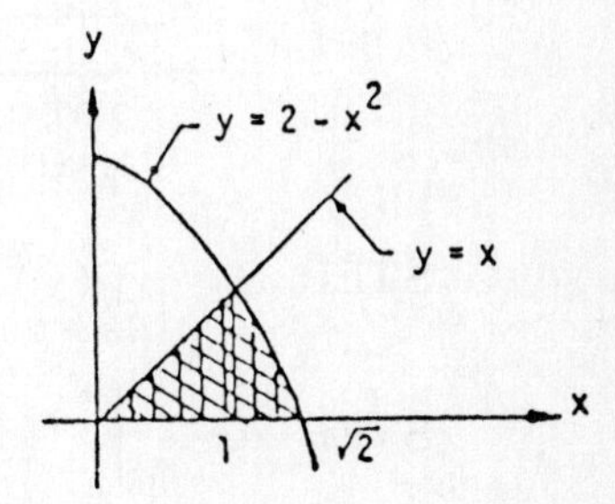

STEP 1: The desired region is shown in the figure at the right. Notice that region has boundaries with changing formulas.

STEP 2: The two curves $y = x$ and $y = 2 - x^2$ intersect when $x = 2 - x^2$ or

$$x^2 + x - 2 = 0.$$

Thus $x =$ _____ or $x =$ _____.

---

1. 1, 4, $1 - (1 - x^{-2})$, x^{-2}, $\int_1^4 x^{-2}\,dx$, $-x^{-1}$, -1, $\frac{3}{4}$

Since $x \geq 0$ in the first quadrant we must pick $x =$ ______ . Thus we partition the region into two subregions and sum the integrations over the intervals $[0,1]$ and $[1,\sqrt{2}]$.

STEP 3: For the interval $[0,1]$: $f(x) - g(x) =$ ____________ .
For the interval $[1,\sqrt{2}]$: $f(x) - g(x) =$ ____________ .

STEP 4: The desired area is given by

$$A = \int_0^1 \underline{\qquad\qquad} dx + \int_1^{\underline{\quad}} (2 - x^2)\,dx$$

$$= \underline{\qquad\qquad}\Big]_0^1 + \underline{\qquad\qquad\qquad}\Big]_1^{\sqrt{2}}$$

$$= \frac{1}{2} + \left(2\sqrt{2} - \frac{1}{3}(\sqrt{2})^3\right) - \left(\underline{\qquad\qquad}\right) = \frac{4\sqrt{2}}{3} - \frac{7}{6} \approx 0.719.$$

OBJECTIVE B: Find the area of the region bounded on the left by the continuous curve $x = g(y)$ and on the right by the continuous curve $x = f(y)$ as y varies from $y = a$ to $y = b$. The calculation of a and b may be required.

3. Find the area bounded by the curves $y^2 = x$ and $y = 6 - x$.

Solution.

STEP 1: The curve on the left is $x = y^2$ and the curve on the right is $x = 6 - y$. The two curves and the region trapped between them are depicted in the figure at the right.

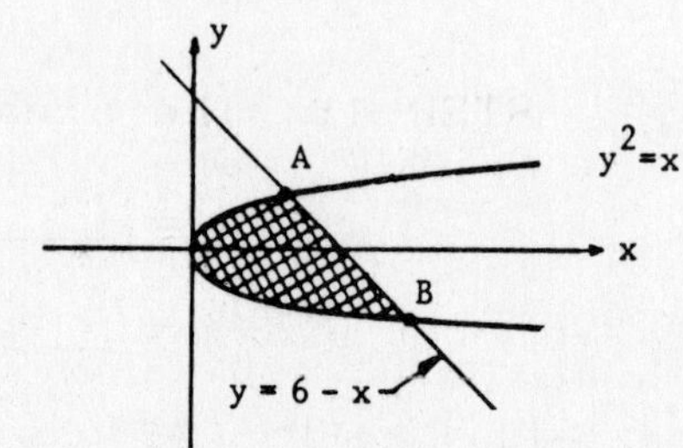

STEP 2: Next we find the y-coordinates of the points A and B of intersection of the two curves. The intersection occurs when $x = y^2$ is the same as $x = 6 - y$ or ______________ . Equivalently, $y^2 + y - 6 = 0$, or $y =$ ______ or $y =$ ______ . The lower limit of integration is $a =$ ______ and the upper limit is $b =$ ______ .

STEP 3: $f(y) - g(y) =$ ________________ .

STEP 4: $\int_a^b [f(y) - g(y)]\,dy = \int_{-3}^{2} (6 - y - y^2)\,dy$

---

2. 1, -2, 1, x - 0, $(2 - x^2) - 0$, x, $\int_1^{\sqrt{2}}$, $\frac{1}{2}x^2$, $2x - \frac{1}{3}x^3$, $2 - \frac{1}{3}$

$= \underline{\hspace{4cm}}\Big]_{-3}^{2}$

$= (12 - 2 - \frac{8}{3}) - (\underline{\hspace{2cm}}) = \underline{\hspace{1.5cm}} \approx \underline{\hspace{2cm}}$.

5.2 FINDING VOLUMES BY SLICING

OBJECTIVE: Use the method of slicing to calculate the volume of a solid whose base is given and whose cross sectional areas are specified.

4. Find the volume of the solid whose base is a triangle cut from the first quadrant by the line $x + 5y = 5$ and whose cross sections perpendicular to the x-axis are semicircles.

Solution.

STEP 1: A typical cross section of the solid is shown at the right.

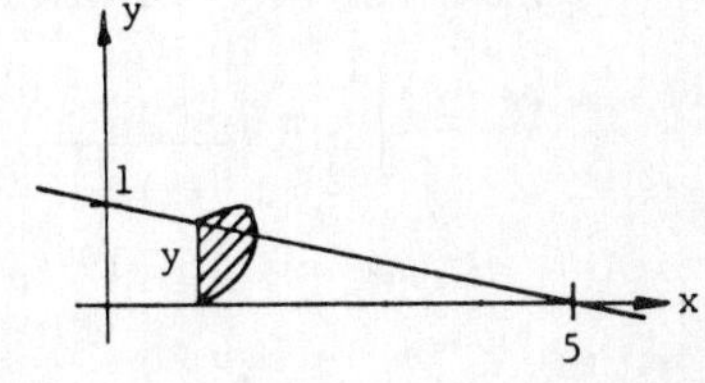

STEP 2: From the equation $x + 5y = 5$ we find that $y = \underline{\hspace{2cm}}$. Since y is the diameter of the semicircle, the cross sectional area given by the area of the semicircle is $A(x) = \frac{\pi}{2}\left(\underline{\hspace{1.5cm}}\right)^2 = \underline{\hspace{2cm}}$.

STEP 3: The semicircles go from $x = 0$ to $x = 5$.

STEP 4: The volume of the solid is given by

$$V = \int_0^5 A(x)\,dx = \int_0^5 \frac{1}{8}\pi y^2\,dx = \int_0^5 \frac{\pi}{8}\left(\underline{\hspace{3cm}}\right)dx$$

$$= \frac{\pi}{8}\left[\frac{x^3}{75} - \frac{x^2}{5} + x\right]_0^5 = \frac{\pi}{8}\left(\underline{\hspace{3cm}}\right)$$

$= \underline{\hspace{1.5cm}} \approx \underline{\hspace{2.5cm}}$ cubic units.

5.3 VOLUMES OF SOLIDS OF REVOLUTIONS -- DISKS AND WASHERS

OBJECTIVE A: Calculate the volume of a solid of revolution generated by the graph $y = f(x)$ of a continuous function over $a \le x \le b$ rotated about the x-axis.

---

3. $y^2 = 6 - y$, 2, -3, -3, 2, $(6 - y) - y^2$, $6y - \frac{1}{2}y^2 - \frac{1}{3}y^3$, $-18 - \frac{9}{2} + 9$, $20\frac{5}{6}$, 20.83333

4. $-\frac{1}{5}x + 1$, $\frac{y}{2}$, $\frac{1}{8}\pi y^2$, $\frac{x^2}{25} - \frac{2x}{5} + 1$, $\frac{5}{3} - 5 + 5$, $\frac{5\pi}{24}$, 0.655

5. Calculate the volume of the solid generated when $y = x^2 - 6x$ is rotated about the x-axis for $0 \le x \le 6$.

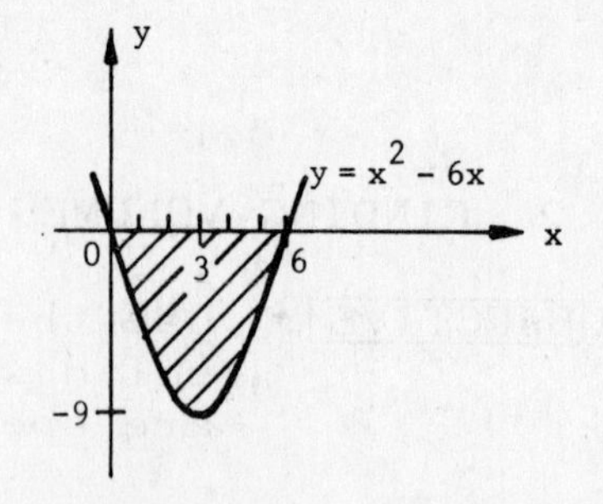

Solution. A graph is shown in the figure at the right. Thus, the volume generated is given by

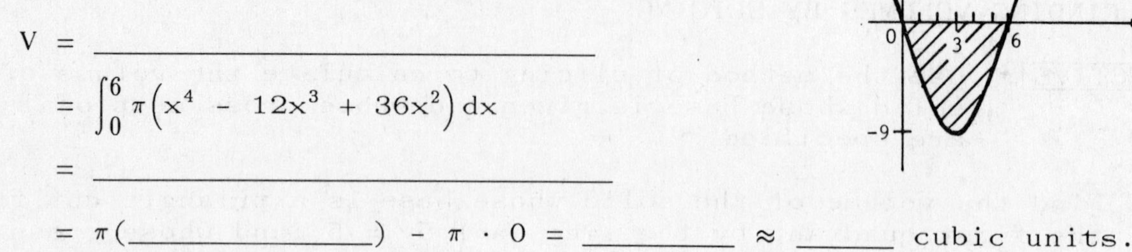

$V =$ ____________________

$= \int_0^6 \pi\left(x^4 - 12x^3 + 36x^2\right) dx$

$=$ ____________________

$= \pi($__________$) - \pi \cdot 0 =$ ______ $\approx$ ______ cubic units.

6. For the volume generated by the graph of $y = |x|$ rotated about the x-axis from $x = -2$ to $x = 1$,

$V = \int_{-2}^{1} \pi($______$)^2\, dx = \int_{-2}^{1}$ ____________

$= \pi[$______$]_{-2}^{1} = \pi($__________$) =$ ______ $\approx$ ______ cubic units.

OBJECTIVE B: Use the method of washers to find the volume generated when a given planar region is rotated about a specified axis or line.

7. The region bounded by the graph of $y^2 = x^2 - 9$, the x-axis and the line $x = 5$ is rotated about the y-axis. Find the volume of the solid.

Solution.

STEP 1: The region is shown in the figure at the right.

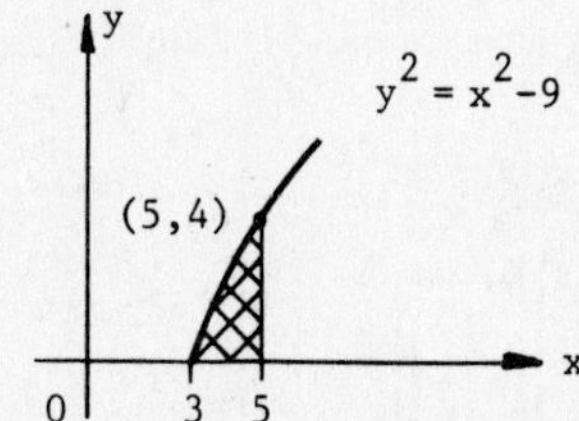

STEP 2: When line segments are drawn through the region and perpendicular to the y-axis, the axis of revolution, we see that the region is intersected from $y = 0$ to $y = 4$. Thus the limits of integration are $c = 0$ and $d = 4$.

STEP 3: As a line segment rotates about the y-axis it sweeps out a washer of inner radius $r = x =$ ________, and outer radius $R =$ ______ . The area of the face of such a washer is $\pi R^2 - \pi r^2 = \pi($__________$)$.

---

5. $\int_0^6 \pi(x^2 - 6x)^2\, dx$, $\pi[\frac{1}{5}x^5 - 3x^4 + 12x^3]_0^6$, $\frac{1}{5}(6)^5 - 3(6)^4 + 12(6)^3$, $\frac{1296\pi}{5}$, 814.3

6. $|x|$, $\pi x^2\, dx$, $\frac{1}{3}x^3$, $\frac{1}{3} - \left(\frac{-8}{3}\right)$, 3π, 9.42

STEP 4: The volume of the solid is

$$V = \int_0^4 \_\_\_\_\_\_\_\_\_\_ \, dy = \pi[\_\_\_\_\_\_\_\_\_\_]_0^4 = \_\_\_\_\_\_\_\_\_\_ \text{ cubic units.}$$

8. Consider the volume generated by rotating the planar region bounded by $y = x^2$ and $y = |x| + 2$ about the y-axis. A sketch of the region is shown at the right. The two curves intersect at the points where $y = x^2$ is the same as $y = |x| + 2$. For $x > 0$, this occurs when $x^2 = x + 2$ or $x =$ __________ . From symmetry it follows that the points of intersection are ________ and ________ . Imagine the solid to be cut into thin slices by planes perpendicular to the y-axis. As y varies from $y = 0$ to $y = 2$, each slice is a thin washer of thickness Δy, inner radius 0, and outer radius $x =$ ________ . As y varies from $y = 2$ to $y = 4$, each slice is a thin washer of thickness Δy, inner radius $x =$ ________ and outer radius $x = \sqrt{y}$. Therefore, the volume is given by

$$V = \int_0^2 \pi(y - \_\_\_\_\_\_) \, dy + \int_2^4 \pi[\_\_\_\_\_\_\_\_\_\_] \, dy$$

$$= \frac{\pi y^2}{2}\Big]_0^2 + \pi\Big(\_\_\_\_\_\_\_\_\_\_\Big)\Big]_2^4 = 2\pi + \pi\left[\left(8 - \frac{8}{3}\right) - \left(\_\_\_\_\_\_\_\_\_\_\right)\right]$$

$$= \_\_\_\_\_\_\_\_\_\_ \approx 16.76 \text{ cubic units.}$$

5.4 CYLINDRICAL SHELLS

OBJECTIVE: Use the method of cylindrical shells to find the volume generated when a given planar region is rotated about a specified axis.

9. Compute the volume generated by rotating about the y-axis the region bounded by $y = 2x$ and $y = x^2 - 2x$.

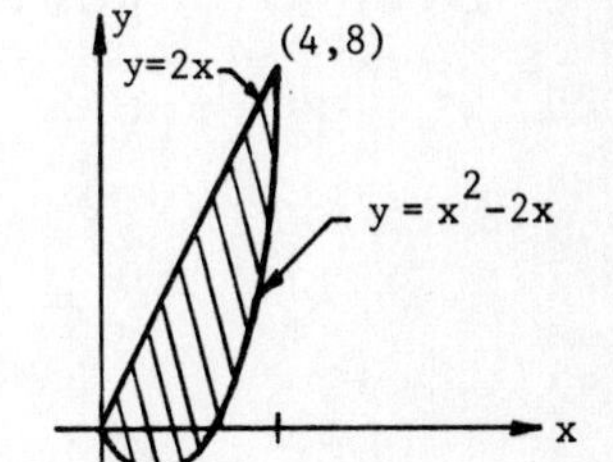

Solution.

STEP 1: A graph of the plane region to be rotated is shown in the figure at the right. We begin with a vertical strip of width Δx, so we integrate with respect to x. A typical height of this strip is $y - 2x$ minus $y =$ __________ . The volume generated

---

7. $\sqrt{y^2 + 9}$, 5, $16 - y^2$, $\pi\left(16 - y^2\right)$, $16y - \frac{1}{3}y^3$, $\frac{128\pi}{3}$

8. 2, (-2,4), (2,4), $\sqrt{y}$, $y - 2$, 0 $y - (y - 2)^2$, $\frac{y^2}{2} - \frac{(y-2)^3}{3}$, $2 - 0$, $\frac{16\pi}{3}$

by revolving this strip about the y-axis is a hollow cylindrical shell of inner circumference ________, inner height ________, and thickness ________ .

STEP 2: For the limits of integration we see that x runs from a = ________ to b = ________ .

STEP 3: We integrate to find the volume:

$$V = \int_0^4 \_\_\_\_\_\_\_\_ \, dx = 2\pi \int_0^4 \_\_\_\_\_\_\_\_ \, dx$$

$$= 2\pi[\_\_\_\_\_\_\_\_]_0^4 = 128\pi\left(\frac{4}{3} - \_\_\_\_\_\_\_\_\right)$$

$$= \_\_\_\_\_\_\_\_ \approx 134.04 \text{ cubic units.}$$

10. The region bounded by the curves $y = x^2$ and $y^2 = x$ is rotated about the x-axis to generate a solid. Find the volume of the solid.

Solution.

STEP 1: The region is shown in the figure at the right. Using the method of shells, we begin with a horizontal strip of width Δy above the x-axis. Thus we integrate with respect to y.

y
y=x²
y²=x
(1,1)
x

A typical length of this strip is $x = \sqrt{y}$ minus $x = y^2$, or in terms of y, ________ . The volume generated by revolving this strip about the x-axis is a hollow cylindrical shell of inner circumference ________, inner length ________, and wall thickness ________ .

STEP 2: For the limits of integration we see that y varies from c = ________ to d = ________ .

STEP 3: We integrate to find the volume:

$$V = \int_0^1 \_\_\_\_\_\_\_\_ \, dy = 2\pi \int_0^1 (\_\_\_\_\_\_\_\_) \, dy$$

$$= 2\pi[\_\_\_\_\_\_\_\_]_0^1 = 2\pi\left(\frac{2}{5} - \frac{1}{4}\right) = \_\_\_\_\_\_\_\_ \approx 0.94248$$

cubic units.

---

9. $x^2 - 2x$, $2\pi x$, $2x - (x^2 - 2x)$, Δx, 0, 4, $2\pi x(4x - x^2)$, $4x^2 - x^3$, $\frac{4}{3}x^3 - \frac{1}{4}x^4$, 1, $\frac{128\pi}{3}$

10. $\sqrt{y} - y^2$, $2\pi y$, $\sqrt{y} - y^2$, Δy, 0, 1, $2\pi y(\sqrt{y} - y^2)$, $y^{3/2} - y^3$, $\frac{2}{5}y^{5/2} - \frac{1}{4}y^4$, $\frac{3\pi}{10}$

5.5 LENGTHS OF PLANE CURVES

OBJECTIVE: Find the length of a smooth curve between two specified points, when the curve is defined by an equation that gives y as a continuously differentiable function of x, or gives x as a continuously differentiable function of y.

11. Find the length of the curve

$$y = \frac{2}{3}(x - 1)^{3/2}, \quad 1 \le x \le 5 .$$

Solution. We use the length equation

$$L = \int_a^b \sqrt{1 + \left(\frac{dy}{dx}\right)^2}\, dx,$$

with a = _____ and b = _____ . Now, $\frac{dy}{dx}$ = ________ , so that $1 + \left(\frac{dy}{dx}\right)^2$ = ________ . The arc length is given by

$$L = \int_1^5 \sqrt{1 + \left(\frac{dy}{dx}\right)^2}\, dx = \int_1^5 \underline{\qquad\qquad}\, dx$$

$$= \underline{\qquad}\Big]_1^5 = \underline{\qquad\qquad} \approx 6.78689 \text{ units.}$$

12. Let the curve be defined by the equation $y^2 = x^3$. To find the arc length from the point (0,0) to the point (4,8), we first calculate $\frac{dy}{dx}$. Differentiating the equation for the curve implicitly gives $2y\frac{dy}{dx}$ = _____ so that $1 + \left(\frac{dy}{dx}\right)^2 = 1 + \frac{\underline{\qquad}}{4y^2}$ = ________ . Therefore, the arc length is given by

$$L = \int_{\_\_}^{\_\_} \underline{\qquad\qquad}\, dx = \frac{4}{9} \cdot \frac{2}{3} \Big[\underline{\qquad\qquad}\Big]_0^4$$

$$= \frac{8}{27} (\underline{\qquad\qquad}) \approx 9.073 \text{ units.}$$

5.6 AREAS OF SURFACES OF REVOLUTION

OBJECTIVE: Find the area of a surface generated by rotating the portion of a smooth curve $y = f(x)$ between $x = a$ and $x = b$ about a specified axis or line parallel to an axis. (The function f is assumed to have a derivative that is continuous on the interval.)

---

11. 1, 5, $\sqrt{x - 1}$, x, $\sqrt{x}$, $\frac{2}{3}x^{3/2}$, $\frac{2}{3}(5\sqrt{5} - 1)$

12. $3x^2$, $9x^4$, $1 + \frac{9x}{4}$, $\int_0^4 \sqrt{1 + \frac{9x}{4}}\, dx$, $\left(1 + \frac{9x}{4}\right)^{3/2}$, $10\sqrt{10} - 1$

13. Find the surface area when the arc $y = 2\sqrt{x}$ from $x = 0$ to $x = 8$ is rotated about the x-axis.

Solution. We evaluate the formula

$$S = \int_a^b 2\pi y \sqrt{1 + \left(\frac{dy}{dx}\right)^2}\, dx$$

with $a = 0$, $b = 1$, $y = 2\sqrt{x}$, and $\frac{dy}{dx} =$ ________ .
Thus,

$$S = \int_{\_\_}^{\_\_} \_\_\_\_\_\_\_\_\_\_\_\_ \, dx = \int_{\_\_}^{\_\_} \_\_\_\_\_\_\_\_\_\_\_\_ \, dx$$

$$= 4\pi(\_\_\_\_\_\_\_\_\_\_)]_0^8 = \frac{8\pi}{3}(\_\_\_\_\_ - 1)$$

$$= \_\_\_\_\_\_\_\_ \approx 217.81709 \text{ sq. units.}$$

14. Find the surface area generated when the arc $y = x^{1/3}$ from $x = 0$ to $x = 1$ is rotated about the y-axis.

Solution. We evaluate the formula

$$S = \int_c^d 2\pi x \sqrt{1 + \left(\frac{dx}{dy}\right)^2}\, dy$$

since the rotation is about the y-axis. To find the limits of integration we note that at $x = 0$, $y = 0$ and at $x = 8$, $y = 2$. Thus, $c =$ _____ and $d =$ _____ . Since $x = y^3$,

$\frac{dx}{dy} =$ __________ and $\sqrt{1 + \left(\frac{dx}{dy}\right)^2}\, dy =$ ____________ dy. Thus,

$$S = 2\pi \int_{\_\_}^{\_\_} \_\_\_\_\_\_\_\_\_\_\_\_ \, dy$$

$$= 2\pi\left(\frac{1}{36} \cdot \frac{2}{3}\right)(\_\_\_\_\_\_\_\_\_\_\_\_)]_{\_\_}^{\_\_} = \frac{\pi}{27}(\_\_\_\_\_\_\_\_\_\_)$$

$$\approx 203.0436 \text{ sq. units.}$$

5.7 MOMENTS AND CENTERS OF MASS

OBJECTIVE A: Find the moment, mass, and center of mass of a thin rod or strip of specified density lying along the x-axis.

15. Find the center of mass of a thin rod of density $\delta(x) = 1 + x$ lying along the interval $0 \le x \le 2$.

Solution.

$$M_0 = \int_0^2 x(1 + x)\, dx = \int_0^2 (x + x^2)\, dx = \_\_\_\_\_\_\_\_\_\_\_\_]_0^2 = \frac{14}{3},$$

---

13. $x^{-1/2}$, $\sqrt{1 + x^{-1}}$, $\int_0^8 4\pi\sqrt{x}\,\sqrt{1 + x^{-1}}\, dx$, $\int_0^8 4\pi\sqrt{x + 1}\, dx$, $\frac{2}{3}(x + 1)^{3/2}$, $9^{3/2}$, $\frac{208\pi}{3}$

14. 0, 2, x, $3y^2$, $\sqrt{1 + 9y^4}$, $\int_0^2 y^3\sqrt{1 + 9y^4}\, dy$, $\left(1 + 9y^4\right)^{3/2}]_0^2$, $(145)^{3/2} - 1$

$M =$ ________________ $= \left(x + \frac{x^2}{2}\right]_0^2 =$ ______ ,

$\overline{x} =$ __________ $=$ ________ .

OBJECTIVE B: Find the center of mass of a thin plate of given density δ covering a given region of the xy-plane.

16. Find the center of mass $(\overline{x}, \overline{y})$ of the thin triangular plate of constant density δ formed in the first quadrant by the coordinate axes and the line $x + 2y = 2$.

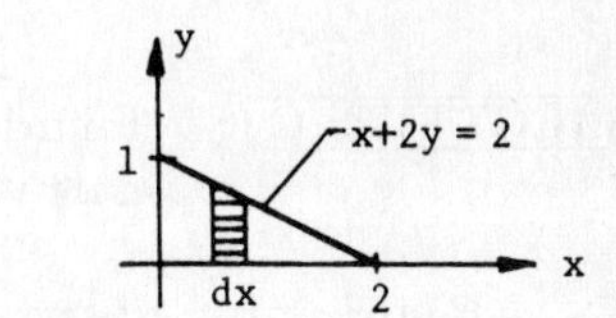

Solution. The plate is depicted in the figure at the right. Let's find $(\overline{x}, \overline{y})$ using vertical strips of width dx parallel to the y-axis. A representative strip is shown in the figure. The mass of this vertical strip is approximately $dm = \delta\, dA = \delta(y$ ____ $) = \delta($______$)\, dx$. The center of mass of the vertical strip is $(\tilde{x}, \tilde{y}) = (x, \frac{y}{2}) =$ (________). Thus,

$$\overline{x} = \frac{\int \tilde{x}\, dm}{\int dm} = \frac{\int_0^2 x \cdot (\_\_\_\_\_\_)\, dx}{\int_0^2 \delta\left(1 - \frac{x}{2}\right) dx} = \frac{\int_0^2 (\_\_\_\_\_\_)\, dx}{\int_0^2 \left(1 - \frac{x}{2}\right) dx}$$

$$= \frac{\_\_\_\_\_\_\Big]_0^2}{x - \frac{1}{4}x^2\Big]_0^2} = \frac{\_\_\_\_}{2 - 1} = \_\_\_\_\_\_ .$$

Likewise,

$$\overline{y} = \frac{\int \tilde{y}\, dm}{\int dm} = \frac{\int_0^2 (\_\_\_\_\_\_)\, \delta\left(1 - \frac{x}{2}\right) dx}{\int_0^2 \delta\left(1 - \frac{x}{2}\right) dx} = \frac{\delta \int_0^2 (\_\_\_\_\_\_\_\_)\, dx}{\delta(2 - 1)}$$

$$= \frac{1}{2}x - \frac{x^2}{4} + \frac{x^3}{8}\Big]_0^2 = \_\_\_\_\_\_ .$$

Alternatively we could find $(\overline{x}, \overline{y})$ just as easily (in this problem) using horizontal strips. A representative strip of width dy is shown in the figure at the right. The mass of this horizontal strip is approximately

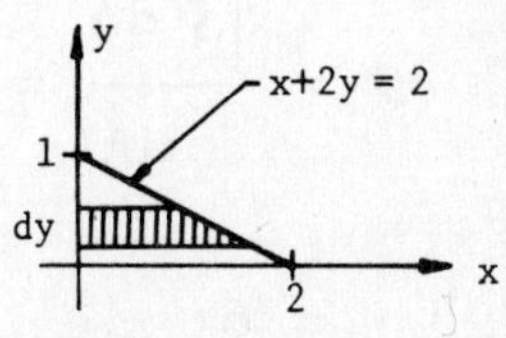

$$dm = \delta\, dA = \delta(\_\_\_\_ dy) = \delta(\_\_\_\_\_\_)\, dy.$$

The center of mass of the horizontal strip is $(\tilde{x}, \tilde{y}) = \left(\frac{x}{2}, y\right) =$ (________). Thus, to compute $\overline{y}$ for instance, we find

---

15. $\frac{x^2}{2} + \frac{x^3}{3}$, $\int_0^2 (1 + x)\, dx$, 4, $\frac{M_0}{M}$, $\frac{7}{6}$

$$\bar{y} = \frac{\int \tilde{y}\,dm}{\int dm} = \frac{\int_0^1 y \cdot (\_\_\_\_\_)\,dy}{\int_0^1 \delta(2 - 2y)\,dy} = \frac{\int_0^1 (\_\_\_\_\_)\,dy}{\int_0^1 (1 - y)\,dy}$$

$$= \frac{\_\_\_\_\_\Big]_0^1}{2y - y^2\Big]_0^1} = \frac{\_\_\_\_}{2 - 1} = \frac{1}{3}.$$

This calculation for $\bar{y}$ agrees with our result using vertical strips. The ease of finding the integrals using vertical or horizontal strips was about the same in this problem, but that may not always be the case for more complicated planar regions.

OBJECTIVE C: Find the centroid of the planar region bounded by given curves and lines.

17. Find the centroid of the region bounded by the parabola $y = x^2$ and the line $y = x$.

<u>Solution</u>. The region is illustrated at the right. To calculate $\bar{x}$, divide the region into vertical strips of width dx. The area of a typical strip is given by $dA = (\_\_\_\_\_\_)\,dx$ and the center of mass of the vertical strip is $(\tilde{x}, \tilde{y}) = \left(x, \frac{x - x^2}{2}\right)$. Hence,

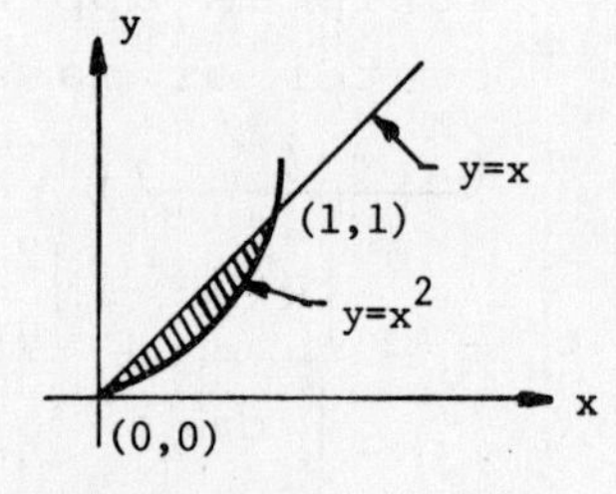

$$\bar{x} = \frac{\int \tilde{x}\,dA}{\int dA} = \frac{\int_0^1 \_\_\_\_\_\_\,dx}{\int_0^1 (x - x^2)\,dx} \qquad (\delta = \text{constant } 1)$$

$$= \frac{\_\_\_\_\_\_\Big]_0^1}{\frac{1}{2}x^2 - \frac{1}{3}x^3\Big]_0^1} = \frac{\_\_\_\_\_\_}{\frac{1}{2} - \frac{1}{3}} = \frac{\_\_\_\_\_}{\frac{1}{6}} = \_\_\_\_\_\_ .$$

To find $\bar{y}$, divide the region into horizontal strips of width dy. The area of a typical strip is $dA = (\_\_\_\_\_\_)\,dy$ and the center of mass of the horizontal strip is $(\tilde{x}, \tilde{y}) = \left(\frac{\sqrt{y} - y}{2}, y\right)$. Then,

$$\bar{y} = \frac{\int \tilde{y}\,dA}{\int dA} = \frac{\_\_\_\_\_\_\_\_}{\frac{1}{6}} = \frac{\_\_\_\_\_\_\_\Big]_0^1}{\frac{1}{6}}$$

$$= 6(\_\_\_\_\_\_\_) = 6(\_\_\_\_\_) = \_\_\_\_\_ .$$

---

16. dx, $1 - \frac{x}{2}$, $\left(x, \frac{1}{2} - \frac{x}{4}\right)$, $\delta\left(1 - \frac{x}{2}\right)$, $x - \frac{1}{2}x^2$, $\frac{1}{2}x^2 - \frac{1}{6}x^3$, $\frac{2}{3}$, $\frac{2}{3}$, $\frac{1}{2} - \frac{x}{4}$, $\frac{1}{2} - \frac{x}{2} + \frac{x^2}{8}$, $\frac{1}{3}$, x, $2 - 2y$, $(1-y, y)$, $\delta(2 - 2y)$, $2(y - y^2)$, $y^2 - \frac{2}{3}y^3$, $1 - \frac{2}{3}$

17. $x - x^2$, $x(x - x^2)$, $\frac{1}{3}x^3 - \frac{1}{4}x^4$, $\frac{1}{3} - \frac{1}{4}$, $\frac{1}{12}$, $\frac{1}{2}$, $\sqrt{y} - y$, $\int_0^1 y\left(\sqrt{y} - y\right)dy$, $\frac{2}{5}y^{5/2} - \frac{1}{3}y^3$, $\frac{2}{5} - \frac{1}{3}$, $\frac{1}{15}$, $\frac{2}{5}$

5.8 WORK

OBJECTIVE A: For a mechanical spring of given natural length L, if a specified force is required to stretch or compress the spring by a certain given amount, calculate (a) the "spring constant" c and (b) the amount of work done in stretching or compressing the spring from a specified length $L = a$ to a specified length $L = b$.

18. Suppose an unstretched spring is 3 feet long. When the spring is used to suspend a 4-pound weight, it stretches to a length of 5 feet. Therefore, the equation $F = cs$ becomes ________, or $c =$ ______ is the value of the spring constant. To calculate the work required to stretch the spring from a length of 4 feet to a length of 6 feet, we find

$$W = \int F\,ds = \int_1^{\_\_} (\_\_\_\_\_\_)\,dx = \_\_\_\_\_\_ \text{ foot-pounds.}$$

OBJECTIVE B: Find the work done in pumping all the water out of a specified container to a given height above the container.

19. A cylindrical tank of radius 5 feet and height 10 feet stands on a platform so that its bottom is 50 feet above the surface of a lake. Water is pumped directly up from the surface of the lake into the water tank. Find the depth d of water in the tank when half the necessary work has been done to fill the tank.

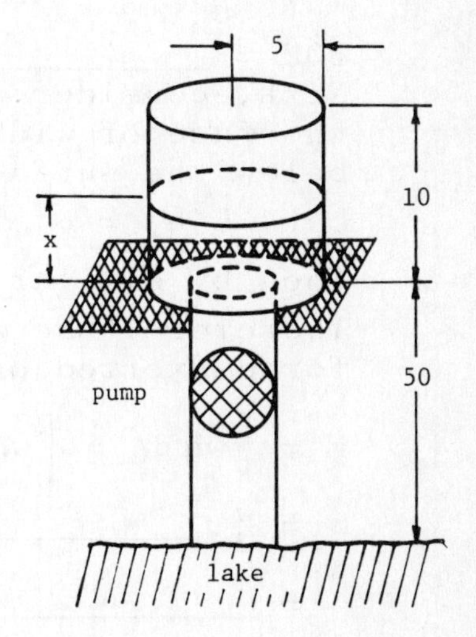

Solution. Let the distance from the bottom of the tank to the surface of the water in the tank be x feet. The situation is depicted in the figure at the right. Consider a slice of water in the tank of thickness dx cut by two planes parallel to the surface of the lake and platform. The volume of this slice is $dV = A\,dx =$ ______. Therefore, its weight is $w\,dV =$ ______, where w is the weight of a cubic foot of water. The total distance this slice is lifted is $50 + x$ feet, where x is its typical height from the bottom of the tank. Therefore the total work required to fill the tank is

$$W_1 = \int \text{distance} \cdot w\,dV = \int_0^{10} (\_\_\_\_\_\_\_\_)\,dx$$

$$= 25\pi w(\_\_\_\_\_\_\_\_)\Big]_0^{10} = 25\pi w(\_\_\_\_\_\_).$$

On the other hand, the work required to fill the tank to the depth d is given by

---

18. $4 = c(5 - 3)$, 2, $\int_1^3 2x\,dx$, 8

$$W_2 = \int_0^d (\underline{\qquad\qquad}) \, dx = 25\pi w(\underline{\qquad\qquad\qquad}).$$

We want to find d when $W_2 = \frac{1}{2} W_1$. This translates into

$$25\pi w(\underline{\qquad\qquad}) = \frac{1}{2}(25\pi w)(550) \quad \text{or} \quad \underline{\qquad\qquad} = 550.$$

Solving this quadratic equation for d (which must be positive) gives

$$d = \frac{-100 \pm \sqrt{10000 + \underline{\qquad\qquad}}}{2}, \quad \text{or} \quad d \approx \underline{\qquad\qquad} \text{ feet.}$$

5.9 FLUID PRESSURES AND FORCES

OBJECTIVE: Find the total force exerted on a given planar region placed vertically under the surface of an incompressible fluid of constant weight density w.

20. Suppose the face of a dam has the shape of an isosceles trapezoid of altitude 20 feet with an upper base of 30 feet and a lower base of 20 feet. A figure illustrating the face of the dam is shown at the right. Let us find the force due to water pressure on the dam when the surface of the water is level with the top of the dam.

From the figure and similar triangles, we see that $\frac{20}{H} = \frac{30}{\underline{\qquad}}$ or

$30H = \underline{\qquad\qquad}$; thus $H = \underline{\qquad}$ feet.

Next, consider a horizontal strip of the face of width dh at a depth h below the surface of the water (see figure).

Let $L(h) = \ell$ denote the horizontal length of this strip.

Then by similar triangles, $\frac{\ell}{60 - h} = \frac{30}{\underline{\qquad}}$, or $\ell = \underline{\qquad\qquad}$.

Therefore, the area of the strip is $dA = \ell dh$, so the total force exerted on the face of the dam is

$$F = \int wh \, dA = \int wh\ell \, dh = \int_0^{20} \underline{\qquad\qquad} = \frac{w}{2} \int_0^{20} (\underline{\qquad\qquad}) \, dh$$

$$= \frac{w}{2} [\underline{\qquad\qquad}]_0^{20} = 200w\left(30 - \underline{\qquad}\right)$$

$$= \underline{\qquad\qquad\qquad} \approx 291{,}667 \text{ pounds using } w = 62.5.$$

---

19. $25\pi\, dx$, $25\pi w\, dx$, $25\pi w(50 + x)$, $50x + \frac{1}{2}x^2$, 550, $25\pi w(50 + x)$, $50d + \frac{1}{2}d^2$, $50d + \frac{1}{2}d^2$, $100d + d^2$, 2200, 5.227

20. $20 + H$, $400 + 20H$, 40, 60, $\frac{1}{2}(60 - h)$, $\frac{w}{2}h(60 - h)\,dh$, $60h - h^2$, $30h^2 - \frac{1}{3}h^3$, $\frac{20}{3}$, $\frac{14000w}{3}$

5.10 THE BASIC PATTERN AND OTHER MODELING APPLICATIONS

OBJECTIVE A: Given a continuous function $v = f(t)$ representing the velocity v of a moving body as a function of time t, find the time intervals in which the velocity is positive and in which it is negative, and calculate the total distance traveled by the body during the specified time interval $t = a$ to $t = b$.

21. Consider the velocity given by $v = \sin(3t - 1)$ for $0 \le t \le 2$. A graph of the velocity is shown in the figure below. From an analysis of the sine curve we observe that the velocity is of periodicity __________, amplitude __________, and horizontal shift __________. Thus, the velocity v is negative for $0 \le t \le \frac{1}{3}$, it is positive for __________, and negative again for __________. Therefore, the total distance traveled by the moving body is

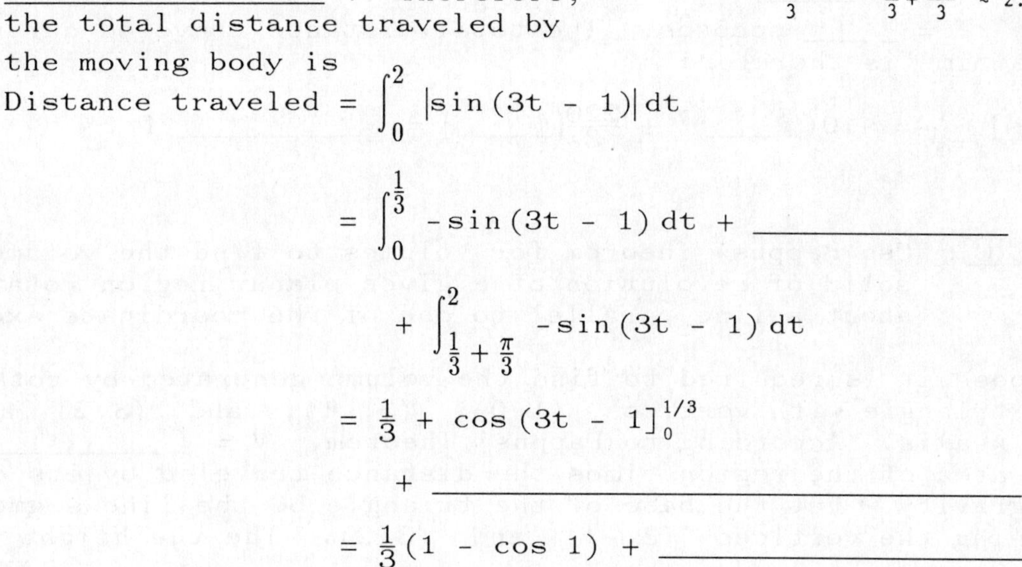

$$\text{Distance traveled} = \int_0^2 |\sin(3t - 1)|\,dt$$

$$= \int_0^{\frac{1}{3}} -\sin(3t - 1)\,dt + \_\_\_\_\_\_\_\_\_\_$$

$$+ \int_{\frac{1}{3}+\frac{\pi}{3}}^{2} -\sin(3t - 1)\,dt$$

$$= \frac{1}{3} + \cos(3t - 1]_0^{1/3}$$

$$+ \_\_\_\_\_\_\_\_\_\_$$

$$= \frac{1}{3}(1 - \cos 1) + \_\_\_\_\_\_\_\_\_\_$$

$$= \_\_\_\_\_\_\_\_\_\_ \approx 1.24779.$$

The shift in the body's position is

$$\text{Position shift} = \int_0^2 \sin(3t - 1)\,dt$$

$$= -\frac{1}{3}\cos(3t - 1)\Big|_0^2 = -\frac{1}{3}\cos 5 + \frac{1}{3}\cos(-1)$$

$$\approx 0.086 \quad \text{units.}$$

---

21. $\frac{2\pi}{3}$, 1, $\frac{1}{3}$, $\frac{1}{3} < t < \frac{1}{3} + \frac{\pi}{3}$, $\frac{1}{3} + \frac{\pi}{3} < t \le 2$, $\int_{\frac{1}{3}}^{\frac{1}{3}+\frac{\pi}{3}} \sin(3t - 1)\,dt$,

$-\frac{1}{3}\cos(3t - 1)]_{\frac{1}{3}}^{\frac{1}{3}+\frac{\pi}{3}} + \frac{1}{3}\cos(3t - 1)]_{\frac{1}{3}+\frac{\pi}{3}}^{2}$, $\frac{2}{3} + \frac{1}{3}(1 + \cos 5)$, $\frac{4}{3} + \frac{1}{3}\cos 5 - \frac{1}{3}\cos 1$

OBJECTIVE B: Given a continuous function $a = f(t)$ representing the acceleration a of a moving body as a function of time t, and given v_0 as its velocity at time $t = 0$, find the total distance traveled by the body during the specified time interval $t = 0$ to $t = b$.

22. Suppose the brakes are applied to a car traveling 50 mph and the brakes give the car a constant negative acceleration of 20 ft/sec^2. How far will the car travel before stopping?

Solution. To make the units consistent in the problem, we convert miles per hour to feet per second. Since 15 mph is equivalent to 22 ft/sec, it follows that 50 mph is equivalent to ________ ft/sec. The problem gives $a =$ _____, so $v = \int a\,dt =$ ______ + C. Since $v_0 = \frac{220}{3}$ when $t = 0$, it follows that C = _____ . Therefore, v = ___________ . Next, the total distance traveled after T seconds is given by $s = \int_0^T \left(-20t + \frac{220}{3}\right)dt =$ ____________________ . When the car stops, v = ______, or $-20T + \frac{220}{3} = 0$, and this occurs at time T = _____ seconds. The total distance traveled during this time is therefore

$$s\Big]_{T=\frac{11}{3}} = -10(\_\_\_\_\_)^2 + \frac{220}{3}(\_\_\_\_\_) = \_\_\_\_\_\_\_\_\_\_ \text{ ft.}$$

OBJECTIVE C: Use Pappus' Theorem for Volumes to find the volume of a solid of revolution of a given planar region rotated about a line parallel to one of the coordinate axes.

23. Suppose it is required to find the volume generated by rotating the triangle with vertices (1,0), (3,-1), and (3,2) about the x-axis. According to Pappus' Theorem, V = __________ is the area of the region times the distance traveled by its center of gravity. Let the base of the triangle be the line segment joining the vertices (3,-1) and (3,2). The the height of the triangle equals 3 - _____ = _____ units, so the area is $A = \frac{1}{2}(2 - \_\_\_\_\_)(2) =$ ______ square units. Next, $\overline{x}$ is located $\frac{2}{3}$ of the distance along the median from the vertex (1,0). That is, $\overline{x} = 1 + \frac{2}{3}(\_\_\_\_\_\_) =$ _______ . Therefore, $V = 2\pi(\_\_\_\_\_\_\_\_)(3) =$ ________ cubic units.

---

22. $\frac{50 \cdot 22}{15} = \frac{220}{3}$, -20, -20t, $\frac{220}{3}$, $-20t + \frac{220}{3}$, $-10T^2 + \frac{220}{3}T$, 0, $\frac{11}{3}$, $\frac{11}{3}$, $\frac{11}{3}$, $\frac{1210}{9} \approx 134.44$

23. $2\pi\overline{x}A$, 1, 2, -1, 3, 3 - 1, $\frac{7}{3}$, $\frac{7}{3}$, 14π

CHAPTER 5 SELF-TEST

1. Find the area between the curves $y = \frac{x^3}{3} - x^2 + 2$ and $y = \frac{x}{3} + 1$.

2. Find the area of the planar region bounded by the curves $x = y^{1/3}$ and $y^2 = x$.

3. Find the volume generated by rotating the ellipse $\frac{x^2}{a^2} + \frac{y^2}{b^2} = 1$ about the x-axis.

4. The base of a certain solid is the region between the planar curves $y = 4 + x^2$ and $y = 12 - x^2$, and the cross sections by planes perpendicular to the x-axis are circles with diameters extending from one curve to the other. Find the volume of the solid.

5. Find the volume generated when the planar region between the lines $y = x$, $y = 3x$ and $x + y = 4$ is rotated about the x-axis.

6. Find the volume generated when the planar region between the curve $y^2 = 9x$ and the lines $x = 4$, $y = 0$ is rotated about the line $x = 5$.

7. Find the length of the curve $y = \frac{2}{3}x^{3/2} - \frac{1}{2}x^{1/2}$ over $0 \leq x \leq 1$.

8. Find the surface area generated by revolving about the x-axis the arc $y = \frac{1}{2\sqrt{2}} x\sqrt{1 - x^2}$, $0 \leq x \leq 1$.

9. If a force of 90 pounds stretches a 10-foot spring by 1 foot, find the work done in stretching the spring from 10 feet to 15 feet.

10. A conical tank is 16 feet across the top, and 12 feet deep. It contains water to a depth of 8 feet. Find the work required to pump all the water to a height of 2 feet above the top of the tank.

11. A water pipe is in the shape of a cylinder placed horizontally in the ground. Its radius is 4 inches. If the pipe is half full of water, find the force on the face of the vertical circular plate that closes off the pipe. Assume the weight-density of the water is $w = 62.5$ pounds per cubic foot.

12. Find the centroid of the planar region bounded by the curves $y = x + 2$, $y = 2x$, and the y-axis.

13. The velocity of a body moving along a coordinate line in meters per second is given by $v(t) = t^3 - \frac{7}{2}t^2 + \frac{3}{2}t$. Find the total distance traveled by the body during the time $0 \leq t \leq 4$ and the shift in the body's position.

SOLUTIONS TO CHAPTER 5 SELF-TEST

1. A figure depicting the region is shown at the right. The two curves intersect when

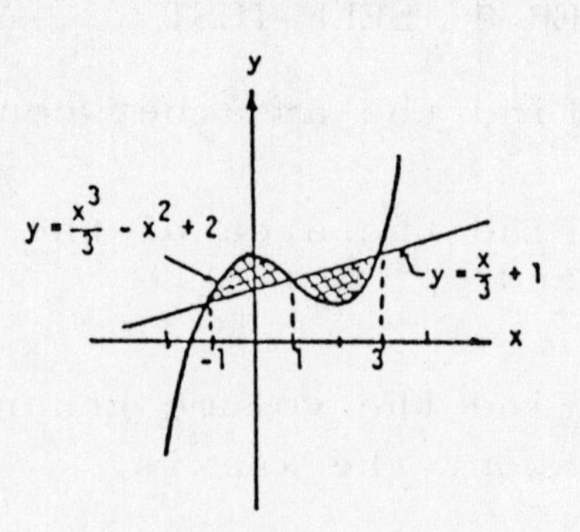

$$\frac{1}{3}x^3 - x^2 + 2 = \frac{1}{3}x + 1$$

or

$$x^3 - 3x^2 - x + 3 = 0$$

or

$$(x + 1)(x - 1)(x - 3) = 0.$$

The points of intersection are $\left(-1, \frac{2}{3}\right)$, $\left(1, \frac{4}{3}\right)$ and $(3, 2)$. The area trapped between the curves is

$$\text{Area} = \int_{-1}^{1}\left[\left(\frac{x^3}{3} - x^2 + 2\right) - \left(\frac{x}{3} + 1\right)dx\right] + \int_{1}^{3}\left[\left(\frac{x}{3} + 1\right) - \left(\frac{x^3}{3} - x^2 + 2\right)\right]dx$$

$$= \left[\frac{x^4}{12} - \frac{x^3}{3} - \frac{x^2}{6} + x\right]_{-1}^{1} + \left[-\frac{x^4}{12} + \frac{x^3}{3} + \frac{x^2}{6} - x\right]_{1}^{3}$$

$$= \left[\left(\frac{1}{12} - \frac{1}{3} - \frac{1}{6} + 1\right) - \left(\frac{1}{12} + \frac{1}{3} - \frac{1}{6} - 1\right)\right]$$

$$+ \left[\left(-\frac{81}{12} + \frac{27}{3} + \frac{9}{6} - 3\right) - \left(-\frac{1}{12} + \frac{1}{3} + \frac{1}{6} - 1\right)\right]$$

$$= \frac{4}{3} + \frac{4}{3} = \frac{8}{3}.$$

2. The planar region is shown at the right. The points of intersection of the two graphs are $(0, 0)$ and $(1, 1)$. Thus the area is given by

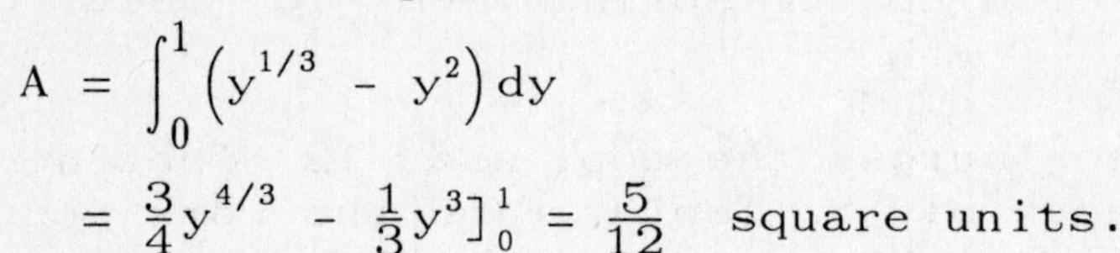

$$A = \int_0^1 \left(y^{1/3} - y^2\right)dy$$

$$= \frac{3}{4}y^{4/3} - \frac{1}{3}y^3\Big]_0^1 = \frac{5}{12} \text{ square units.}$$

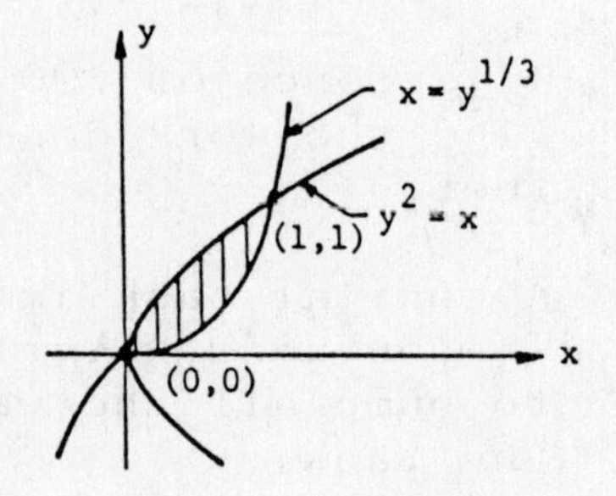

Alternatively,

$$A = \int_0^1 \left(\sqrt{x} - x^3\right)dx$$

$$= \frac{2}{3}x^{3/2} - \frac{1}{4}x^4\Big]_0^1 = \frac{5}{12} \text{ square units.}$$

3. $$V = \int_{-a}^{a} \pi y^2\, dx = \int_{-a}^{a} \pi \cdot \frac{b^2}{a^2}(a^2 - x^2)\, dx$$

$$= \frac{\pi b^2}{a^2}\left[a^2x - \frac{1}{3}x^3\right]_{-a}^{a} = \frac{4}{3}\pi b^2 a.$$

This generalizes the formula for the volume of a sphere, when $a = b = r$ is its radius.

4. The two curves intersect when $4 + x^2 = 12 - x^2$ or $x = \pm 2$. (See the figure at the right depicting the base of the solid.) The cross-sectional area of a typical slice is

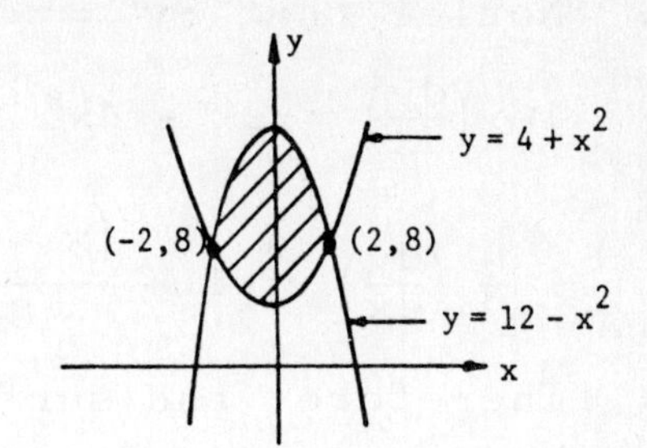

$$\pi\left(\frac{12 - x^2 - (4 + x^2)}{2}\right)^2 = \pi\left(4 - x^2\right)^2$$

$$= \pi(16 - 8x^2 + x^4).$$

Thus the volume is given by

$$V = \int A(x)\,dx = \int_{-2}^{2} \pi(16 - 8x^2 + x^4)\,dx$$

$$= \pi[16x - \tfrac{8}{3}x^3 + \tfrac{1}{5}x^5]_{-2}^{2} = \frac{512\pi}{15} \approx 107.2 \text{ cubic units.}$$

5. The planar region to be rotated about the x-axis is shown in the figure at the right. We slice the region into horizontal slices and use the method of cylindrical shells. Thus, from the figure we see that

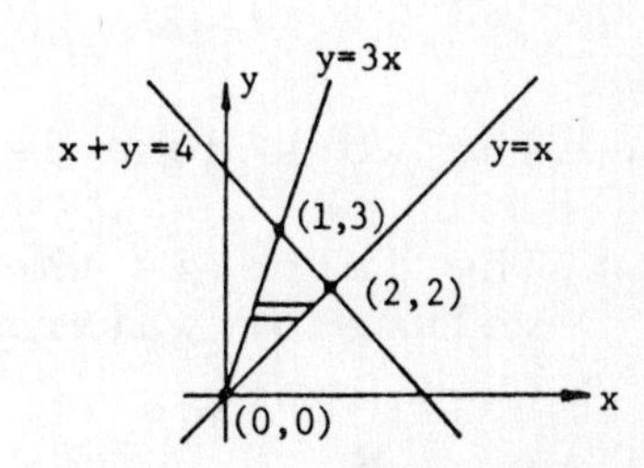

$$V = \int_0^2 2\pi y(y - \tfrac{y}{3})\,dy + \int_2^3 2\pi y(4 - y - \tfrac{y}{3})\,dy$$

$$= 2\pi[\tfrac{2}{9}y^3]_0^2 + 2\pi[2y^2 - \tfrac{4}{9}y^3]_2^3 = \frac{20\pi}{3} \approx 20.9 \text{ cubic units.}$$

6. The planar region to be rotated is shown in the figure at the right. We slice the region into horizontal slices and use the method of washers. Thus, from the figure we find the volume is,

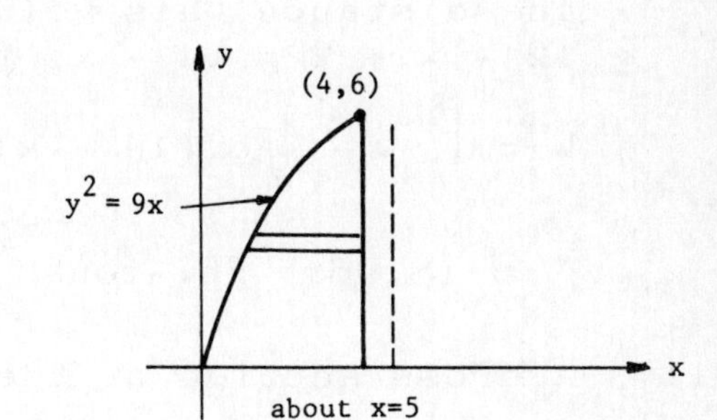

$$V = \int_0^6 \pi\left[\left(5 - \frac{y^2}{9}\right)^2 - 1^2\right]dy$$

$$= \pi\int_0^6 \left(24 - \tfrac{10}{9}y^2 + \tfrac{1}{81}y^4\right)dy$$

$$= \pi[24y - \tfrac{10}{27}y^3 + \tfrac{1}{405}y^5]_0^6 = 83.2\pi \approx 261.4 \text{ cubic units.}$$

7. $y = \frac{2}{3}x^{3/2} - \frac{1}{2}x^{1/2}$ and $\frac{dy}{dx} = x^{1/2} - \frac{1}{4}x^{-1/2}$

$$1 + \left(\frac{dy}{dx}\right)^2 = \left(x - \tfrac{1}{2} + \tfrac{1}{16}x^{-1}\right) + 1 = \left(x^{1/2} + \tfrac{1}{4}x^{-1/2}\right)^2,$$

so the length of the plane curve is given by

$$L = \int_0^1 \left(x^{1/2} + \tfrac{1}{4}x^{-1/2}\right)dx = \tfrac{2}{3}x^{3/2} + \tfrac{1}{2}x^{1/2}]_0^1 = \tfrac{7}{6} \text{ units.}$$

8. Notice that $8y^2 = x^2 - x^4$. Differentiating implicitly,

$16y\left(\frac{dy}{dx}\right) = 2x - 4x^3$ or $\frac{dy}{dx} = \frac{x - 2x^3}{8y}$. Thus,

$$1 + \left(\frac{dy}{dx}\right)^2 = 1 + \frac{\left(x - 2x^3\right)^2}{64y^2} = 1 + \frac{\left(x - 2x^3\right)^2}{8\left(x^2 - x^4\right)} = \frac{\left(3 - 2x^2\right)^2}{8\left(1 - x^2\right)}.$$

Therefore, the surface area is given by

$$S = \int 2\pi y\, ds = \int_0^1 2\pi \frac{x}{2\sqrt{2}} \sqrt{1 - x^2} \cdot \frac{3 - 2x^2}{2\sqrt{2}\sqrt{1 - x^2}}\, dx$$

$$= \frac{\pi}{4}\int_0^1 \left(3 - 2x^2\right) x\, dx = \frac{\pi}{4}\left[\frac{3}{2}x^2 - \frac{1}{2}x^4\right]_0^1 = \frac{\pi}{4} \text{ square units.}$$

9. $F = kx$ so $90 = k \cdot 1$ or $k = 90$. Thus, the work done is given by $W = \int F\, dx = \int_0^5 90x\, dx = 45x^2\Big]_0^5 = 1125$ ft.-lb.

10. The work done is $W = \int F\, ds$.

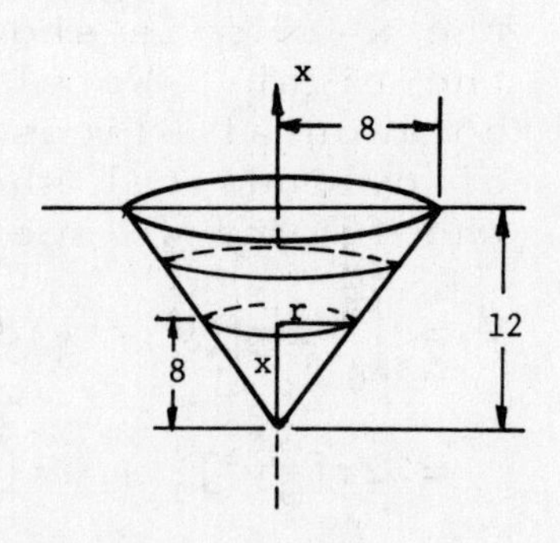

The force is the weight of a typical volume of water. From the figure at the right,

$\frac{r}{x} = \frac{8}{12}$ or $r = \frac{2}{3}x$.

Thus, a typical volume element in a horizontal slice of the water is

$dV = \pi r^2\, dx = \frac{4\pi}{9} x^2\, dx$.

The distance this volume element is lifted is $12 - x + 2 = 14 - x$ feet. Thus,

$$W = \int_0^8 w \cdot \frac{4\pi}{9} x^2 (14 - x)\, dx = \frac{4\pi w}{9}\left(\frac{14}{3}x^3 - \frac{1}{4}x^4\right)\Big]_0^8 = \frac{4\pi w \cdot 8^4}{27}$$

$\approx 18.96\pi$ ft.-tons (assuming $w = 62.5$).

11. A cross section of the half-filled pipe is shown at the right. Here $r = 4$ inches $= \frac{1}{3}$ foot. By the Theorem of Pythagoras,

$\left(\frac{\ell}{2}\right)^2 + h^2 = r^2$ so that

$\ell = \frac{2}{3}\sqrt{1 - 9h^2}$.

The force F is given by

$$F = \int wh\, dA = \int_0^{1/3} wh\ell\, dh = \frac{2w}{3}\int_0^{1/3} h\sqrt{1 - 9h^2}\, dh$$

$$= \left(\frac{2w}{3}\right)\left(\frac{2}{3}\right)\left(-\frac{1}{18}\right)\left(1 - 9h^2\right)^{3/2}\Big]_0^{1/3} = \frac{2w}{81} \approx 1.54 \text{ pounds.}$$

12. The planar region is shown at the right. Using vertical strips,

$$dA = [(x + 2) - 2x]\,dx$$

and

$$(\tilde{x}, \tilde{y}) = \left(x, \frac{2 - x}{2}\right)$$

so that

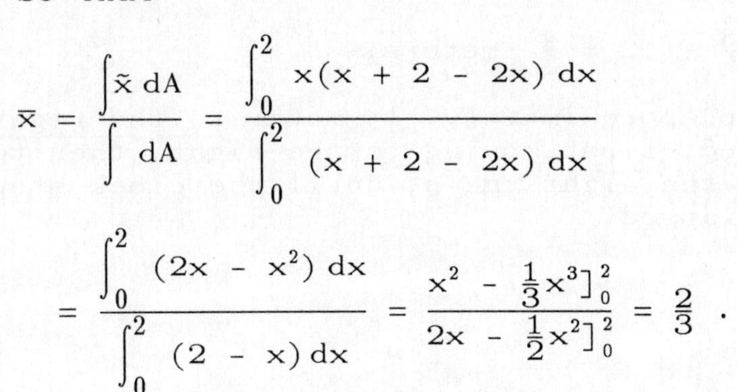

$$\bar{x} = \frac{\int \tilde{x}\,dA}{\int dA} = \frac{\int_0^2 x(x + 2 - 2x)\,dx}{\int_0^2 (x + 2 - 2x)\,dx}$$

$$= \frac{\int_0^2 (2x - x^2)\,dx}{\int_0^2 (2 - x)\,dx} = \frac{x^2 - \frac{1}{3}x^3\big]_0^2}{2x - \frac{1}{2}x^2\big]_0^2} = \frac{2}{3}\,.$$

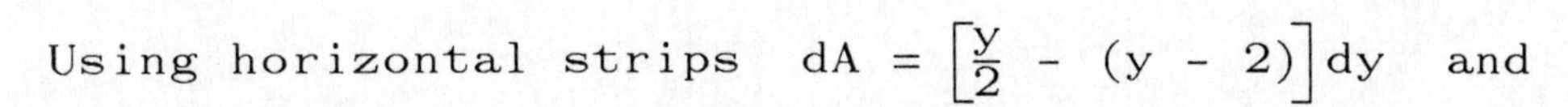

Using horizontal strips $dA = \left[\frac{y}{2} - (y - 2)\right]dy$ and

$(\tilde{x}, \tilde{y}) = \left(\frac{4 - y}{4}, y\right)$. Thus,

$$\bar{y} = \frac{\int \tilde{y}\,dA}{\int dA} = \frac{\int_0^2 y\left(\frac{y}{2}\right)dy + \int_2^4 y\left[\frac{y}{2} - (y - 2)\right]dy}{2}$$

$$= \frac{1}{2}\left[\left(\frac{1}{6}y^3\right)\right]_0^2 + \left(y^2 - \frac{1}{6}y^3\right)\Big]_2^4 = \frac{1}{2}\left(\frac{4}{3} + 16 - \frac{32}{3} - 4 + \frac{4}{3}\right) = 2.$$

13. $v(t) = t^3 - \frac{7}{2}t^2 + \frac{3}{2}t = t\left(t - \frac{1}{2}\right)(t - 3)$

Thus, $v \geq 0$ when $0 \leq t \leq \frac{1}{2}$ or $t \geq 3$; and

$v < 0$ when $\frac{1}{2} < t < 3$.

Hence, we find

$$\text{Distance traveled} = \int_0^{1/2} \left(t^3 - \frac{7}{2}t^2 + \frac{3}{2}t\right)dt$$

$$- \int_{1/2}^3 \left(t^3 - \frac{7}{2}t^2 + \frac{3}{2}t\right)dt + \int_3^4 \left(t^3 - \frac{7}{2}t^2 + \frac{3}{2}t\right)dt$$

$$= \left[\frac{1}{4}t^4 - \frac{7}{6}t^3 + \frac{3}{4}t^2\right]_0^{1/2} - \left[\frac{1}{4}t^4 - \frac{7}{6}t^3 + \frac{3}{4}t^2\right]_{1/2}^3$$

$$+ \left[\frac{1}{4}t^4 - \frac{7}{6}t^3 + \frac{3}{4}t^2\right]_3^4$$

$$= \frac{11}{96} + 9 + \frac{4}{3} \approx 10.45 \text{ meters.}$$

The position shift is simply the integral of the velocity function. Thus,

$$\text{Position shift} = \int_0^4 \left(t^3 - \frac{7}{2}t^2 + \frac{3}{2}t\right)dt$$

$$= \left(\frac{1}{4}t^4 - \frac{7}{6}t^3 + \frac{3}{4}t^2\right]_0^4$$

$$= 64 - \frac{448}{6} + 12 = \frac{4}{3} \quad \text{meters.}$$

That is, the body ends up approximately 1 meter to the right of where it initially started, first moving to the right, then to the left, and finally to the right again, until the clock stops after 4 seconds have elapsed.

NOTES.

CHAPTER 6 TRANSCENDENTAL FUNCTIONS

6.1 INVERSE FUNCTIONS

1. If f and g are inverse functions on suitably restricted domains, then $g(f(x)) =$ ______ and $f(g(y)) =$ ______ . That is, the composite of g and f or of f and g is the ______ function.

2. Given a function $y = f(x)$, to find a formula for the inverse function f^{-1}, solve the equation $y = f(x)$ for ______ in terms of ______ . Interchange the letters x and y. The resulting formula is the inverse ______ .

OBJECTIVE: Use the derivative rule for inverse functions to calculate the derivative of the inverse for a specified function.

3. If f and f^{-1} are inverse functions on suitably restricted domains, then $\frac{df^{-1}}{dx}(b) = \frac{1}{f'(a)}$ where a and b are related by ______ and ______ .

4. Let $f(x) = -6x + 2$ and let f^{-1} denote the inverse of f. We wish to calculate the derivative $\frac{df^{-1}}{dx}(14)$. First, $-6x + 2 = 14$ implies $x =$ ______ . Thus, $f(-2) = 14$ so $a =$ ______ $b =$ ______ in Problem 3. Then, $f'(x) =$ ______ so that $\frac{df^{-1}}{dx}(14) = \frac{1}{f'(\_\_\_)} =$ ______ .

5. To calculate the inverse of $y = -6x + 2$, interchange the letters x and y obtaining ______ . Solving the resultant equation for y yields ______, or $f^{-1}(x) =$ ______ is the inverse function of $f(x) = -6x + 2$. Calculating the derivative $\frac{df^{-1}}{dx}(14)$ directly from the formula for $f^{-1}(x)$ gives $-\frac{1}{6}$ as before.
   Remark. The advantage of the derivative formula for the inverse function given by Theorem 1 of the text is that it provides for the calculation of the derivative $\frac{df^{-1}}{dx}$ even though a formula for the inverse function f^{-1} is not known.

---

1. x, y, identity   2. x, y, $y = f^{-1}(x)$ 3. $b = f(a)$ and $a = f^{-1}(b)$

4. -2, $a = -2$ and $b = 14$, -6, -2, $-\frac{1}{6}$ 5. $x = -6y + 2$, $y = -\frac{1}{6}(x - 2)$, $f^{-1}(x) = -\frac{1}{6}(x - 2)$

6. Let f^{-1} be the inverse of $f(x) = x^2 + 4x - 3$ for $x > -2$. To find $\frac{df^{-1}}{dx}(-6)$, first set $f(x) = x^2 + 4x - 3$ equal to ______ and solve the quadratic equation yielding $x = -3$ or $x =$ ______ . We reject $x = -3$ because -3 is outside the allowable interval $x > -2$. Thus, $a =$ ______ and $b = f(a) =$ ______ . Now $\frac{d}{dx}(x^2 + 4x - 3) =$ ______ so $f'(-1) =$ ______ . Thus, $\frac{df^{-1}}{dx}(-6) = \frac{1}{f'(\_\_\_)} =$ ______ . Notice that we did not need a formula for the inverse function f^{-1} itself.

7. If $y = 4 - 7x$, then the inverse function is $f^{-1}(x) =$ ______ .

6.2 NATURAL LOGARITHMS

8. The natural logarithm is defined by $\ln x =$ ______ for x satisfying ______ .

9. By the first part of the Fundamental Theorem of Calculus, $\frac{d}{dx} \ln x =$ ______ , so the natural logarithm is a continuous function for $x > 0$ because it is ______ .

OBJECTIVE A: Use the four properties of the natural logarithm to rewrite a logarithmic expression as a sum, difference, or multiple of logarithms.

10. $\ln ax =$ ______ for $a > 0$ and $x > 0$.

11. $\ln \frac{a}{x} =$ ______ for $a > 0$ and $x > 0$.

12. $\ln x^n =$ ______ for $x > 0$ and n rational.

13. $\ln \sqrt[3]{\frac{x^2}{a^4}} = \ln\left(\frac{x^2}{a^4}\right)^{\_\_} = (\_\_\_) \ln\left(\frac{x^2}{a^4}\right) = \frac{1}{3}(\ln x^2 - \_\_\_\_\_)$
$= \frac{2}{3} \ln x -$ ______ .

14. $\ln(b^3 \cdot \sqrt{x}) = \ln b^3 +$ ______ $= 3$ ______ $+ \ln x^{1/2} =$ ______ .

15. $\ln(x^2 + 2x + 1) = \ln(x + 1)^{\_\_} =$ ______ for $x > -1$.

---

6. -6, -1, -1, -6, $2x + 4$, 2, -1, $\frac{1}{2}$
7. $\frac{1}{7}(4 - x)$
8. $\int_1^x \frac{1}{t}\,dt$, $x > 0$
9. $\frac{1}{x}$, differentiable
10. $\ln a + \ln x$
11. $\ln a - \ln x$
12. $n \ln x$
13. $\frac{1}{3}$, $\frac{1}{3}$, $\ln a^4$, $\frac{4}{3} \ln a$
14. $\ln \sqrt{x}$, $\ln b$, $3 \ln b + \frac{1}{2} \ln x$
15. 2, $2 \ln(x + 1)$

OBJECTIVE B: Summarize the characteristics of the graph of $y = \ln x$, and graph functions involving the natural logarithm.

16. The domain of $y = \ln x$ is the set __________, and its range is the set __________ .

17. The graph of $y = \ln x$ is increasing ____________________ . It is concave downward ____________________ .

18. Since $y = \ln x$ is differentiable for $x > 0$, it is a ______________ function of x.

19. $\lim_{x\to\infty} \ln x =$ ________ and $\lim_{x\to 0^+} \ln x =$ __________ .

20. Consider the curve $y = x - \ln x$. The derivative $\frac{dy}{dx} =$ ________ so that $\frac{dy}{dx} = 0$ implies $\frac{1}{x} =$ ______ or $x =$ ______. Notice that the domain of y is the set __________ . The second derivative $\frac{d^2y}{dx^2} =$ _______ is always positive. Therefore, the critical point $x = 1$ gives a relative ____________ value of $y(1) =$ _______ . As $x \to 0$, $y \to$ ______ . To examine the curve as x increases, notice that $\frac{dy}{dx} = 1 - \frac{1}{x} > 0$ whenever $x > 1$. Thus the graph is everywhere increasing. Sketch a graph of y at the right.

OBJECTIVE C: Differentiate functions whose expressions involve the natural logarithmic function.

21. $$\frac{d}{dx} \ln\left(5 + 2x^3\right)^4 = \frac{1}{\left(5 + 2x^3\right)^4} \frac{d}{dx}\left(\_\_\_\_\_\_\_\_\right)$$
$$= \frac{1}{\left(5 + 2x^3\right)^4}\left[\_\_\_\_\_\_\_\_\_\_\right]\frac{d}{dx}\left(5 + 2x^3\right)$$
$$= \frac{\_\_\_\_\_\_}{\left(5 + 2x^3\right)^4} = \_\_\_\_\_\_\_\_\_\_ .$$

---

16. $x > 0$, $-\infty < y < +\infty$ 17. for all x in the domain, for all $x > 0$ 18. continuous

19. $+\infty$, $-\infty$ 20. $1 - \frac{1}{x}$, 1, 1, $x > 0$, $\frac{1}{x^2}$, minimum, 1, $+\infty$, $\frac{1}{x}$, 1, 1, x

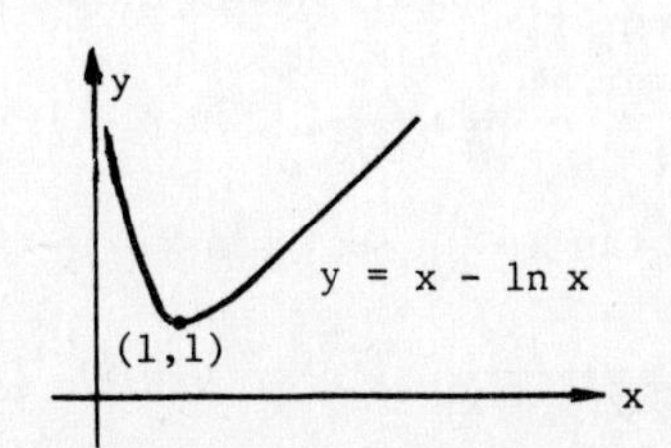

21. $\left(5 + 2x^3\right)^4$, $4\left(5 + 2x^3\right)^3$, $24x^2\left(5 + 2x^3\right)^3$, $\frac{24x^2}{5 + 2x^3}$

22. $\frac{d}{dx}\left[\ln(\sin x)\right]^2 = \underline{\hspace{3cm}} \frac{d}{dx} \ln (\sin x)$
$= 2 \ln(\sin x) \cdot \underline{\hspace{2cm}} \cdot \frac{d}{dx}(\sin x)$
$= 2 \csc x \ln(\sin x) \cdot \underline{\hspace{2cm}} = \underline{\hspace{3cm}}$.

23. $\frac{d}{dx} x^2 \ln \sqrt{x} = 2x \ln \sqrt{x} + \underline{\hspace{3cm}} \frac{d}{dx} \ln \sqrt{x}$
$= 2x \ln \sqrt{x} + \underline{\hspace{3cm}} \frac{d}{dx} \sqrt{x}$
$= 2x \ln \sqrt{x} + \underline{\hspace{4cm}} = \frac{x}{2}(\underline{\hspace{3cm}})$.

OBJECTIVE D: Use the method of logarithmic differentiation to calculate derivatives.

24. Find $\frac{dy}{dx}$ if $y = x^{\tan x}$, $x > 0$.

Solution.
$\ln y = \ln(x^{\tan x}) = \underline{\hspace{4cm}}$ so that
$\frac{1}{y}\frac{dy}{dx} = \sec^2 x \ln x + \underline{\hspace{4cm}}$, or
$\frac{dy}{dx} = x^{\tan x}(\underline{\hspace{4cm}})$.

25. Find $\frac{dy}{dx}$ if $y = \sqrt{\frac{1 - x}{1 + x}}$, $-1 < x < 1$.

Solution.
$\ln y = \ln \sqrt{\frac{1 - x}{1 + x}} = \frac{1}{2} \ln (1 - x) - \underline{\hspace{4cm}}$ so that
$\frac{1}{y}\frac{dy}{dx} = -\frac{1}{2(1 - x)} - \underline{\hspace{3cm}} = \frac{-(1 + x) - (\underline{\hspace{1.5cm}})}{2(1 - x)(1 + x)}$
$\frac{dy}{dx} = -y(1 - x)^{-1} \cdot \underline{\hspace{3cm}} = \underline{\hspace{4cm}}$.

26. Find $\frac{dy}{dx}$ if $y = \left(x^r\right)^x$, $x > 0$.
Solution.
$\ln y = \ln \left(x^r\right)^x = \underline{\hspace{3cm}} = \underline{\hspace{3cm}}$.
$\frac{1}{y}\frac{dy}{dx} = r \ln x + \underline{\hspace{3cm}} = r(\underline{\hspace{3cm}})$ so that
$\frac{dy}{dx} = \underline{\hspace{4cm}}$.

---

22. $2 \ln (\sin x)$, $\frac{1}{\sin x}$, $\cos x$, $2 \cot x \ln (\sin x)$

23. x^2, $x^2 \cdot \frac{1}{\sqrt{x}}$, $x^2 \cdot \frac{1}{\sqrt{x}} \cdot \frac{1}{2\sqrt{x}}$, $2 \ln x + 1$ or, $4 \ln \sqrt{x} + 1$

24. $\tan x \cdot \ln x$, $\tan x \cdot \frac{1}{x}$, $\sec^2 x \ln x + \frac{1}{x} \tan x$

25. $\frac{1}{2} \ln (1 + x)$, $\frac{1}{2(1 + x)}$, $1 - x$, $(1 + x)^{-1}$, $-(1 - x)^{-1/2}(1 + x)^{-3/2}$

26. $x \ln x^r$, $rx \ln x$, $rx \cdot \frac{1}{x}$, $\ln x + 1$, $r(x^r)^x (\ln x + 1)$

OBJECTIVE E: Integrate functions whose antiderivatives involve the natural logarithm function.

27. $\int \frac{x\,dx}{x^2+4}$

Let $u = x^2 + 4$. Then $du =$ ______ so $x\,dx =$ ______ . Thus the integral becomes

$$\int \frac{x\,dx}{x^2+4} = \int \frac{du}{\_\_\_\_} = \_\_\_\_\_\_ + C = \_\_\_\_\_\_ .$$

28. $\int \frac{3x+1}{x}\,dx = \int (3 + \_\_\_\_)\,dx = \int 3\,dx + \_\_\_\_$

$= 3x +$ ______ $+ C$.

29. $\int \frac{dx}{x \ln \sqrt{x}}$

Let $u = \ln \sqrt{x}$. Then $\frac{du}{dx} = \_\_\_\_ \frac{d}{dx}(\_\_\_\_) = \_\_\_\_$.

Hence, $2\,du =$ ______ dx. Thus the integral becomes

$$\int \frac{dx}{x \ln \sqrt{x}} = \int \frac{2\,du}{\_\_\_\_} = \_\_\_\_\_\_ + C = \_\_\_\_\_\_ .$$

6.3 THE EXPONENTIAL FUNCTION

OBJECTIVE A: Use the equivalent equations $y = e^x$ and $x = \ln y$ to simplify logarithms of exponentials, and exponentials of logarithms.

30. The equation $y = e^{\ln x}$ is equivalent to $\ln y =$ ______ . Since the logarithm is one - one, the last equation is equivalent to ______; that is, $e^{\ln x} =$ ______ . In other words, the exponential "undoes" the natural logarithm.

31. The equation $y = \ln(e^x)$ is equivalent to $e^y =$ ______ . Since the exponential function is one - one, the last equation is equivalent to ______; that is, $\ln(e^x) =$ ______ .

32. $e^{-2\ln(x+1)} = e^{\ln \_\_\_\_} =$ ______ by Problem 30.

33. $\ln(\sqrt{x}e^{2-x}) = \ln \sqrt{x} +$ ______ $=$ ______ .

---

27. $2x\,dx$, $\frac{1}{2}\,du$, $2u$, $\frac{1}{2}\ln|u|$, $\frac{1}{2}\ln(x^2+4) + C$

28. $\frac{1}{x}$, $\int \frac{dx}{x}$, $\ln|x|$

29. $\frac{1}{\sqrt{x}}$, $\sqrt{x}$, $\frac{1}{2x}$, $\frac{1}{x}$, u, $2\ln|u|$, $2\ln(\ln\sqrt{x}) + C$

30. $\ln x$, $y = x$, x

31. e^x, $y = x$, x

32. $(x+1)^{-2}$, $(x+1)^{-2}$

33. $\ln(e^{2-x})$, $\ln\sqrt{x} + (2 - x)$

OBJECTIVE B: Differentiate functions whose expressions involve exponential functions.

34. $\frac{d}{dx} e^{\sqrt{1-x^2}} = e^{\sqrt{1-x^2}} \cdot \frac{d}{dx}$ (________)

$= e^{\sqrt{1-x^2}} \cdot \frac{1}{2\sqrt{1-x^2}} \cdot \frac{d}{dx}$ (________)

$=$ ____________________ .

35. $\frac{d}{dx} e^{x \ln x} =$ ________ $\cdot \frac{d}{dx} (x \ln x) = e^{x \ln x}$ (________).

36. $\frac{d}{dx} \ln(e^x + 1) = \frac{1}{e^x + 1} \cdot \frac{d}{dx}$ (________) $=$ ________ .

37. $\frac{d}{dx} \sin \sqrt{e^x} = \cos \sqrt{e^x} \cdot \frac{d}{dx}$ (________)

$= \cos \sqrt{e^x} \cdot \left(\frac{1}{2\sqrt{e^x}}\right) \cdot \frac{d}{dx}$ (________)

$=$ ____________________ .

OBJECTIVE C: Integrate functions whose antiderivatives involve exponential functions.

38. $\int x^2 e^{-x^3} dx$

Let $u = -x^3$. Then $du =$ ________ so that $x^2\,dx =$ ________ . Thus the integral becomes,

$\int x^2 e^{-x^3} dx = \int$ ________ $du =$ ________ $+ C =$ ________ .

39. $\int (e^x - 2)^4 e^x\,dx$

Let $u = e^x - 2$. Then $du =$ ________ and the integral becomes,

$\int (e^x - 2)^4 e^x\,dx = \int$ ________ $du =$ ________ $+ C =$ ________ .

40. $\int \frac{e^x}{1 + e^{2x}} dx$

Let $u = e^x$. Then $du =$ ________ and the integral becomes,

$\int \frac{e^x\,dx}{1 + e^{2x}} = \int \frac{du}{\_\_\_\_} =$ ________ $+ C =$ ________ .

---

34. $\sqrt{1-x^2}$, $1 - x^2$, $\frac{-x}{\sqrt{1-x^2}} e^{\sqrt{1-x^2}}$

35. $e^{x \ln x}$, $\ln x + 1$

36. $e^x + 1$, $\frac{e^x}{e^x + 1}$

37. $\sqrt{e^x}$, e^x, $\frac{1}{2}\sqrt{e^x} \cos \sqrt{e^x}$

38. $-3x^2\,dx$, $-\frac{1}{3}\,du$, $-\frac{1}{3}e^u$, $-\frac{1}{3}e^u$, $-\frac{1}{3}e^{-x^3} + C$

39. $e^x\,dx$, u^4, $\frac{1}{5}u^5$, $\frac{1}{5}(e^x - 2)^5 + C$

40. $e^x\,dx$, $1 + u^2$, $\tan^{-1} u$, $\tan^{-1}(e^x) + C$

6.4 a^x AND $\log_a x$

41. The function $y = a^x$ is defined by $a^x =$ __________ and it is well-defined whenever __________ .

42. The definition in Problem 41 is equivalent to saying that $\ln a^x =$ __________ .

43. If $x > 0$, the number x^n is defined for <u>any</u> real number n and means

$$x^n = \_\_\_\_\_\_\_\_\_\_, \quad x > 0 .$$

44. If $x > 0$, then

$$\frac{d}{dx}(x^n) = \_\_\_\_\_\_\_\_\_\_ .$$

OBJECTIVE A: Differentiate functions whose expressions involve an exponential function a^u, where u is a differentiable function of x.

45. The derivative of a^u, where u is a differentiable function of x, is given by $\frac{d}{dx} a^u =$ __________ .

46. $\frac{d}{dx} 2^{\sec x} =$ __________ $\cdot \frac{d}{dx} \sec x =$ __________ .

47. $\frac{d}{dx} x^2 3^x = 2x \cdot$ __________ $+ x^2 \cdot$ __________ $= 3^x$ (__________).

48. Find $\frac{dy}{dx}$ if $y = x^{\tan x}$, $x > 0$.

<u>Solution</u>.

$\ln y = \ln(x^{\tan x}) =$ __________ so that

$\frac{1}{y}\frac{dy}{dx} = \sec^2 x \ln x +$ __________, or

$\frac{dy}{dx} = x^{\tan x}$ (__________).

49. Find $\frac{dy}{dx}$ if $y = \left(x^r\right)^x$, $x > 0$, r any real number.

<u>Solution</u>.

$\ln y = \ln\left(x^r\right)^x =$ __________ $=$ __________ .

---

41. $e^{x \ln a}$, $a > 0$
42. $x \ln a$
43. $e^{n \ln x}$
44. nx^{n-1}
45. $a^u \cdot \frac{du}{dx} \cdot \ln a$
46. $2^{\sec x} \cdot \ln 2$, $(\sec x \tan x)\, 2^{\sec x} \cdot \ln 2$
47. 3^x, $3^x \ln 3$, $2x + x^2 \ln 3$
48. $\tan x \cdot \ln x$, $\tan \cdot \frac{1}{x}$, $\sec^2 x \ln x + \frac{1}{x} \tan x$

$\frac{1}{y}\frac{dy}{dx} = r \ln x +$ __________ $= r($__________$)$ so that

$\frac{dy}{dx} =$ ____________________ .

OBJECTIVE B: Graph functions whose expressions involve an exponential function a^u.

50. Consider the curve $y = (0.2)^{x+1} + 2$.
$\frac{dy}{dx} =$ __________ $\frac{d}{dx}(x + 1) =$ _______________ .
Therefore, $\frac{dy}{dx}$ is of constant ________ sign, since $\ln(.2) < 0$, and the curve is everywhere ___________ .
Calculation of the second derivative gives $\frac{d^2y}{dx^2} =$ ___________ which is of constant ________ sign; hence, the curve is everywhere concave ________ .
As $x \to \infty$, $(.2)^{x+1} \to$ _____ so $y \to$ _____ . As $x \to -\infty$, $(.2)^{x+1} \to$ _____ so $y \to$ _____ .
Finally, the points $(-1,$ _____$)$, $(0,$ _____$)$, $(1,$ _____$)$ lie on the curve. You can now sketch the graph of the curve in the coordinate system provided at the right.

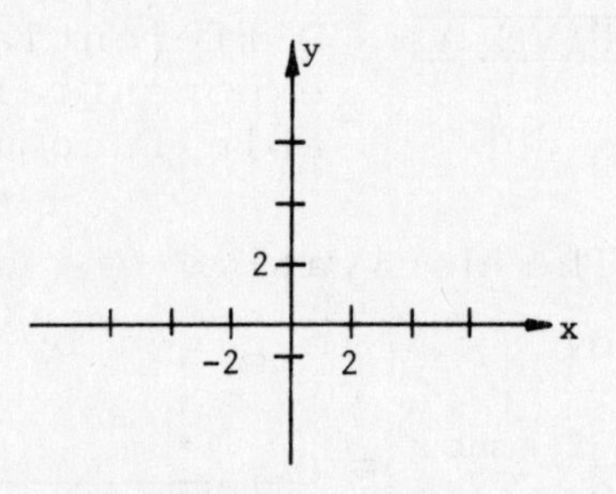

OBJECTIVE C: Use the definition of $\log_a x$ to evaluate simple expressions.

51. If $a > 0$ and $a \neq 1$, then $y = \log_a x$ is defined and equivalent to $a^y =$ _______ whenever x is positive. If a or x is negative, $\log_a x$ is __________ .

52. In terms of natural logarithms, $\log_a x =$ _____________ .

---

49. $x \ln x^r$, $rx \ln x$, $rx \cdot \frac{1}{x}$, $\ln x + 1$, $r(x^r)^x (\ln x + 1)$

50. $(0.2)^{x+1} \ln(.2)$, $(0.2)^{x+1} \ln(.2)$, negative, decreasing, $(0.2)^{x+1}[\ln(.2)]^2$, positive, upward, 0, 2, $+\infty$, $+\infty$, 3, 2.2, 2.04

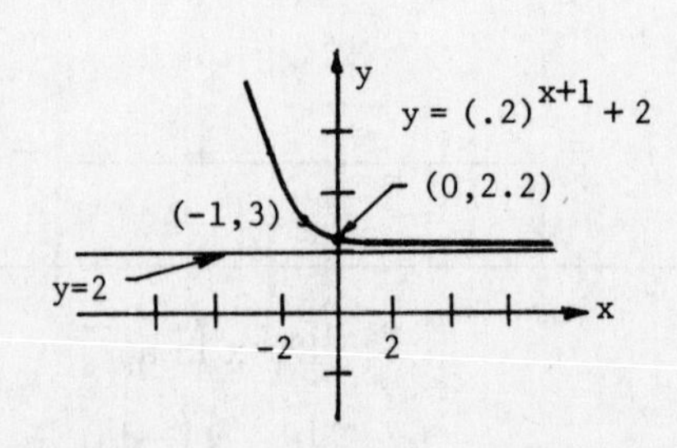

51. x, defined

52. $\frac{\ln x}{\ln a}$

53. $\log_8 4$

If $y = \log_8 4$, then $8^y =$ ______ or $2^{\_\_} = 2^2$. Thus, $3y =$ ______ or $y =$ ______ . Therefore, $\log_8 4 =$ ______ .

54. $\log_{.75} \frac{27}{64}$

If $y = \log_{.75} \frac{27}{64}$, then $\frac{27}{64} =$ ______ . Now, $\left(\frac{3}{4}\right)^3 =$ ______ so that $\log_{.75} \frac{27}{64} =$ ______ .

OBJECTIVE D: Solve exponential and logarithmic equations.

55. $5^{x^2} = 7^x$

Taking the natural logarithm of both sides gives x^2 (______) $= x \ln 7$ or x(______) $= 0$. Hence, $x = 0$ and $x =$ ______ both solve the equation.

56. $\log_7 (x^2 - 6x) = 1$ is equivalent to $x^2 - 6x =$ ______ or, $0 =$ ______ $= (x - 7)$(______). Therefore, $x =$ ______ and $x =$ ______ solve the logarithmic equation.

57. $5^{\log_5 2} =$ ______ because $y = 5^u$ and $y = \log_5 u$ are ______ functions of each other.

58. In general, $a^{\log_a u} =$ ______ and $\log_a a^x =$ ______ .

OBJECTIVE E: Differentiate functions involving a logarithmic function $\log_a u$.

59. $\frac{d}{dx} \log_a u = \frac{d}{dx} \frac{\_\_\_\_\_\_}{\ln a} =$ ______ $\cdot \frac{du}{dx}$.

60. $\frac{d}{dx} \log_{10} (x^2 - e^x) = \frac{1}{\_\_\_\_\_\_} \frac{d}{dx}(x^2 - e^x) =$ ______ .

61. Find $\frac{dy}{dx}$ if $y = \left(1 + \sqrt{x}\right)^{\log_2 x}$

Solution.

$\ln y = \ln \left(1 + \sqrt{x}\right)^{\log_2 x} =$ ______ .

---

53. 4, 3y, 2, $\frac{2}{3}$, $\frac{2}{3}$

54. $\left(\frac{3}{4}\right)^y$, $\frac{27}{64}$, 3

55. ln 5, x ln 5 - ln 7, $\frac{\ln 7}{\ln 5}$

56. 7^1, $x^2 - 6x - 7$, $x + 1$, -1, 7

57. 2, inverse

58. u, x

59. ln u, $\frac{1}{u \ln a}$

60. $(x^2 - e^x) \ln 10$, $\frac{2x - e^x}{(x^2 - e^x) \ln 10}$

$$\frac{1}{y}\frac{dy}{dx} = \_\_\_\_\_\_\_\_ \ln(1 + \sqrt{x}) + \log_2 x \cdot \frac{1}{1 + \sqrt{x}} \cdot \_\_\_\_\_\_\_\_$$

$$\frac{dy}{dx} = \left(1 + \sqrt{x}\right)^{\log_2 x} [\_\_\_\_\_\_\_\_\_\_\_\_\_\_\_\_].$$

OBJECTIVE F: Integrate functions whose antiderivatives involve an exponential function a^u.

62. $\int \frac{dx}{2^x} = \int 2^{-x}\, dx$

Let $u = -x$ so that $du =$ ______, and the integral becomes

$\int 2^{-x}\, dx = \int$ ______ $du =$ ________ $+ C =$ __________ .

63. $\int_1^2 x\, 10^{x^2-1}\, dx$

Let $u = x^2 - 1$. Then $du =$ ________ so that $x\, dx =$ ________ .

$\int x\, 10^{x^2-1}\, dx = \int$ ______ $du =$ ________ $+ C$. Hence,

$\int_1^2 x\, 10^{x^2-1}\, dx =$ ________ $]_1^2 = \frac{1}{2 \ln 10}$ (________)

$=$ ________ .

6.5 GROWTH AND DECAY

OBJECTIVE: Solve exponential growth and decay initial value problems:

Differential equation: $\frac{dy}{dt} = ky$

Initial condition: $y = y_0$ when $t = 0$.

64. The Law of Exponential Change asserts that the dependent variable y is functionally related to the independent variable t according to the rule:

________________ .

If $k > 0$ the rule gives exponential ________; if ______ the rule gives exponential decay.

---

61. $\log_2 x \cdot \ln(1 + \sqrt{x})$, $\frac{1}{x \ln 2}$, $\frac{1}{2\sqrt{x}}$, $\frac{1}{x \ln 2} \ln(1 + \sqrt{x}) + \frac{\ln x}{2 \ln 2 \cdot \sqrt{x}(1 + \sqrt{x})}$

62. $-dx$, -2^u, $-\frac{1}{\ln 2} 2^u$, $-\frac{1}{2^x \ln 2} + C$

63. $2x\, dx$, $\frac{1}{2} du$, $\frac{1}{2} 10^u$, $\frac{1}{2 \ln 10} 10^u$, $\frac{1}{2 \ln 10} 10^{x^2-1}$, $10^3 - 1$, $\frac{999}{2 \ln 10}$

64. $y = y_0 e^{kt}$, growth, $k < 0$

65. The half-life of a radioactive substance is the length of time it takes for ________ of a given amount of the substance to disintegrate through radiation. The half-life of the carbon isotope C^{14} is about 5700 years.

66. Assume that the amount x of C^{14} present in a dead organism decays exponentially from the time of death. Then, $x = x_0e^{kt}$, where x_0 is the original amount present. To find the constant k in the case of carbon C^{14},

$$\left(\frac{1}{2}\right)x_0 = x_0\, e^{\_\_\_\_ k} \quad \text{or} \quad \ln\frac{1}{2} = \_\_\_\_\_\_\_\_ .$$

Thus, $k = \dfrac{-\ln 2}{\_\_\_\_} \approx -1.22 \times 10^{-4}$.

Suppose we want to determine the amount of C^{14} present after 10,000 years. Then, the percentage is given by the ratio

$$\frac{x_0\, e^{k\cdot 10^4}}{x_0} \approx e^{\_\_\_\_\_\_} \approx 0.2964,$$

or approximately 29.64 percent of the amount of C^{14} remains after 10,000 years.

67. The "1470" skull found in Kenya by Richard Leakey is reputed to be 2,500,000 years old. The percentage of C^{14} remaining is given by

$$\frac{x_0\, e^{k\cdot 2.5 \times 10^6}}{x_0} \approx e^{\_\_\_\_} .$$

However, if $x < -21$ then $e^x < 10^{-9}$ so the percentage of C^{14} left in the skull would be negligible. The current reliable limit for C^{14} dating is about 40,000 years, so another method for dating the skull had to be found.

68. Suppose an object is immersed in a surrounding fluid having constant temperature T_s. If $T(t)$ is the temperature of the object at time t, then Newton's Law of Cooling asserts that

$$\frac{dT}{dt} = \_\_\_\_\_\_\_\_\_\_\_\_\_\_ ,$$

where $k > 0$ if the object is cooling; if $k < 0$ the object is ________ . If $T = T_0$ is the temperature of the object at time $t = 0$, the solution to the differential equation (also called Newton's Law of Cooling) is

______________________ .

69. A thermometer is taken from a room where the air temperature is 70°F to the outside where the temperature is 20°F. Write an initial value problem modeling the situation, assuming $T(t)$ is the temperature of the thermometer at time t.

---

65. half 66. 5700, 5700k, 5700, $-\frac{\ln 2}{57} \times 10^2$ 67. -304

68. $-k(T - T_s)$, warming, $T - T_s = (T_0 - T_s)e^{-kt}$

Differential equation: ____________________

Initial condition: ____________________ .

The solution to this initial value problem is

____________________ .

As $t \to \infty$, the temperature of the thermometer approaches ______ .

70. Continuing with Problem 69, to evaluate the constant k we need additional information. Suppose then that after 2/3 minute the thermometer reads 45°. From the solution in Problem 69, this condition means

$$\frac{45 - 20}{50} = \_\_\_\_\_\_\_\_ .$$

Applying the natural logarithm function to both sides to solve for k gives $\ln \frac{1}{2} = \_\_\_\_\_\_\_\_$, or

$$k = \_\_\_\_\_\_\_\_ \approx 1.04 .$$

Thus, after 1 minute the temperature of the thermometer will be

$$T = 20 + \_\_\_\_\_\_\_\_ \approx 37.7°F.$$

We used a calculator to evaluate $e^{-\left(\frac{3}{2}\ln 2\right)} \approx 0.354$.

6.6 L'HÔPITAL'S RULE

OBJECTIVE A: Use l'Hôpital's rule to find limits of indeterminate forms of type 0/0 or ∞/∞.

71. $\lim_{x\to\infty} xe^{-x} = \lim_{x\to\infty} \frac{x}{e^x}$ is an indeterminate form of type ______ so l'Hôpital's rule applies. Thus, $\lim_{x\to\infty} \frac{x}{e^x} = \lim_{x\to\infty} \_\_\_\_\_\_ = \_\_\_\_\_\_$.

72. $\lim_{x\to 0} \frac{e^{2x} - 1}{x}$ is an indeterminate form of type _____ so l'Hôpital's rule applies. Thus,

$$\lim_{x\to 0} \frac{e^{2x} - 1}{x} = \lim_{x\to 0} \frac{\_\_\_\_}{1} = \_\_\_\_\_\_ .$$

---

69. $\frac{dT}{dt} = k(T - 20)$, $T(0) = 70$, $T - 20 = (70 - 20)e^{-kt}$, 20°F

70. $e^{-k(2/3)}$, $-\frac{2}{3}k$, $\frac{3}{2}\ln 2$, $50e^{-\left(\frac{3}{2}\ln 2\right)1}$

71. ∞/∞, $\frac{1}{e^x}$, 0

72. 0/0, $2e^{2x}$, 2

73. $\lim_{\theta \to \frac{\pi}{2}^-} \dfrac{\ln (\tan \theta)}{\sec \theta}$

As $\theta \to \frac{\pi}{2}^-$, $\tan \theta \to \infty$ and $\sec \theta \to$ ______ . Hence $\ln (\tan \theta) \to$ ______ and this is an indeterminate form of type ________ . Therefore, l'Hôpital's rule applies:

$$\lim_{\theta \to \frac{\pi}{2}^-} \frac{\ln (\tan \theta)}{\sec \theta} = \lim_{\theta \to \frac{\pi}{2}^-} \frac{\_\_\_\_\_\_}{\sec \theta \tan \theta} = \lim_{\theta \to \frac{\pi}{2}^-} \frac{\_\_\_\_\_\_}{\sin^2 \theta} = \_\_\_\_\_\_ .$$

74. $\lim_{x \to 0^+} x^2 \ln x$

As $x \to 0^+$, $x^2 \to 0$ and $\ln x \to$ ______ so this is an indeterminate form $0 \cdot \infty$. Writing the limit as

$\lim_{x \to 0^+} x^2 \ln x = \lim_{x \to 0^+} \dfrac{\ln x}{\_\_\_\_}$, we see this form is now of the type

∞/∞. Therefore l'Hôpital's rule applies:

$$\lim_{x \to 0^+} \frac{\ln x}{1/x^2} = \lim_{x \to 0^+} \_\_\_\_\_\_ = \lim_{x \to 0^+} \_\_\_\_\_\_ = \_\_\_\_\_\_ .$$ Hence,

$\lim_{x \to 0^+} x^2 \ln x =$ ________ .

OBJECTIVE B: Use l'Hôpital's rule to find limits of indeterminate forms of type 1^∞, 0^0, and ∞^0.

75. $\lim_{x \to 0} (1 - x)^{1/\sqrt{x}}$ is of the form ________ . We let $f(x) = (1 - x)^{1/\sqrt{x}}$ and find $\lim_{x \to 0} \ln f(x)$. Now,

$\ln f(x) =$ ________ $\cdot \ln (1 - x)$, so l'Hôpital's rule gives

$$\lim_{x \to 0} \ln f(x) = \lim_{x \to 0} \frac{\_\_\_\_\_\_}{\sqrt{x}}, \quad (0/0 \text{ form})$$

$$= \lim_{x \to 0} \frac{-1/(1 - x)}{\_\_\_\_\_\_} = \lim_{x \to 0} \frac{2\sqrt{x}}{x - 1} = \_\_\_\_\_\_ .$$

Therefore, $\lim_{x \to 0} f(x) =$ ________ .

76. $\lim_{x \to 0^+} x^{\sin x}$ is of the form ______ . Let $y = x^{\sin x}$.

Then $\ln y =$ ____________ . Thus $\ln y = \dfrac{\ln x}{\_\_\_\_}$.

Then $\lim_{x \to 0^+} \dfrac{\ln x}{\csc x}$ is an indeterminate form of type ________ and

---

73. ∞, ∞, $\frac{\infty}{\infty}$, $\sec^2 \theta / \tan \theta$, $\cos \theta$, 0

74. $-\infty$, $\frac{1}{x^2}$, $\frac{1/x}{-2/x^3}$, $-\frac{1}{2}x^2$, 0, 0

75. 1^∞, $\frac{1}{\sqrt{x}}$, $\ln (1 - x)$, $\frac{1}{2\sqrt{x}}$, 0, $e^0 = 1$

l'Hôpital's rule applies. Hence,

$$\lim_{x \to 0^+} \frac{\ln x}{\csc x} = \lim_{x \to 0^+} \frac{1/x}{-\csc x \text{ ctn } x} = \lim_{x \to 0^+} \frac{\_\_\_\_\_\_}{-x \cos x}$$

$$= \lim_{x \to 0^+} \frac{\_\_\_\_\_\_}{x \sin x - \cos x} = \_\_\_\_\_\_ .$$

Since $\ln y \to 0$ as $x \to 0^+$, $y \to$ ________ and the required limit is ________ .

6.7 RELATIVE RATES OF GROWTH

77. If $\lim_{x \to \infty} \frac{f(x)}{g(x)} = \infty$ we say that f grows ________ than g, or g grows ________ than f as $x \to \infty$. Equivalently, $\lim_{x \to \infty} \frac{g(x)}{f(x)} =$ ________ .

78. If f and g grow at the same rate as $x \to \infty$, then $\lim_{x \to \infty} \frac{f(x)}{g(x)} =$ ________, where L is ________ .

OBJECTIVE A: Given two functions f(x) and g(x), compare the rates at which they grow as $x \to \infty$.

79. Let's compare the rates at which x^3 and e^{2x} grow as $x \to \infty$. Thus we compute

$$\lim_{x \to \infty} \frac{x^3}{e^{2x}} = \lim_{x \to \infty} \frac{3x^2}{2e^{2x}} = \_\_\_\_\_\_ = \lim_{x \to \infty} \frac{6}{8e^{2x}} = \_\_\_\_\_\_ .$$

We conclude that x^3 grows ________ than e^{2x}.

80. $\lim_{x \to \infty} \frac{5^x}{2^x} = \lim_{x \to \infty} \left(\frac{5}{2}\right)^x =$ ________ .

We conclude that 5^x grows ________ than 2^x.

81. $\lim_{x \to \infty} \frac{x^3 + x^2}{2x^3 - x^2} = \lim_{x \to \infty} \frac{1 + \frac{1}{x}}{\_\_\_\_\_\_} =$ ________ .

We conclude that $x^3 + x^2$ and $2x^3 - x^2$ grow ________ ________ .

---

76. 0^0, $\sin x \cdot \ln x$, $\csc x$, ∞/∞, $\sin^2 x$, $2 \sin x \cos x$, 0, 1, 1

77. faster, slower, 0

78. $L \neq 0$, nonzero and finite

79. $\lim_{x \to \infty} \frac{6x}{4e^{2x}}$, 0, slower

80. ∞, faster

81. $2 - \frac{1}{x}$, $\frac{1}{2}$, at the same rate

OBJECTIVE B: Use the "little-oh" and "big-oh" notation and terminology for comparing two functions as $x \to \infty$.

82. The notation $f = o(g)$ as $x \to \infty$ is another way of saying ________________ as $x \to \infty$, or $\lim_{x\to\infty} \frac{f(x)}{g(x)} =$ ______ .

83. The notation $f = O(g)$ means there is a positive number M such that for x sufficiently large,

________________ .

84. $\lim_{x\to\infty} \frac{x^3}{e^{2x}} = 0$ from Problem 79. Thus $x^3 =$ ________ as $x \to \infty$. Also, x^3 is ________________ because $f = o(g)$ implies $f = O(g)$.

85. $\lim_{x\to\infty} \frac{x + \ln x}{3x} = \lim_{x\to\infty} \frac{1 + \frac{1}{x}}{\_\_\_\_} =$ ______ . Thus $x + \ln x =$ ______ as $x \to \infty$.

6.8 INVERSE TRIGONOMETRIC FUNCTIONS

OBJECTIVE A: Find the values of inverse trigonometric functions at selected points without the use of tables or a calculator.

86. $y = \sin^{-1} x$ is equivalent to ______ , where ____ $\le x \le$ ____ and ____ $\le y \le$ ____ .

87. $y = \tan^{-1} x$ is equivalent to ______ , where ____ $< x <$ ____ and ____ $< y <$ ____ .

88. Let $y = \sin^{-1}\left(-\frac{\sqrt{2}}{2}\right)$; then $\sin y =$ ______ , so $y =$ ______ .

That is, $\sin^{-1}\left(-\frac{\sqrt{2}}{2}\right) =$ ______ .

89. If $\alpha = \tan^{-1}\left(-\frac{\sqrt{3}}{3}\right)$, then $\tan \alpha =$ ____ , so $\alpha =$ ______ .

Hence, $\sin \alpha =$ ______ and $\cos \alpha =$ ______ .

---

82. f grows slower than g, 0

83. $\frac{f(x)}{g(x)} \le M$

84. $o(e^x)$, "big-oh" of e^x

85. 3, $\frac{1}{3}$, $O(3x)$

86. $x = \sin y$, -1, 1, $-\frac{\pi}{2}$, $\frac{\pi}{2}$

87. $x = \tan y$, $-\infty$, ∞, $-\frac{\pi}{2}$, $\frac{\pi}{2}$

88. $-\frac{\sqrt{2}}{2}$, $-\frac{\pi}{4}$, $-\frac{\pi}{4}$

89. $-\frac{\sqrt{3}}{3}$, $-\frac{\pi}{6}$, $-\frac{1}{2}$, $\frac{\sqrt{3}}{2}$

OBJECTIVE B: Evaluate expressions involving inverse trigonometric functions and trigonometric functions.

90. To find $\sin\left(\cos^{-1}\left(-\frac{\sqrt{3}}{2}\right)\right)$, let $y = \cos^{-1}\left(-\frac{\sqrt{3}}{2}\right)$. Then, $\cos y =$ ______ so $y =$ ______ . Hence, $\sin y =$ ______ . Alternatively since $\sin^2 y + \cos^2 y = 1$ holds,

$$\sin\left(\cos^{-1}\left(-\frac{\sqrt{3}}{2}\right)\right) = \sqrt{1 - \cos^2\left(\cos^{-1}\left(-\frac{\sqrt{3}}{2}\right)\right)}$$

$$= \left[1 - \cos^2\left(\cos^{-1}\left(-\frac{\sqrt{3}}{2}\right)\right)\right]^{1/2}$$

$$= \sqrt{1 - \underline{\qquad}} = \underline{\qquad} .$$

6.9 DERIVATIVES OF INVERSE TRIGONOMETRIC FUNCTIONS; INTEGRALS

OBJECTIVE A: Differentiate functions whose expressions involve inverse trigonometric functions.

91. $\frac{d}{dx}(\sin^{-1} u) =$ ______________ .

92. $\frac{d}{dx}(\tan^{-1} u) =$ ______________ .

93. $\frac{d}{dx}(\sec^{-1} u) =$ ______________ .

94. $\frac{d}{dx}\left(\sin^{-1}\frac{x}{5}\right)^2 = \left(2\sin^{-1}\frac{x}{5}\right)\left(\frac{1}{\sqrt{1 - (x/5)^2}}\right)\frac{d}{dx}(\underline{\qquad})$

$= \left(2\sin^{-1}\frac{x}{5}\right)\left(\frac{5}{\sqrt{25 - x^2}}\right)(\underline{\qquad}) =$ ______________ .

95. $\frac{d}{dx}\left(\sec^{-1}\frac{1}{x}\right) = \frac{1}{\frac{1}{|x|}\sqrt{\underline{\qquad} - 1}} \cdot (\underline{\qquad}) = \frac{-1}{\sqrt{\underline{\qquad}}}$.

96. $\frac{d}{dx}(\tan^{-1}\sqrt{x - 1}) = \frac{1}{1 + (x - 1)}\frac{d}{dx}(\underline{\qquad}) =$ ______________ .

---

90. $-\frac{\sqrt{3}}{2}$, $\frac{5\pi}{6}$, $\frac{1}{2}$, $\frac{3}{4}$, $\frac{1}{2}$

91. $\frac{du/dx}{\sqrt{1 - u^2}}$

92. $\frac{du/dx}{1 + u^2}$

93. $\frac{du/dx}{|u|\sqrt{u^2 - 1}}$

94. $\frac{x}{5}$, $\frac{1}{5}$, $\frac{2}{\sqrt{25 - x^2}}\sin^{-1}\frac{x}{5}$

95. $\frac{1}{x^2}$, $-\frac{1}{x^2}$, $1 - x^2$

96. $\sqrt{x - 1}$, $\frac{1}{2x\sqrt{x - 1}}$

97. Differentiating $\tan^{-1}\frac{x}{y} = \frac{1}{2}$ implicitly, we find

$$0 = \frac{d}{dx}(\tan^{-1}\frac{x}{y}) = \frac{1}{1 + (x/y)^2}\cdot\frac{d}{dx}(\_\_\_\_) = \frac{y^2}{x^2+y^2}(\_\_\_\_\_\_\_\_)$$

$$= \frac{y - \_\_\_\_}{x^2 + y^2}; \quad \text{thus } \frac{dy}{dx} = \_\_\_\_\_\_ .$$

OBJECTIVE B: Evaluate integrals leading to inverse trigonometric functions.

98. $\int_0^1 \frac{x\,dx}{\sqrt{1 - x^4}}$. Let $u = x^2$, then $du = \_\_\_\_\_\_$ so that $x\,dx = \_\_\_\_\_\_$, and the indefinite integral becomes,

$$\int \frac{x\,dx}{\sqrt{1-x^4}} = \frac{1}{2}\int \frac{du}{\_\_\_\_\_\_} = \_\_\_\_\_\_\_\_ + C. \quad \text{Thus,}$$

$$\int_0^1 \frac{x\,dx}{\sqrt{1-x^4}} = \_\_\_\_\_\_\_\_\Big]_0^1 = \frac{1}{2}(\sin^{-1} 1 - \_\_\_\_) = \_\_\_\_\_\_ .$$

99. $\int_0^{\pi/2} \frac{\cos x\,dx}{1 + \sin^2 x}$. Let $u = \sin x$, then $du = \_\_\_\_\_\_\_\_$, and the indefinite integral becomes

$$\int \frac{\cos x\,dx}{1+\sin^2 x} = \int \frac{du}{\_\_\_\_\_\_} = \_\_\_\_\_\_\_\_ + C. \quad \text{Thus,}$$

$$\int_0^{\pi/2} \frac{\cos x\,dx}{1+\sin^2 x} = \_\_\_\_\_\_\_\_\Big]_0^{\pi/2} = \tan^{-1} 1 - \_\_\_\_\_\_$$

$$= \_\_\_\_ - 0 = \_\_\_\_ .$$

OBJECTIVE C: Use l'Hôpital's rule to find limits of indeterminate forms involving inverse trigonometric functions.

100. $\lim_{x\to 0} \frac{\sin^{-1} x}{\sin x}$ is of the form 0/0, so l'Hôpital's rule applies.

Thus, $\lim_{x\to 0} \frac{\sin^{-1} x}{\sin x} = \lim_{x\to 0} \frac{\_\_\_\_}{\cos x} = \frac{\_\_\_\_}{1} = \_\_\_\_\_\_ .$

---

97. $\frac{x}{y}$, $\frac{y - x\frac{dy}{dx}}{y^2}$, $x\frac{dy}{dx}$, $\frac{y}{x}$

98. $2x\,dx$, $\frac{1}{2}du$, $\sqrt{1 - u^2}$, $\frac{1}{2}\sin^{-1}u$, $\frac{1}{2}\sin^{-1}x^2$, $\sin^{-1}0$, $\frac{\pi}{4}$

99. $\cos x\,dx$, $1 + u^2$, $\tan^{-1}u$, $\tan^{-1}(\sin x)$, $\tan^{-1}0$, $\frac{\pi}{4}$, $\frac{\pi}{4}$

100. $\frac{1}{\sqrt{1 - x^2}}$, 1, 1

101. $\lim\limits_{x\to+\infty} \dfrac{\frac{\pi}{2} - \tan^{-1} x}{1/x} = \lim\limits_{t\to 0^+} \dfrac{\frac{\pi}{2} - \tan^{-1} \frac{1}{t}}{\underline{\qquad\qquad}}$

$= \lim\limits_{t\to 0^+} \left(\underline{\qquad\qquad}\right)\dfrac{d}{dt}\left(\dfrac{1}{t}\right) = \lim\limits_{t\to 0^+} \left(-\dfrac{t^2}{t^2+1}\right)\left(\underline{\qquad\qquad}\right)$

$= \lim\limits_{t\to 0^+} \underline{\qquad\qquad} = \underline{\qquad\qquad}$.

6.10 HYPERBOLIC FUNCTIONS

OBJECTIVE A: Define the six hyperbolic functions and graph them.

102. $\cosh x =$ ________________ .
Sketch the graph at the right.

103. $\sinh x =$ ________________ .
Sketch the graph at the right on the same coordinate system as the $\cosh x$.

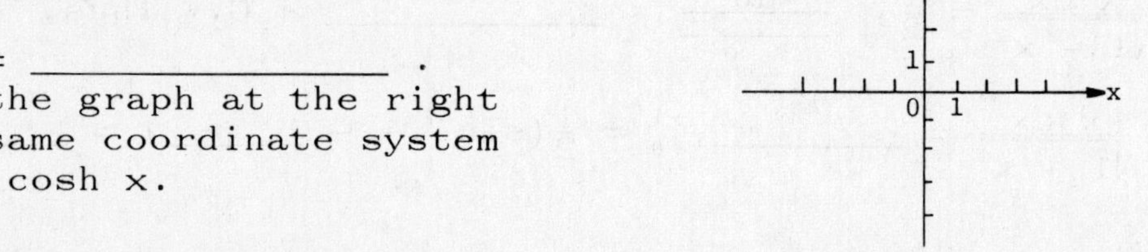

104. $\tanh x =$ ________________ .

105. $\coth x =$ ________________ .
Sketch the graphs at the right.

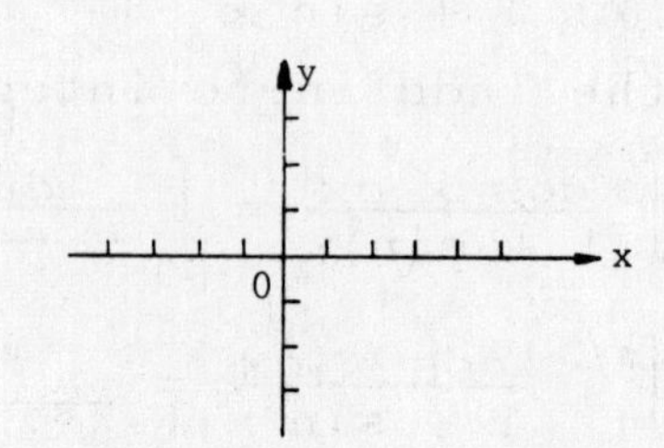

---

101. t, $-\dfrac{1}{1+(1/t)^2}$, $-\dfrac{1}{t^2}$, $\dfrac{1}{t^2+1}$, 1

102. $\dfrac{e^x + e^{-x}}{2}$ 103. $\dfrac{e^x - e^{-x}}{2}$ 104. $\dfrac{\sinh x}{\cosh x}$ 105. $\dfrac{\cosh x}{\sinh x}$

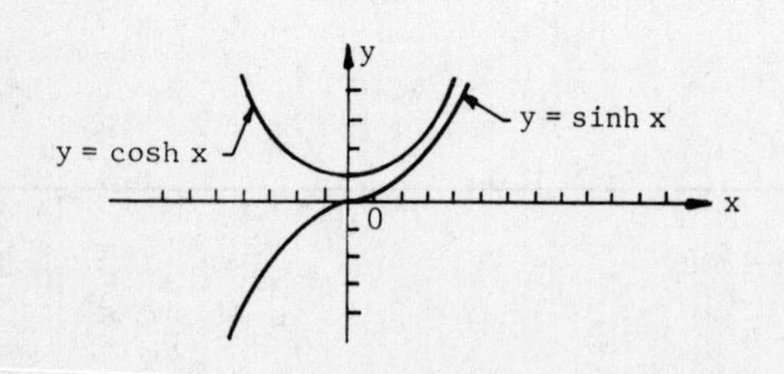

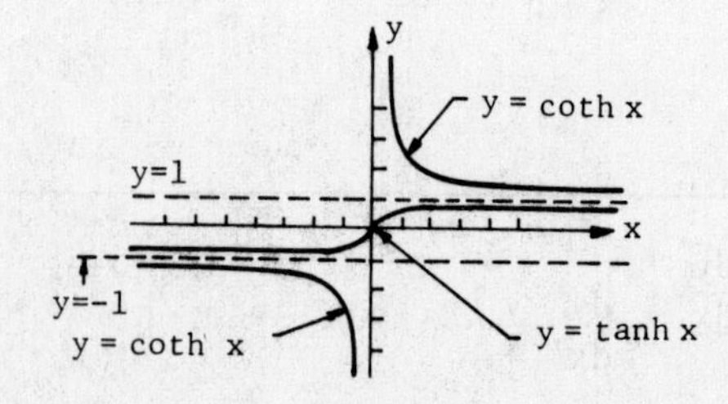

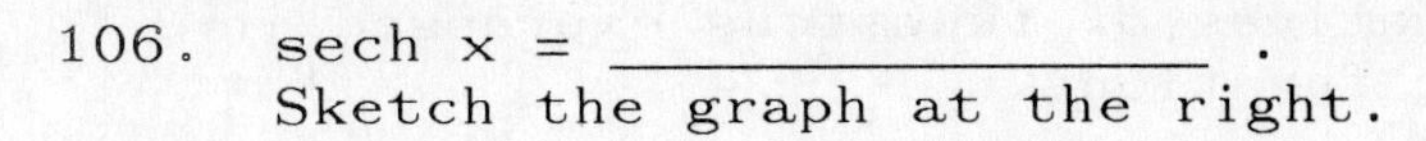

106. sech x = ________________ .
Sketch the graph at the right.

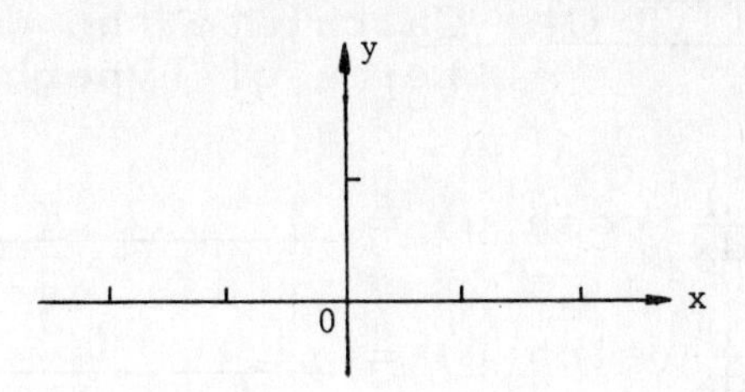

107. csch x = ________________ .

OBJECTIVE B: Given the value for one of the six hyperbolic functions at a point, determine the values of the remaining five at that point. Also use hyperbolic identities.

108. Suppose $\tanh x = -\frac{\sqrt{3}}{2}$. Then,
$\text{sech}^2 x = 1 -$ ____________ = ________ or sech x = ________ .
Thus, $\cosh x = \frac{1}{\text{sech } x} =$ ________ .

109. Continuing Problem 108, $\sinh^2 x = \cosh^2 x -$ ________ = ________ .
Since tanh x is negative and cosh x is positive, it follows that sinh x = ________ . Then, csch x = ________ and coth x = ________ .

110. cosh(-x) = ________________ .

111. sinh(-x) = ________________ .

112. sinh(x + y) = ____________________ .

113. cosh(x + y) = ____________________ .

114. sinh 2x = ____________________ .

115. cosh 2x = ____________________ .

116. cosh 2x - 1 = ________________ .

---

106. $\frac{1}{\cosh x}$

107. $\frac{1}{\sinh x}$

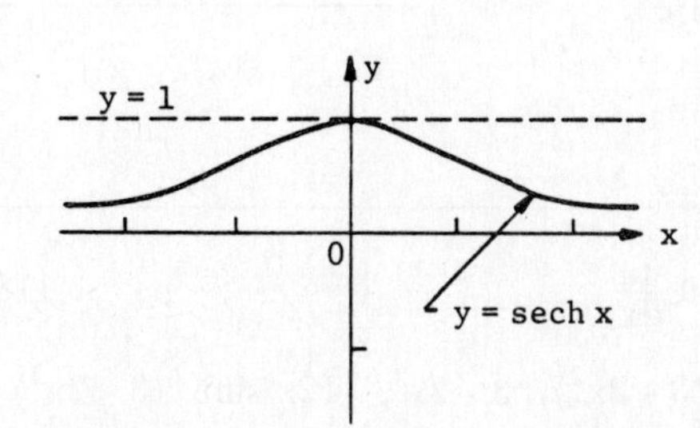

108. $\tanh^2 x$, $\frac{1}{4}$, $\frac{1}{2}$, 2

109. 1, 3, $-\sqrt{3}$, $-\frac{1}{\sqrt{3}}$, $\frac{-2}{\sqrt{3}}$

110. cosh x

111. -sinh x

112. sinh x cosh y + cosh x sinh y

113. cosh x cosh y + sinh x sinh y

114. 2 sinh x cosh x

115. $\cosh^2 x + \sinh^2 x$

116. $2 \sinh^2 x$

OBJECTIVE C: Calculate the derivatives of functions expressed in terms of hyperbolic functions.

117. $\frac{d}{dx}(\cosh u) =$ ________________ .

118. $\frac{d}{dx}(\sinh u) =$ ________________ .

119. $\frac{d}{dx}(\tanh u) =$ ________________ .

120. $y = \sinh^3(3 - 2x^2)$

$\frac{dy}{dx} = 3 \sinh^2(3 - 2x^2) \cdot \frac{d}{dx}$ ________________

$= 3 \sinh^2(3 - 2x^2) \cosh(3 - 2x^2) \frac{d}{dx}$ ____________

$=$ ________________________ .

121. $y = e^x \tanh 2x$

$\frac{dy}{dx} = e^x \frac{d}{dx}$ (______________) $+ e^x \tanh 2x$

$= e^x$ __________ $\frac{d}{dx}$ (_____) $+ e^x \tanh 2x$

$=$ ________________ .

122. $y = x^{\sinh x}, \quad x > 0$

$\frac{dy}{dx} = \frac{d}{dx}(e^{\sinh x \cdot \ln x}) = e^{\sinh x \cdot \ln x} \frac{d}{dx}$ (________________)

$= e^{\sinh x \cdot \ln x}(\sinh x \cdot \frac{d}{dx} \ln x + \ln x \cdot \frac{d}{dx}$ ____________)

$= e^{\sinh x \cdot \ln x}$ (________________________)

$= x^{\sinh x - 1}$ (________________________).

123. $e^y = \operatorname{sech} x$

Differentiating implicitly, $\frac{d}{dx}(e^y) = \frac{d}{dx} \operatorname{sech} x$, or

_________ $= - \operatorname{sech} x \tanh x$. Thus,

$\frac{dy}{dx} = -e^{-y}$ (________________________) $=$ ________________ .

---

117. $\sinh u \frac{du}{dx}$ 118. $\cosh u \frac{du}{dx}$ 119. $\operatorname{sech}^2 u \frac{du}{dx}$

120. $\sinh(3 - 2x^2)$, $3 - 2x^2$, $-12x \sinh^2(3 - 2x^2) \cosh(3 - 2x^2)$

121. $\tanh 2x$, $\operatorname{sech}^2 2x$, $2x$, $e^x(2 \operatorname{sech}^2 2x + \tanh 2x)$

122. $\sinh x \cdot \ln x$, $\sinh x$, $\frac{\sinh x}{x} + \cosh x \cdot \ln x$, $\sinh x + x \cosh x \ln x$

123. $e^y \frac{dy}{dx}$, $\operatorname{sech} x \tanh x$, $-\tanh x$

OBJECTIVE D: Integrate functions whose expressions involve hyperbolic functions.

124. $\int x \cosh(x^2 + 3)\, dx$

Let $u = x^2 + 3$, then $du =$ ________, and the integral becomes

$\int x \cosh(x^2 + 3)\, dx = \int$ ________ $du =$ ________ $+ C$

$=$ ________ .

125. $\int \sinh^2 x\, dx$

From the identities $\cosh 2x = \sinh^2 x + \cosh^2 x$ and $\cosh^2 x - \sinh^2 x = 1$, we have $\cosh 2x =$ ________ or $\sinh^2 x = \frac{1}{2}($________$)$.

Thus, $\int \sinh^2 x\, dx =$ ________ $+ C$.

126. $\int \tanh x \ln(\cosh x)\, dx$

Let $u = \ln(\cosh x)$, so $du =$ ________ and the integral becomes

$\int \ln(\cosh x) \tanh x\, dx = \int$ ____ $du =$ ____ $+ C$

$=$ ________ .

OBJECTIVE E: Define and use the six inverse hyperbolic functions. Be able to calculate their derivatives.

127. $y = \sinh^{-1} x$ means ________ . Thus $x = \frac{e^y - \text{\_\_\_\_\_}}{2}$ or $2xe^y =$ ________ or $e^{2y} -$ ________ $- 1 = 0$. Solution of this quadratic equation by the quadratic formula gives,

$e^y = \frac{2x \pm \sqrt{\text{\_\_\_\_\_\_}}}{2}$. Since $e^y > 0$ we must have

$e^y =$ ________ or $\sinh^{-1} x = y =$ ________ .

128. $\text{sech}^{-1} x =$ ________ .

129. $\coth^{-1} x =$ ________ .

---

124. $2x\, dx$, $\frac{1}{2}\cosh u$, $\frac{1}{2}\sinh u$, $\frac{1}{2}\sinh(x^2 + 3) + C$

125. $2\sinh^2 x + 1$, $\cosh 2x - 1$, $\frac{1}{4}\sinh 2x - \frac{1}{2}x$

126. $\frac{1}{\cosh x} \cdot \sinh x\, dx$, u, $\frac{1}{2}u^2$, $\frac{1}{2}\ln^2(\cosh x) + C$

127. $x = \sinh y$, e^{-y}, $e^{2y} - 1$, $2xe^y$, $4x^2 + 4$, $x + \sqrt{x^2 + 1}$, $\ln\left(x + \sqrt{x^2 + 1}\right)$

128. $\cosh^{-1} \frac{1}{x}$

129. $\tanh^{-1} \frac{1}{x}$

130. $\operatorname{csch}^{-1}x =$ ______________ .

131. We can use the formula found in Problem 127 to calculate $\frac{dy}{dx}$ for $y = \sinh^{-1}x$:

$$\frac{d}{dx}\ln\left(x + \sqrt{x^2 + 1}\right) = \frac{1}{x + \sqrt{x^2 + 1}} \cdot \frac{d}{dx}(\_\_\_\_\_\_\_\_)$$

$$= \frac{1}{x + \sqrt{x^2 + 1}} \cdot \left(\_\_\_\_\_\_\_\_\right)$$

$$= \frac{1}{x + \sqrt{x^2 + 1}} \cdot \left(\frac{\_\_\_\_\_\_\_\_}{\sqrt{x^2 + 1}}\right) = \_\_\_\_\_ .$$

132. An alternate way to calculate the derivative of $y = \sinh^{-1}x$ is as follows: Differentiate $x = \sinh y$ implicitly: $\frac{d}{dx}x = \frac{d}{dx}\sinh y$ or $1 =$ ______________ . Thus,

$$\frac{dy}{dx} = \frac{1}{\_\_\_\_\_} = \frac{1}{\sqrt{1 + \_\_\_\_\_}} = \_\_\_\_\_\_\_\_ .$$

The positive square root is taken in the penultimate step because $\cosh y$ is always ______________ .

OBJECTIVE F: Calculate the derivatives of functions expressed in terms of inverse hyperbolic functions.

133. $\frac{d}{dx}(\tanh^{-1}e^x) = \frac{1}{\_\_\_\_\_} \cdot \frac{d}{dx}e^x =$ ______________ .

134. $\frac{d}{dx}\ln(\sinh^{-1}x) = \frac{1}{\sinh^{-1}x} \cdot \frac{d}{dx}$ ______________ $=$ ______________ .

135. $\frac{d}{dx}\sqrt{\coth^{-1}x} = \frac{1}{2}(\coth^{-1}x)^{-1/2} \cdot \frac{d}{dx}$ ______________

$=$ ______________ .

136. $\frac{d}{dx}\cosh^{-1}\frac{3}{x^2} =$ ______________ $\cdot \frac{-6}{x^3} =$ ______________ .

---

130. $\sinh^{-1}\frac{1}{x}$

131. $x + \sqrt{x^2 + 1}$, $1 + \frac{2x}{2\sqrt{x^2 + 1}}$, $\sqrt{x^2 + 1} + x$, $\frac{1}{\sqrt{x^2 + 1}}$

132. $\cosh y \cdot \frac{dy}{dx}$, $\cosh y$, $\sinh^2 y$, $\frac{1}{\sqrt{1 + x^2}}$, positive

133. $1 - e^{2x}$, $\frac{e^x}{1 - e^{2x}}$

134. $\sinh^{-1}x$, $\frac{1}{\sinh^{-1}x \cdot \sqrt{1 + x^2}}$

135. $\coth^{-1}x$, $\frac{1}{2(1 - x^2)\sqrt{\coth^{-1}x}}$

136. $\frac{1}{\sqrt{\frac{9}{x^4} - 1}}$, $\frac{-6}{x\sqrt{9 - x^4}}$

OBJECTIVE G: Evaluate integrals using integration formulas for inverse hyperbolic functions.

137. $\int_{-3}^{-2} \frac{dx}{\sqrt{x^2 + 1}}$

Since $\int \frac{dx}{\sqrt{x^2 + 1}} =$ ______ + C = ln (______) + C (Problem 127)

$\int_{-3}^{-2} \frac{dx}{\sqrt{x^2 + 1}} =$ ______$\Big]_{-3}^{-2}$

$= \ln(\sqrt{5} - 2) -$ ______ ≈ 0.375.

138. $\int_{0.5}^{0.9} \frac{dx}{x\sqrt{1 - x^2}}$

For $0 < x < 1$, $\int \frac{dx}{x\sqrt{1 - x^2}} =$ ______ + C.

Now, $\text{sech}^{-1} x = \cosh^{-1}$ ______ $= \ln\left(\frac{1}{x} + \sqrt{\frac{1}{x^2} - 1}\right) = \ln\left(\frac{\quad}{x}\right)$.

Thus,

$\int_{0.5}^{0.9} \frac{dx}{x\sqrt{1 - x^2}} =$ ______$\Big]_{0.5}^{0.9}$

$= -\ln\left(\frac{1 + \sqrt{1 - .81}}{.9}\right) +$ ______

$= -\ln\left(\frac{10 + \sqrt{19}}{9}\right) +$ ______

≈ 0.850.

139. $\int \frac{dx}{16 - x^2} = \frac{1}{16} \int$ ______

Let $u = \frac{x}{4}$, du = ______, and the integral becomes

$\int \frac{dx}{16 - x^2} = \int \frac{4\ du}{\quad} = \frac{1}{4}($______$) + C$

= ______ + C.

---

137. $\sinh^{-1} x$, $x + \sqrt{x^2 + 1}$, $\ln\left(x + \sqrt{x^2 + 1}\right)$, $\ln(\sqrt{10} - 3)$

138. $-\text{sech}^{-1} x$, $\frac{1}{x}$, $1 + \sqrt{1 - x^2}$, $-\ln\left(\frac{1 + \sqrt{1 - x^2}}{x}\right)$, $\ln\left(\frac{1 + \sqrt{1 - .25}}{.5}\right)$, $\ln(2 + \sqrt{3})$

139. $\frac{dx}{1 - (x^2/16)}$, $\frac{1}{4}dx$, $16(1 - u^2)$, $\frac{1}{2}\ln\left|\frac{1 + u}{1 - u}\right|$, $\frac{1}{8}\ln\left|\frac{4 + x}{4 - x}\right|$

6.11 FIRST ORDER DIFFERENTIAL EQUATIONS

140. A first order differential equation is a relation

_______________ .

141. A function $y = y(x)$ is said to be a _______ of a differential equation if the latter is satisfied when _______ and its _______ are replaced throughout by _______ and its corresponding derivative.

OBJECTIVE A: Show that a given function is a solution to a specified first order differential equation.

142. Consider the differential equation

$$3xy' - y = \ln x + 1, \quad x > 0,$$

and the function

$$y = y(x) = Cx^{1/3} - \ln x - 4,$$

where C is any constant. Then,

$$\frac{dy}{dx} = y'(x) = \_\_\_\_\_\_\_\_,$$

and

$$3xy' = \_\_\_\_\_\_\_\_ .$$

Thus,

$$3xy' - y = (Cx^{1/3} - 3) - (Cx^{1/3} - \ln x - 4)$$
$$= \_\_\_\_\_\_\_\_ .$$

Therefore, the function $y(x) = Cx^{1/3} - \ln x - 4$, and its derivative, satisfy the differential equation. We have verified that $y = y(x)$ is a solution.

OBJECTIVE B: Solve first order differential equations in which the variables can be separated. If initial conditions are prescribed, determine the value of the constant of integration.

143. Solve the differential equation

$$(xy - x)\,dx + (xy + y)\,dy = 0 .$$

Solution. We separate the variables and integrate:

$$x(y - 1)\,dx + y(x + 1)\,dy = 0 ,$$

$$\frac{y\,dy}{y - 1} = \_\_\_\_\_\_\_\_, \quad y \neq 1.$$

---

140. $\frac{dy}{dx} = f(x, y)$

141. solution, $y(x)$, derivative, $y(x)$

142. $\frac{1}{3}Cx^{-2/3} - \frac{1}{x}$, $Cx^{1/3} - 3$, $\ln x + 1$

Then,

$$\left(1 + \frac{1}{y - 1}\right) dy = \underline{\hspace{3cm}}$$

$$y + \ln|y - 1| = \underline{\hspace{3cm}} + \ln C.$$

We introduce $\ln C$, $C > 0$, as the constant of integration in order to simplify the form of the solution. Thus, by algebra,

$$x + y = \ln C\left|\frac{x + 1}{y - 1}\right| \quad \text{or,} \quad |y - 1|e^{x+y} = \underline{\hspace{3cm}}, \quad y \neq 1.$$

144. Solve the differential equation

$$x^2yy' = e^y \ ; \quad \text{when} \quad x = 2, \quad y = 0 \ .$$

Solution: We change to differential form, separate the variables, and integrate:

$$x^2y\,dy = e^y dx \quad \text{or,} \quad ye^{-y}dy = \underline{\hspace{3cm}} .$$

In Chapter 7 you will see that the left side integrates to $-e^{-y}(y + 1)$. Thus,

$$-e^{-y}(y + 1) = \underline{\hspace{2.5cm}},$$

or simplifying algebraically,

$$x(y + 1) = \underline{\hspace{2.5cm}} .$$

Using the initial condition $x = 2$ and $y = 0$ gives,

$$2(0 + 1) = \underline{\hspace{2.5cm}}, \quad \text{or} \quad C = \underline{\hspace{2.5cm}} .$$

Thus, the solution is given by $x(y + 1) = \left(1 + \frac{x}{2}\right)e^y$.

OBJECTIVE C: Determine if a differential equation of first order is linear, and if it is, solve it.

145. A differential equation of first order, which is linear in the dependent variable y, can always be put in the standard form

$$\underline{\hspace{3cm}},$$

where P and Q are functions of x.

146. Assuming that P and Q are continuous functions of x, we can solve a linear differential equation $y' + Py = Q$ by finding an integrating factor,

---

143. $-\frac{x\,dx}{x + 1}$, $-\left(1 - \frac{1}{x + 1}\right) dx$, $-x + \ln|x + 1|$, $C|x + 1|$

144. $x^{-2}\,dx$, $-\frac{1}{x} + C$, $(1 - Cx)e^y$, $1 - 2C$, $-\frac{1}{2}$

145. $\frac{dy}{dx} + Py = Q$

$$v(x) = \underline{\hspace{3cm}},$$

providing a solution equation $v(x) \cdot y = \underline{\hspace{4cm}}$.

147. Let us solve the equation $x\frac{dy}{dx} + (x - 2)y = 3x^3e^{-x}$. In standard form,

$$\frac{dy}{dx} + \left(1 - \frac{2}{x}\right)y = 3x^2e^{-x} .$$

Here $P = \underline{\hspace{2cm}}$ and $Q = \underline{\hspace{2cm}}$, and the differential equation is linear. An integrating factor is given by

$$v(x) = e^{\int P\,dx} = e^{\int (1-\frac{2}{x})\,dx} = e^{\underline{\hspace{2cm}}} = x^{\underline{\hspace{1cm}}} e^x .$$

Hence a solution is given by

$$x^{-2}e^x y = \int \underline{\hspace{3cm}}\, dx + C = \int \underline{\hspace{3cm}}\, dx + C$$

$$= \underline{\hspace{3cm}} .$$

Thus,

$$y = \underline{\hspace{5cm}} .$$

148. The differential equation $y' = x - 4xy$ can be written in standard form as

$$y' + 4xy = x ,$$

so it is linear. The equation may also be written in the form

$$\frac{dy}{1 - 4y} = \underline{\hspace{2cm}},$$

so it is separable in the variables x and y. Thus we have a choice of methods of solution. As a separable equation, we integrate the last equation, and find

$$-\frac{1}{4}\ln|1 - 4y| = \frac{1}{2}x^2 + \ln C ,$$

or

$$|1 - 4y| = C_1 \underline{\hspace{2cm}}, \quad \text{where } C_1 = C^{-4} .$$

If we consider the differential equation as linear, an integrating factor is

$$v(x) = e^{\int 4x\,dx} = \underline{\hspace{3cm}},$$

---

146. $e^{\int P(x)dx}$, $\int v(x)\,Q(x)\,dx + C$

147. $1 - \frac{2}{x}$, $3x^2e^{-x}$, $x - 2\ln x$, -2, $x^{-2}e^x \cdot 3x^2e^{-x}$, 3, $3x + C$, $(3x^3 + Cx^2)e^{-x}$

from which we get

$$ye^{2x^2} = \int \_\_\_\_\_\_\_\_ dx + C_2 = \_\_\_\_\_\_\_\_ + C_2$$

or

$$4y = \_\_\_\_\_\_\_\_\_\_\_\_ .$$

If $1 - 4y < 0$, we choose $4C_2 = C_1$, and if $1 - 4y \geq 0$, we choose $4C_2 = -C_1$. Thus both solution forms agree.

6.12 EULER'S NUMERICAL METHOD; SLOPE FIELDS

OBJECTIVE: Find the first three approximations y_1, y_2, y_3 using the Euler approximation for an initial value problem $y' = f(x,y)$, $y(x_0) = y_0$.

149. Consider the initial value problem $y' = f(x,y)$ and $y(x_0) = y_0$. Then Euler's method allows you to approximate the solution by stepping along the tangent lines in increments dx according to the formulas

$$x_n = \_\_\_\_\_\_\_\_\_\_\_\_ \quad \text{and} \quad y_n = \_\_\_\_\_\_\_\_\_\_\_\_ .$$

150. For the initial value problem $y' = x^2 - y$, $y(1) = 2$ we find using the increment size $dx = 0.1$ that

$$x_0 = \_\_\_\_\_\_\_\_ \quad \text{and} \quad y_1 = 2 + (\_\_\_\_\_\_\_\_)(0.1) = 1.9 .$$

Next,

$$x_1 = \_\_\_\_ + 0.1 \quad \text{and} \quad y_2 = \_\_\_\_ + [(1.1)^2 - 1.9](0.1) = 1.831 .$$

At the third step we obtain,

$$x_2 = \_\_\_\_ \quad \text{and} \quad y_3 = \_\_\_\_\_\_\_\_\_\_\_\_\_\_\_\_ = 1.7919 .$$

---

148. $x\,dx$, e^{-2x^2}, e^{2x^2}, xe^{2x^2}, $\frac{1}{4}e^{2x^2}$, $1 + 4C_2e^{-2x^2}$

149. $x_{n-1} + dx$, $y_{n-1} + f(x_{n-1}, y_{n-1})\,dx$

150. 1, $1^2 - 2$, 1, 1.9, 1.3, $1.831 + [(1.2)^2 - 1.831](0.1)$

CHAPTER 6 SELF-TEST

In Problems 1-14 calculate the derivative $\frac{dy}{dx}$.

1. $y = \tan\left(\cos \frac{2}{x}\right)$
2. $y = \cos^{-1}\left(\frac{1 - x^2}{1 + x^2}\right), \quad x > 0$
3. $y = 5^x \log_5 x$
4. $y = 2^{\sin^{-1} x}$
5. $y = 2x \tan^{-1} 2x - \ln \sqrt{1 + 4x^2}$
6. $y = \frac{\sec 3\sqrt{x}}{\sqrt{x}}$
7. $y = (x)^{\sqrt{x}}, \quad x > 0$
8. $e^y + e^x = e^{x+y}$
9. $y = \tanh(\sin x)$
10. $y = \coth^{-1}(\ln x)$
11. $y = \sqrt{\cosh^{-1} x^2}$
12. $y = \ln(\sinh x^3)$
13. $y = x^{-1} \tanh^{-1} x^2$
14. $y = \sinh^{-1}(\tan x)$

In Problems 15-23 calculate the indicated integrals.

15. $\int_{-5/4}^{5/4} \frac{dx}{25 + 16x^2}$
16. $\int \frac{e^{-x}}{\sqrt{1 - e^{-2x}}} dx$
17. $\int_1^e \frac{2^{\ln x}}{3x} dx$
18. $\int \frac{4\, dx}{(e^x - e^{-x})^2}$
19. $\int \frac{\sinh(\ln x)\, dx}{x}$
20. $\int \sqrt{1 + \cosh x}\, dx$
21. $\int_3^7 \frac{dx}{\sqrt{x^2 - 1}}$
22. $\int_0^{1/2} \frac{\cosh x\, dx}{1 - \sinh^2 x}$
23. $\int_0^1 \frac{dx}{\sqrt{e^{2x} + 1}}$

In Problems 24-31 evaluate the limits uisng l'Hôpital's rule.

24. $\lim_{x \to 0} \frac{\sin x - x \cos x}{x^3}$
25. $\lim_{x \to \infty} \frac{x^2 - 5}{2x^2 + 3x}$
26. $\lim_{x \to \infty} \frac{\sin(3/x)}{2/x}$
27. $\lim_{x \to \frac{\pi}{2}^-} \left(x \tan x - \frac{\pi}{2} \sec x\right)$
28. $\lim_{x \to 0} \csc x \sin^{-1} x$
29. $\lim_{x \to 1} x^{1/(1-x)}$
30. $\lim_{x \to 0^+} \frac{e^x - \cos x}{x \sin x}$
31. $\lim_{x \to 0} \frac{4^x - 2^x}{x}$

32. Determine the following values.

(a) $\sin^{-1}\left(-\frac{\sqrt{3}}{2}\right)$

(b) $\tan^{-1} \frac{\sqrt{3}}{3}$

(c) $\tan^{-1}\left(\cos \frac{\pi}{2}\right)$

(d) $\cos^{-1}\left(\sin \frac{\pi}{6}\right)$

33. Simplify the expression $\dfrac{\log_3 243}{\log_2 \sqrt[4]{64} + \log_8 8^{-10}}$.

34. Let $f(x) = \log_a x$, $f(5) = 1.46$, $f(2) = 0.63$, $f(7) = 1.77$. Use the properties of logarithms to find,
(a) $f(10)$ (b) $f(49)$ (c) $f\left(\frac{5}{7}\right)$ (d) $f(1.4)$

35. Solve the following equations for x.
(a) $3^{-8x+6} = 27^{-x-8}$ (b) $\log_5 (5x - 1) = -2$
(c) $e^x = 10^{x+1}$

36. Graph the curve $y = \dfrac{2}{1 + 3e^{-2x}}$.

37. Sketch the graph of $y = \dfrac{\ln x}{x^2}$.

38. Define the hyperbolic function $y = \operatorname{csch} x$, and sketch its graph.

39. Given that $\cosh x = 2$, $x < 0$, find the values of the remaining hyperbolic functions at x.

40. Find the length of the catenary $y = 3 \cosh \frac{x}{3}$ from $x = 0$ to $x = 3$.

41. Suppose that a dose d mg of a drug is injected into the bloodstream. Assume that the drug leaves the blood and enters the urine at a rate proportional to the amount of the drug present in the blood. Assume that at the end of 1 hour the amount of drug in the urine is $\frac{1}{2}d$. Find the time at which 10 percent of the original dose is in the bloodstream.

42. Suppose that the number of bacteria in a yeast culture grows at a rate proportional to the number present. If the population of a colony of yeast bacteria doubles in one hour, find the number of bacteria present at the end of 3.5 hours.

43. Solve the differential equation $\dfrac{dy}{dx} = x^3e^x + \dfrac{2y}{x} - 1$, $x > 0$.

44. Solve the differential equation $e^x(y - 1)\,dx + 2(e^x + 4)\,dy = 0$.

SOLUTIONS TO CHAPTER 6 SELF-TEST

1. $\dfrac{dy}{dx} = \sec^2\left(\cos \frac{2}{x}\right) \cdot \left(-\sin \frac{2}{x}\right) \cdot \left(-\frac{2}{x^2}\right)$

2. $$\frac{dy}{dx} = \frac{-1}{\sqrt{1 - \left(\frac{1 - x^2}{1 + x^2}\right)^2}} \cdot \left[\frac{(1 + x^2)(-2x) - (1 - x^2)(2x)}{(1 + x^2)^2}\right]$$

$$= \frac{-(1 + x^2)}{\sqrt{(1 + x^2)^2 - (1 - x^2)^2}} \cdot \frac{-4x}{(1 + x^2)^2} = \frac{4x}{(1 + x^2)\sqrt{4x^2}}$$

$$= \frac{2x}{(1 + x^2)|x|} = \frac{2}{1 + x^2}$$

3. $$\frac{dy}{dx} = 5^x \ln 5 \cdot \log_5 x + \frac{5^x}{x \ln 5} = 5^x\left(\ln x + \frac{1}{x \ln 5}\right)$$

4. $$\frac{dy}{dx} = 2^{\sin^{-1} x}(\ln 2)\frac{1}{\sqrt{1 - x^2}}$$

5. $$\frac{dy}{dx} = 2\tan^{-1} 2x + 2x \cdot \frac{1}{1 + 4x^2} \cdot 2 - \frac{1}{\sqrt{1 + 4x^2}} \cdot \frac{8x}{2\sqrt{1 + 4x^2}}$$

$$= 2\tan^{-1} 2x$$

6. $$\frac{dy}{dx} = \frac{1}{\sqrt{x}}(\sec 3\sqrt{x} \tan 3\sqrt{x}) \cdot \frac{3}{2\sqrt{x}} + \left(-\frac{1}{2}x^{-3/2}\right)\sec 3\sqrt{x}$$

$$= \frac{\sec 3\sqrt{x}}{2x\sqrt{x}}(3\sqrt{x} \tan 3\sqrt{x} - 1)$$

7. $y = (x)^{\sqrt{x}}$ gives $\ln y = \sqrt{x} \ln x$

$$\frac{1}{y}\frac{dy}{dx} = \frac{1}{2\sqrt{x}} \ln x + \sqrt{x} \cdot \frac{1}{x} = \frac{1}{2\sqrt{x}}(\ln x + 2)$$

$$\frac{dy}{dx} = \frac{1}{2\sqrt{x}}(\ln x + 2)(x)^{\sqrt{x}}$$

8. $$e^y \frac{dy}{dx} + e^x = e^{x+y}\frac{d}{dx}(x + y) = e^{x+y}\left(1 + \frac{dy}{dx}\right)$$

Hence, $\left(e^y - e^{x+y}\right)\frac{dy}{dx} = e^{x+y} - e^x$, or from the original expression, $-e^x \frac{dy}{dx} = e^y$. Thus, $\frac{dy}{dx} = -e^{y-x}$.

9. $$\frac{dy}{dx} = \text{sech}^2(\sin x) \cdot \cos x$$

10. $$\frac{dy}{dx} = \frac{1}{[1 - (\ln x)^2]} \cdot \frac{1}{x}, \quad \ln x > 1$$

11. $$\frac{dy}{dx} = \frac{1}{2}\left(\cosh^{-1} x^2\right)^{-1/2} \cdot \frac{d}{dx}\left(\cosh^{-1} x^2\right)$$

$$= \frac{1}{2}\left(\cosh^{-1} x^2\right)^{1/2} \cdot \frac{1}{\sqrt{x^4 - 1}} \cdot \frac{d}{dx}(x^2)$$

$$= \frac{x}{\sqrt{(x^4 - 1)\cosh^{-1} x^2}}, \quad x > 1$$

12. $\frac{dy}{dx} = \frac{1}{\sinh x^3} \cdot \frac{d}{dx}(\sinh x^3) = \frac{1}{\sinh x^3} \cdot \cosh x^3 \cdot \frac{d}{dx}(x^3)$

$= 3x^2 \coth x^3$

13. $\frac{dy}{dx} = -\frac{1}{x^2} \tanh^{-1} x^2 + x^{-1} \cdot \frac{1}{1 - x^4} \cdot 2x$

$= -x^{-2} \tanh^{-1} x^2 + \frac{2}{1 - x^4}, \quad x^4 < 1$

14. $\frac{dy}{dx} = \frac{1}{\sqrt{\tan^2 x + 1}} \cdot \sec^2 x = \frac{\sec^2 x}{\sqrt{\sec^2 x}} = \sec x, \quad \text{if} \quad -\frac{\pi}{2} < x < \frac{\pi}{2}$

15. $\int \frac{dx}{25 + 16x^2} = \frac{1}{25} \int \frac{dx}{1 + \left(\frac{4}{5}x\right)^2} = \frac{1}{25} \cdot \frac{5}{4} \tan^{-1} \frac{4}{5}x + C$

$\int_{-5/4}^{5/4} \frac{dx}{25 + 16x^2} = \frac{1}{20}\left(\tan^{-1} 1 - \tan^{-1}(-1)\right) = \frac{\pi}{40}$

16. $u = e^{-x}, \quad du = -e^{-x}\, dx$

$\int \frac{-du}{\sqrt{1 - u^2}} = \cos^{-1}\left(e^{-x}\right) + C$

17. $u = \ln x, \quad du = \frac{1}{x} dx$

$\int \frac{2^{\ln x}}{3x}\, dx = \frac{1}{3} \int 2^u\, du = \frac{1}{3 \ln 2}\, 2^{\ln x} + C$

$\int_1^e \frac{2^{\ln x}}{3x}\, dx = \frac{1}{3 \ln 2}\left(2^{\ln e} - 1\right) = \frac{1}{3 \ln 2}$

18. $\int \frac{4\, dx}{\left(e^x - e^{-x}\right)^2} = \int \operatorname{csch}^2 x\, dx = -\coth x + C$

19. Let $u = \ln x$, $du = \frac{1}{x} dx$, and the integral becomes

$\int \frac{\sinh(\ln x)\, dx}{x} = \int \sinh u\, du = \cosh u + C = \cosh(\ln x) + C$

20. From the identity $2 \cosh^2 x = \cosh 2x + 1$, we find that

$\sqrt{2} \cosh \frac{x}{2} = \sqrt{\cosh x + 1}$. Thus,

$\int \sqrt{\cosh x + 1}\, dx = \sqrt{2} \int \cosh \frac{x}{2}\, dx = 2\sqrt{2} \sinh \frac{x}{2} + C$

21. $\int_{-3}^{7} \frac{dx}{\sqrt{x^2 - 1}} = \cosh^{-1} x\Big]_3^7 = \ln\left(x + \sqrt{x^2 + 1}\right)\Big]_3^7$

$= \ln(7 + \sqrt{48}) - \ln(3 + \sqrt{8})$

$= \ln\left(\frac{7 + 4\sqrt{3}}{3 + 2\sqrt{2}}\right) \approx 2.39.$

22. Let $u = \sinh x$, $du = \cosh x\,dx$, so that

$$\int_0^{1/2} \frac{\cosh x\,dx}{1 - \sinh^2 x} = \tanh^{-1}(\sinh x)\Big]_0^{1/2} = \frac{1}{2}\ln\left[\frac{1 + \sinh x}{1 - \sinh x}\right]_0^{1/2}$$

$$= \frac{1}{2}\ln\left[\frac{1 + 0.5211}{1 - 0.5211}\right] \approx 0.5778 \text{ by Tables.}$$

23. $\displaystyle\int_0^1 \frac{dx}{\sqrt{e^{2x} + 1}} = \int_0^1 \frac{e^x\,dx}{e^x\sqrt{e^{2x} + 1}}$, which is of the form

$$\int \frac{du}{|u|\sqrt{u^2 + 1}} = -\text{csch}^{-1}u + C, \quad \text{for } u = e^x \neq 0.$$

Thus, $\displaystyle\int_0^1 \frac{dx}{\sqrt{e^{2x} + 1}} = -\text{csch}^{-1}e^x\Big]_0^1 = -\ln\left(\frac{1 + \sqrt{1 + e^{2x}}}{e^x}\right)\Big]_0^1$

$$= -\ln\left(\frac{1 + \sqrt{1 + e^2}}{e}\right) + \ln(1 + \sqrt{2}) \approx 0.52.$$

24. $$\lim_{x\to 0} \frac{\sin x - x\cos x}{x^3} = \lim_{x\to 0} \frac{\cos x - \cos x + x\sin x}{3x^2}$$

$$= \lim_{x\to 0} \frac{\sin x}{3x} = \frac{1}{3}\lim_{x\to 0} \frac{\sin x}{x} = \frac{1}{3}.$$

25. $\displaystyle\lim_{x\to\infty} \frac{x^2 - 5}{2x^2 + 3x}$ is of the form ∞/∞. Applying l'Hôpital's rule,

$$\lim_{x\to\infty} \frac{x^2 - 5}{2x^2 + 3x} = \lim_{x\to\infty} \frac{2x}{4x + 3} \text{ [still } \infty/\infty] = \lim_{x\to\infty} \frac{2}{4} = \frac{1}{2}.$$

26. $\displaystyle\lim_{x\to\infty} \frac{\sin(3/x)}{2/x}$ is of the form $0/0$. Applying l'Hôpital's rule,

$$\lim_{x\to\infty} \frac{\sin(3/x)}{2/x} = \lim_{x\to\infty} \frac{(-3/x^2)\cos(3/x)}{-2/x^2} = \lim_{x\to\infty} \frac{3}{2}\cos\frac{3}{x} = \frac{3}{2}\cos 0$$

$$= \frac{3}{2}.$$

27. $\displaystyle\lim_{x\to\frac{\pi}{2}^-} \left(x\tan x - \frac{\pi}{2}\sec x\right)$ is of the form $\infty - \infty$. However,

$$\lim_{x\to\frac{\pi}{2}^-} \left(x\tan x - \frac{\pi}{2}\sec x\right) = \lim_{x\to\frac{\pi}{2}^-} \frac{x\sin x - \frac{\pi}{2}}{\cos x} \text{ is of the form } 0/0.$$

Applying l'Hôpital's rule,

$$\lim_{x\to\frac{\pi}{2}^-} \frac{x\sin x - \frac{\pi}{2}}{\cos x} = \lim_{x\to\frac{\pi}{2}^-} \frac{\sin x + x\cos x}{-\sin x} = \frac{1 + \left(\frac{\pi}{2}\right)(0)}{(-1)} = -1.$$

28. $$\lim_{x\to 0} \csc x \sin^{-1} x = \lim_{x\to 0} \frac{\sin^{-1}x}{\sin x} \quad \text{(type } 0/0)$$

$$= \lim_{x\to 0} \frac{1/\sqrt{1 - x^2}}{\cos x} = \frac{1}{1} = 1$$

29. Let $y = x^{1/(1-x)}$ so $\ln y = \dfrac{\ln x}{1 - x}$

$$\lim_{x\to 1} \ln y = \lim_{x\to 1} \frac{\ln x}{1 - x} \quad \text{(type 0/0)} \quad = \lim_{x\to 1} \frac{1/x}{-1} = -1$$

By the continuity of the natural logarithm,

$$\lim_{x\to 1} x^{1/(1-x)} = e^{-1}.$$

30. $$\lim_{x\to 0^+} \frac{e^x - \cos x}{x \sin x} \quad \text{(type 0/0)} \quad = \lim_{x\to 0^+} \frac{e^x + \sin x}{x \cos x + \sin x} = +\infty.$$

31. $$\lim_{x\to 0} \frac{4^x - 2^x}{x} \quad \text{(type 0/0)} \quad = \lim_{x\to 0} \frac{4^x \ln 4 - 2^x \ln 2}{1}$$
$$= \ln 4 - \ln 2 = \ln 2.$$

32. (a) $-\frac{\pi}{3}$ (b) $\frac{\pi}{6}$ (c) 0 (d) $\frac{\pi}{3}$

33. $$\frac{\log_3 243}{\log_2 \sqrt[4]{64} + \log_8 8^{-10}} = \frac{5}{\frac{1}{4}(6) - 10(1)} = -\frac{10}{17}$$

34. (a) $\log_a 10 = \log_a 2 + \log_a 5 = 2.09$

(b) $\log_a 49 = 2 \log_a 7 = 3.54$

(c) $\log_a \frac{5}{7} = \log_a 5 - \log_a 7 = -0.31$

(d) $\log_a 1.4 = \log_a 14 - \log_a 10$
$= \log_a 2 + \log_a 7 - \log_a 5 - \log_a 2 = 0.31$

35. (a) $3^{-8x+6} = 3^{-3x-24}$ or $-8x + 6 = -3x - 24$ or $x = 6$.

(b) $5^{-2} = 5x - 1$ or $x = \frac{26}{125}$.

(c) $x = (x + 1) \ln 10$ or $x = \dfrac{\ln 10}{1 - \ln 10} \approx -1.768$.

36. $f(x) = \dfrac{2}{1 + 3e^{-2x}}$, $f'(x) = \dfrac{12e^{-2x}}{\left(1 + 3e^{-2x}\right)^2}$,

$$f''(x) = \frac{24e^{-2x}\left(-1 + 3e^{-2x}\right)}{\left(1 + 3e^{-2x}\right)^3}$$

$f' > 0$ for every x;

$f''(x) = 0$ when $x = \frac{1}{2} \ln 3 \approx 0.55$

$\lim_{x\to +\infty} f(x) = 2$ and $\lim_{x\to -\infty} f(x) = 0$

The graph is sketched at the right.

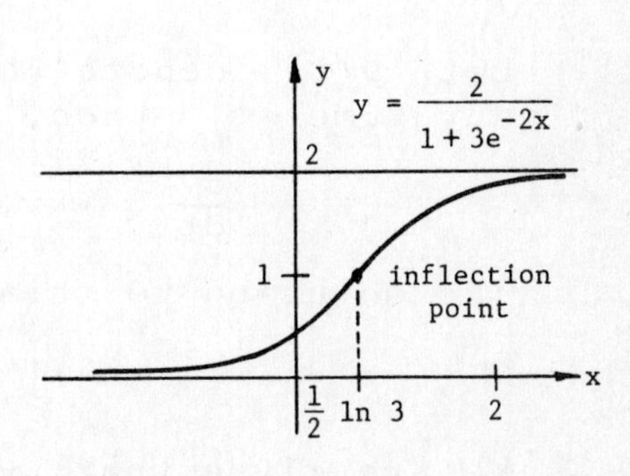

37. $f(x) = \dfrac{\ln x}{x^2}$,

$f'(x) = \dfrac{1 - 2\ln x}{x^3}$ so

$f'(x) = 0$ implies $\ln x = \frac{1}{2}$ or $x = \sqrt{e}$

$f''(x) = \dfrac{-5 + 6\ln x}{x^4}$ so

$f''(x) = 0$ implies $x = e^{5/6} \approx 2.3$

$\lim\limits_{x\to\infty} f(x) = \lim\limits_{x\to\infty} \dfrac{1/x}{2x} = 0$ and

$\lim\limits_{x\to 0^+} f(x) = -\infty$. The graph is sketched at the right.

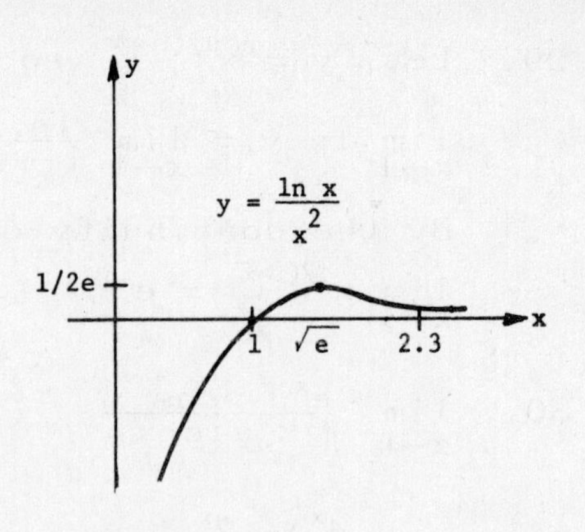

38. $y = \operatorname{csch} x = \dfrac{1}{\sinh x}$,

where $\sinh x = \frac{1}{2}(e^x - e^{-x})$.
The graph is sketched at the right.

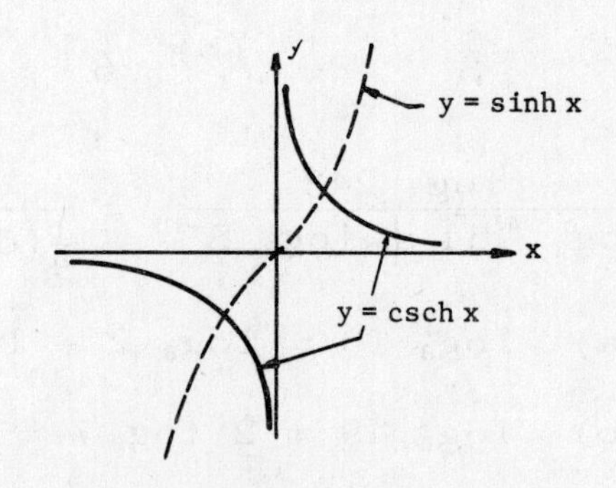

39. $\operatorname{sech} x = \dfrac{1}{\cosh x} = \dfrac{1}{2}$; $\tanh^2 x = 1 - \operatorname{sech}^2 x = \dfrac{3}{4}$ so that

$\tanh x = \dfrac{-\sqrt{3}}{2}$. Since $x < 0$; $\sinh x = \cosh x \tanh x = -\sqrt{3}$;

$\coth x = \dfrac{1}{\tanh x} = -\dfrac{2}{\sqrt{3}}$; and $\operatorname{csch} x = \dfrac{1}{\sinh x} = -\dfrac{1}{\sqrt{3}}$.

40. $s = \displaystyle\int_0^3 \sqrt{1 + \left(\frac{dy}{dx}\right)^2}\, dx = \int_0^3 \sqrt{1 + \sinh^2 \frac{x}{3}}\, dx$

$= \displaystyle\int_0^3 \cosh \frac{x}{3}\, dx = 3 \sinh \frac{x}{3}\Big]_0^3 = 3 \sinh 1 \approx 3.53.$

41. Let $y(t)$ denote the amount of the drug in the bloodstream at any time t. then,

$$\frac{dy}{dt} = -ky, \quad k > 0 \quad \text{and} \quad y(0) = d\ .$$

The solution to this initial value problem is

$$y(t) = de^{-kt}\ .$$

We are given that $y(1) = \frac{1}{2}d$, so

$$\frac{1}{2}d = de^{-k(1)}\ .$$

Then

$$-k = \ln \frac{1}{2} \quad \text{or} \quad k = \ln 2 .$$

Substituting into the original solution,

$$y(t) = de^{-(\ln 2)t} .$$

The desired time $t = T$ is such that $y(T) = \frac{1}{10}d$. Thus,

$$\frac{1}{10}d = de^{-(\ln 2)T}$$

or

$$\ln \frac{1}{10} = (-\ln 2)T .$$

Solving,

$$T = \frac{-\ln 10}{-\ln 2} \approx 3.32 \text{ hr} .$$

42. Let x denote the number of bacteria present at any time t. Then, $\frac{dy}{dx} = kx$ or $x = Ce^{kt}$, for some constant C. If x_0 is the initial number of bacteria at $t = 0$, then $C = x_0$, so $x = x_0e^{kt}$. When $t = 1$, $x = 2x_0$ so that $2 = e^k$ or $k = \ln 2$. Therefore,

$$x = x_0 \, e^{t \ln 2} .$$

Finally, when $t = 3.5$, $x = x_0 e^{3.5 \ln 2} \approx 11.31\, x_0$. Thus, there are 11.31 times the initial number of bacteria present at the end of 3.5 hours.

43. $y' - \frac{2}{x}y = x^3e^x - 1$, $x > 0$ is linear. An integrating factor is

$$v(x) = e^{\int -\frac{2}{x}dx} = e^{-2\ln x} = x^{-2}, \quad x > 0 .$$

Thus,

$$\begin{aligned} x^{-2}y &= \int x^{-2}(x^3e^x - 1)\,dx \\ &= \int xe^{-x}\,dx - \int x^{-2}\,dx \\ &= -(x+1)e^{-x} + x^{-1} + C \end{aligned}$$

Remark: You will learn $\int xe^{-x}\,dx = -(x+1)e^{-x} + C$ in Chapter 7.

Thus,

$$y = Cx^2 + x - x^2(x+1)e^{-x} .$$

44. The variables are separable, and the differential equation can be written as

$$\frac{e^x}{e^x + 4}\, dx + \frac{2}{y - 1}\, dy = 0 .$$

Integration gives,

$$\ln(e^x + 4) + 2\ln|y - 1| = \ln C, \quad \text{or} \quad (y - 1)^2(e^x + 4) = C .$$

NOTES.

CHAPTER 7 TECHNIQUES OF INTEGRATION

7.1 BASIC INTEGRATION FORMULAS

OBJECTIVE: Evaluate indefinite integrals by reducing the integrands to basic forms through algebra and substitution.

1. $\int x(a + x)^{1/3}\, dx$, where a is constant.

Let $u = a + x$. Then $du =$ ______ and $x =$ ______. Thus,

$$\int x(a + x)^{1/3}\, dx = \int \_\_\_\_\_\_\, du = \int u^{4/3}\, du - \int \_\_\_\_\_\_\, du$$

$$= \_\_\_\_\_\_ - \frac{3}{4} a u^{4/3} + C$$

$$= \_\_\_\_\_\_ + C.$$

2. $\int \frac{dx}{x\sqrt{x - 5}}$

Let $u = \sqrt{x - 5}$. Then $du =$ ______ and $x =$ ______.

$$\int \frac{dx}{x\sqrt{x - 5}} = \int \_\_\_\_\_\_\, du = \_\_\_\_\_\_ \int \frac{du}{\left(\frac{u}{\sqrt{5}}\right)^2 + 1}$$

$$= \_\_\_\_\_\_ \tan^{-1}\frac{u}{\sqrt{5}} + C = \_\_\_\_\_\_ + C.$$

3. $\int \frac{\cos x\, dx}{\sqrt{1 - \sin x}}$

Let $u = \sin x$, then $du =$ ______ so that

$$\int \frac{\cos x\, dx}{\sqrt{1 - \sin x}} = \int \_\_\_\_\_\_\, du = \int (1 - u)^{\_\_\_}\, du$$

$$= \_\_\_\_\_\_ + C = \_\_\_\_\_\_ .$$

4. $\int \frac{3^{\tan x}\, dx}{\cos^2 x}$

Let $u = \tan x$. Then $du =$ ______ and the integral is

$$\int \frac{3^{\tan x}\, dx}{\cos^2 x} = \int (\sec^2 x)\, 3^{\tan x}\, dx = \int \_\_\_\_\_\_\, du$$

$$= \_\_\_\_\_\_ + C = \_\_\_\_\_\_ .$$

---

1. dx, $u - a$, $(u - a)u^{1/3}$, $au^{1/3}$, $\frac{3}{7}u^{7/3}$, $3(a + x)^{4/3}\left[\frac{1}{7}(a + x) - \frac{a}{4}\right]$

2. $\frac{dx}{2\sqrt{x - 5}}$, $u^2 + 5$, $\frac{2}{u^2 + 5}$, $\frac{2}{5}$, $\frac{2\sqrt{5}}{5}$, $\frac{2}{\sqrt{5}}\tan^{-1}\sqrt{\frac{x}{5} - 1}$

3. $\cos x\, dx$, $\frac{1}{\sqrt{1 - u}}$, $-\frac{1}{2}$, $-2(1 - u)^{1/2}$, $-2\sqrt{1 - \sin x} + C$

4. $\sec^2 x\, dx$, 3^u, $\frac{1}{\ln 3}3^u$, $\frac{1}{\ln 3}3^{\tan x} + C$

5. $\int \frac{dx}{x + x \ln^2 x}$

Let $u = \ln x$. Then $du =$ ________ so that

$$\int \frac{dx}{x + x \ln^2 x} = \int \frac{dx}{x\left(\underline{\qquad}\right)} = \int \frac{du}{\underline{\qquad}}$$

$$= \underline{\qquad\qquad} + C = \underline{\qquad\qquad} .$$

6. $\int \frac{(x + 1)\,dx}{x^2 + x + 5}$

Complete the square in the denominator to write it in the form

$$x^2 + x + 5 = (x^2 + x + \underline{\quad}) + 5 - \frac{1}{4} = (\underline{\qquad})^2 + \frac{19}{4}.$$

Now, substitute $u = x + \frac{1}{2}$ and $du =$ ______ so the integral becomes

$$\int \frac{(x + 1)\,dx}{x^2 + x + 5} = \int \frac{(x + 1)\,dx}{\left(x + \frac{1}{2}\right)^2 + \frac{19}{4}} = \int \frac{(\underline{\qquad})\,du}{u^2 + \frac{19}{4}} .$$

Next, separate the fraction:

$$\int \frac{\left(u + \frac{1}{2}\right)du}{u^2 + \frac{19}{4}} = \int \frac{\underline{\qquad}}{u^2 + \frac{19}{4}} + \frac{1}{2}\int \frac{du}{u^2 + \frac{19}{4}} .$$

Finally, substitute $z = u^2 + \frac{19}{4}$ and $dz = 2u\,du$ in the first term on the right and then integrate:

$$\int \frac{(x + 1)\,dx}{x^2 + x + 5} = \int \frac{\underline{\qquad}}{z} + \frac{1}{2}\int \frac{du}{u^2 + \frac{19}{4}}$$

$$= \underline{\qquad\qquad} + \frac{1}{2}\sqrt{\frac{4}{19}}\tan^{-1}\frac{u}{\sqrt{19/4}} + C$$

$$= \frac{1}{2}\ln(\underline{\qquad\qquad}) + \frac{1}{\sqrt{19}}\tan^{-1}\frac{2u}{\sqrt{19}} + C$$

$$= \underline{\qquad\qquad\qquad\qquad\qquad} .$$

7. $\int \frac{x^3}{x^2 + 1}\,dx$

First divide the denominator into the numerator, getting a quotient plus a remainder that is a proper fraction (where the degree of the numerator is less than the degree of the denominator):

---

5. $\frac{1}{x}\,dx$, $1 + \ln^2 x$, $1 + u^2$, $\tan^{-1} u$, $\tan^{-1}(\ln x) + C$

6. $\frac{1}{4}$, $x + \frac{1}{2}$, dx, $u + \frac{1}{2}$, $u\,du$, $\frac{1}{2}\,dz$, $\frac{1}{2}\ln|z|$, $u^2 + \frac{19}{4}$, $\frac{1}{2}\ln(x^2 + x + 5) + \frac{1}{\sqrt{19}}\tan^{-1}\frac{2x + 1}{\sqrt{19}} + C$

$$\frac{x^3}{x^2+1} = x - \underline{\hspace{8em}} .$$

Then,

$$\int \frac{x^3}{x^2+1} = \int \underline{\hspace{5em}} - \int \frac{x}{x^2+1}\,dx$$

$$= \underline{\hspace{5em}} - \int \underline{\hspace{5em}}\,du \quad (\text{where } u = x^2+1)$$

$$= \underline{\hspace{14em}}$$

$$= \frac{1}{2}x^2 + \ln\sqrt{x^2+1} + C .$$

7.2 INTEGRATION BY PARTS

OBJECTIVE: Evaluate integrals by the method of integration by parts when the method applies.

8. The method of integration by parts is based on the product rule for differentiation. In terms of integrals,

$$\int u\,dv = \underline{\hspace{8em}} .$$

9. To employ the method of integration by parts successfully we must be able to integrate the part ____ immediately in order to obtain ____ . Also it is desired that the new integral $\int v\,du$ be simpler than the original integral $\int u\,dv$.

10. $\int \frac{xe^x\,dx}{(1+x)^2}$

Let $u = xe^x$, $du = \underline{\hspace{5em}}$, $dv = \frac{dx}{(1+x)^2}$, $v = \underline{\hspace{5em}}$.
Then,

$$\int \frac{xe^x\,dx}{(1+x)^2} = \frac{-xe^x}{1+x} + \int \underline{\hspace{3em}}\,dx$$

$$\int \frac{xe^x\,dx}{(1+x)^2} = e^x\left(\frac{-x}{1+x} + \underline{\hspace{3em}}\right) + C = \underline{\hspace{6em}} .$$

11. $\int \tan^{-1}\sqrt{x}\,dx$

Let $u = \tan^{-1}\sqrt{x}$, $du = \underline{\hspace{5em}}$, $dv = \underline{\hspace{3em}}$, $v = \underline{\hspace{3em}}$.
Then,

$$\int \tan^{-1}\sqrt{x}\,dx = \underline{\hspace{7em}} - \int \frac{\sqrt{x}\,dx}{2(1+x)} .$$

---

7. $\frac{x}{x^2+1}$, $x\,dx$, $\frac{1}{2}x^2$, $\frac{1/2}{u}$, $\frac{1}{2}x^2 + \frac{1}{2}\ln|u| + C$

8. $uv - \int v\,du$

9. dv, v

10. $e^x(1+x)\,dx$, $-\frac{1}{1+x}$, e^x, 1, $\frac{e^x}{1+x} + C$

Let $w = \sqrt{x}$ and $dw =$ ______, and the latter integral becomes

$$\int \frac{\sqrt{x}\,dx}{2(1+x)} = \int \frac{w^2\,dw}{\_\_\_\_} = \int \left(1 - \_\_\_\_\_\_\right) dw$$

$$= w - \_\_\_\_\_\_ + C$$

$$= \_\_\_\_\_\_\_\_$$

Putting this together with our previous result, we find

$$\int \tan^{-1}\sqrt{x}\,dx = \_\_\_\_\_\_\_\_\_\_ .$$

12. $\int_1^e (\ln x)^2\,dx$

Let $u = (\ln x)^2$, $du =$ ________, $dv =$ ________, $v =$ ________ . Then,

$$\int_1^e (\ln x)^2\,dx = x(\ln x)^2]_1^e - 2\int_1^e \_\_\_\_\_\_\_\_$$

$$= (e - \_\_\_\_\_\_) - 2[\_\_\_\_\_\_\_\_]_1^e$$

↑ Example 4 in this section of the text

$$= e - 2(\_\_\_\_\_\_\_\_) = \_\_\_\_\_\_\_\_ .$$

13. $\int x^2 \sin x\,dx$

Let $u = x^2$, $du =$ ______, $dv =$ ______, $v =$ ______ . Then

$$\int x^2 \sin x\,dx = \_\_\_\_\_\_\_\_ + 2\int x \cos x\,dx.$$

To determine the latter integral we integrate by parts again: $U = x$, $dU =$ ______, $dV =$ ______, $V =$ ______ . Then,

$$\int x \cos x\,dx = x \sin x - \int \_\_\_\_\_\_ = \_\_\_\_\_\_\_\_ + C$$

Therefore, putting these results together, we find

$$\int x^2 \sin x\,dx = \_\_\_\_\_\_\_\_\_\_ .$$

---

11. $\frac{dx}{2\sqrt{x}(1+x)}$, dx, x, $x \tan^{-1}\sqrt{x}$, $\frac{dx}{2\sqrt{x}}$, $1 + w^2$, $\frac{1}{1+w^2}$, $\tan^{-1} w$, $\sqrt{x} - \tan^{-1}\sqrt{x}$,

$(x + 1)\tan^{-1}\sqrt{x} - \sqrt{x} + C$

12. $\frac{2}{x}\ln x\,dx$, dx, x, $\ln x\,dx$, 0, $x \ln x - x$, $(e - e + 1)$, $e - 2$

13. $2x\,dx$, $\sin x\,dx$, $-\cos x$, $-x^2 \cos x$, dx, $\cos x\,dx$, $\sin x$, $\sin x\,dx$, $x \sin x + \cos x$,

$(2 - x^2)\cos x + 2x \sin x + C$

14. $\displaystyle\int \frac{\tan^{-1} x \, dx}{x^2}$

Let $u = \tan^{-1} x$, $du =$ ________, $dv =$ ________, $v =$ ________ .
Then,

$$\int \frac{\tan^{-1} x \, dx}{x^2} = \underline{\qquad\qquad} + \int \frac{dx}{x(1 + x^2)}$$
$$= \underline{\qquad\qquad} + \int \frac{dx}{x} + \int \frac{\qquad}{1 + x^2}$$
$$= \underline{\qquad\qquad\qquad\qquad} .$$

15. Evaluate $\int e^{2x} \sin 3x \, dx$ by tabular integration.

<u>Solution</u>.

Let $u = f(x) = e^{2x}$ and $dv = g(x)\,dx = \sin 3x \, dx$. Then complete the following table:

f(x) and its derivatives	signs	g(x) and its integrals
e^{2x}	(+)	$\sin 3x$
____	( )	____
$\int$ ____ ____	( )	____

The "stopping rule" for the last row in the table is that you can integrate the product of the functions in the last row, or the product is a <u>constant</u> times the product of the functions in the first row. Of course, if you obtain a zero in the f(x) column the table is finished because the integral of zero is zero. In this problem, the product of the last row is a constant times the first row. The table reads:

$$+e^{2x}\left(-\frac{1}{3}\cos 3x\right) - 2e^{2x}\left(-\frac{1}{9}\sin 3x\right) + \int 4e^{2x}\left(-\frac{1}{9}\right)\sin 3x \, dx$$

Therefore,

$$\int e^{2x} \sin 3x \, dx = e^{2x}\left(-\frac{1}{3}\cos 3x + \frac{2}{9}\sin 3x\right) - \frac{4}{9}\int e^{2x} \sin 3x \, dx + C .$$

Combining the two integrals gives

$$\int e^{2x} \sin 3x \, dx = \underline{\qquad\qquad\qquad\qquad\qquad\qquad}$$

or, simplifying arithmetically,

$$\int e^{2x} \sin 3x \, dx = \frac{1}{13} e^{2x} (2 \sin 3x - 3 \cos 3x) + C_1 .$$

---

14. $\frac{1}{1 + x^2}\,dx$, $\frac{dx}{x^2}$, $-\frac{1}{x}$, $-\frac{\tan^{-1} x}{x}$, $-\frac{\tan^{-1} x}{x}$, $-x\,dx$, $-\frac{\tan^{-1} x}{x} + \ln|x| - \frac{1}{2}\ln(1 + x^2) + C$

15.

f(x)		g(x)
e^{2x}	(+)	$\sin 3x$
$2e^{2x}$	(-)	$-\frac{1}{3}\cos 3x$
$4e^{2x}$	(+)	$-\frac{1}{9}\sin 3x$

, $\frac{9}{13} e^{2x}\left[\frac{2}{9}\sin 3x - \frac{1}{3}\cos 3x\right] + C_1$ $\left(\text{where } C_1 = \frac{9}{13} C\right)$

7.3 PARTIAL FRACTIONS

OBJECTIVE: Find indefinite integrals of rational functions by the method of partial fractions expansion.

16. The success of separating a rational function $\frac{f(x)}{g(x)}$ into a sum of partial fractions hinges upon two things:
(1) The degree of $f(x)$ must be ______________________ .
If this is not the case, one must first perform ______________, then work with the ____________ term.
(2) The factors of ________ must be known. In practice it may be difficult to perform the factorization.

17. To find A and B in the partial fractions expansion of

$$\frac{3x + 1}{(x + 1)(x + 2)} = \frac{A}{x + 1} + \frac{B}{x + 2}$$

by the method of undetermined coefficients, first clear the equation of fractions:

$3x + 1 =$ ______________________ .

Next collect like powers of x on the right side:

$3x + 1 = ($________$)x + (2A + B)$.

This will be an identity in x if and only if

$A + B =$ ______ and __________ $= 1$.

These equations determine the values of A and B to be $A =$ ______, $B = 5$.

18. From Problem 17,

$$\int \frac{3x + 1}{(x + 1)(x + 2)}\,dx = \int \frac{-2}{x + 1}\,dx + \int \text{\_\_\_\_\_\_\_\_\_\_}$$

$$= \text{\_\_\_\_\_\_\_\_\_\_\_\_\_\_\_\_\_\_\_\_\_\_\_\_} .$$

19. To find A, B, and C in the partial fractions expansion

$$\frac{2x^2 - x + 1}{(x + 1)(x - 3)(x + 2)} = \frac{A}{x + 1} + \frac{B}{x - 3} + \frac{C}{x + 2}$$

by the method of undetermined coefficients, first clear the equation of fractions:

$2x^2 - x + 1 = A(x - 3)(x + 2) + B(x + 1)(x + 2) + C($____$)($____$)$.

Next expand the products on the right and collect like powers of x:

$2x^2 - x + 1 = (A + B + C)x^2 + ($__________$)x + (-6A + 2B - 3C)$.

---

16. less than the degree of $g(x)$, long division, remainder, $g(x)$

17. $A(x + 2) + B(x + 1)$, $A + B$, 3, $2A + B$, -2

18. $\frac{5}{x + 2}\,dx$, $-2 \ln|x + 1| + 5 \ln|x + 2| + C$

This equation is an identity in x if and only if coefficients of like powers of x on the two sides are equal:

$A + B + C =$ ______, $-A + 3B - 2C =$ ______, ____________ $= 1$.

These equations determine the values of A, B and C to be $A = -1$, $B = \frac{4}{5}$ and $C = \frac{11}{5}$. Check that these values satisfy the three equations.

20. To find A, B, and C in the partial fractions expansion

$$\frac{2x^2 - x + 1}{(x + 1)(x - 3)(x + 2)} = \frac{A}{x + 1} + \frac{B}{x - 3} + \frac{C}{x + 2}$$

by the Heaviside technique, the value of A can be found by covering up the factor ______ in the left side and evaluating the result at $x =$ ______:

$$A = \frac{2(-1)^2 - (-1) + 1}{\rule{3cm}{0.4pt}} = \frac{4}{\rule{1cm}{0.4pt}} = \rule{2.5cm}{0.4pt}.$$

$$\text{Similarly,}\quad B = \frac{2(3)^2 - (3) + 1}{\rule{3cm}{0.4pt}} = \frac{16}{\rule{1cm}{0.4pt}} = \rule{2cm}{0.4pt}, \text{ and}$$

$$C = \frac{\rule{3cm}{0.4pt}}{(-2 + 1)(-2 - 3)} = \frac{\rule{1cm}{0.4pt}}{5}.$$

21. $$\int \frac{(2x^2 - x + 1)\,dx}{(x + 1)(x - 3)(x + 2)} = \int \frac{-\,dx}{\rule{1cm}{0.4pt}} + \frac{4}{5}\int \frac{dx}{\rule{1.5cm}{0.4pt}} + \frac{11}{5}\int \frac{dx}{\rule{1.5cm}{0.4pt}}$$

$$= \rule{6cm}{0.4pt} + C.$$

22. $$\int \frac{4\,dx}{x^3 + 4x}$$

To expand $\frac{4}{x^3 + 4x}$ into a sum of partial fractions, first write,

$$\frac{4}{x^3 + 4x} = \frac{4}{x(x^2 + 4)} = \frac{A}{x} + \frac{\rule{2cm}{0.4pt}}{x^2 + 4}.\quad \text{Then,}$$

$4 = A(x^2 + 4) +$ ________ $=$ (______)$x^2 + Cx + 4A$. Thus, equating coefficients of like powers of x, $A + B =$ ______, $C =$ ______, and $4A =$ ______. Solving these equations gives $A =$ ________, $B =$ ________, and $C =$ ________. Hence,

---

19. $(x + 1)(x - 3)$, $-A + 3B - 2C$, 2, -1, $-6A + 2B - 3C$

20. $(x + 1)$, -1, $(-1 - 3)(-1 + 2)$, -4, -1, $(3 + 1)(3 + 2)$, 20, $\frac{4}{5}$, $2(-2)^2 - (-2) + 1$, 11

21. $x + 1$, $x - 3$, $x + 2$, $-\ln|x + 1| + \frac{4}{5}\ln|x - 3| + \frac{11}{5}\ln|x + 2|$

$$\int \frac{4\,dx}{x^3 + 4x} = \int \frac{dx}{\_\_\_\_} - \int \frac{x\,dx}{\_\_\_\_}$$

$$= \ln|x| - \_\_\_\_\_\_\_\_ + C' = \_\_\_\_\_\_\_\_ .$$

23. $\int \frac{dx}{\sin x \cos x}$

Write $\int \frac{dx}{\sin x \cos x} = \int \frac{\sin x\,dx}{\sin^2 x \cos x} = \int \frac{\sin x\,dx}{(1 - \cos^2 x)\cos x}$,

and let $u = \cos x$, $du = \_\_\_\_\_\_$ so that

$\int \frac{dx}{\sin x \cos x} = \int \frac{-du}{\_\_\_\_\_\_}$. Next, let

$$\frac{1}{(1 - u^2)u} = \frac{1}{(1 - u)(1 + u)u} = \frac{A}{u} + \frac{B}{1 - u} + \frac{C}{1 + u}.$$

By the Heaviside technique, $A = \_\_\_\_\_\_$, $B = \_\_\_\_\_\_$, and $C = \_\_\_\_\_\_$. Thus,

$$\int \frac{-du}{(1 - u^2)u} = -\int \frac{du}{\_\_\_\_} - \frac{1}{2}\int \frac{du}{\_\_\_\_} + \frac{1}{2}\int \frac{du}{\_\_\_\_}$$

$$= \_\_\_\_\_\_\_\_\_\_\_\_\_\_\_\_ + C'$$

$$= \ln \frac{1}{|u|}[\_\_\_\_\_\_\_\_\_\_] + C'$$

$$= \ln \frac{\sqrt{1 - \cos^2 x}}{\_\_\_\_\_\_} + C' = \_\_\_\_\_\_\_\_\_\_ .$$

24. $\int \frac{(2x^2 + x + 2)\,dx}{x(x - 1)^2}$

By partial fractions,

$$\frac{2x^2 + x + 2}{x(x - 1)^2} = \frac{A}{x} + \frac{B}{x - 1} + \_\_\_\_\_\_\_\_ .$$

Clearing fractions and equating numerators gives,

$$2x^2 + x + 2 = A(x - 1)^2 + Bx(x - 1) + \_\_\_\_\_\_$$

$$= (A + B)x^2 + (\_\_\_\_\_\_)x + A .$$

Equating coefficients of like powers of x gives the equations $A + B = 2$, $\_\_\_\_\_\_ = 1$, and $\_\_\_\_\_\_ = 2$. Solving, $A = \_\_\_\_\_\_$, $B = \_\_\_\_\_\_$, and $C = 5$. Hence,

$$\int \frac{(2x^2 + x + 2)\,dx}{x(x - 1)^2} = 2\int \frac{dx}{\_\_\_\_} + 5\int \frac{dx}{\_\_\_\_} = \_\_\_\_\_\_\_\_ .$$

---

22. $Bx + C$, $x(Bx + C)$, $A + B$, 0, 0, 4, 1, -1, 0, x, $x^2 + 4$, $\frac{1}{2}\ln(x^2 + 4)$, $\ln \frac{|x|}{\sqrt{x^2 + 4}} + C'$

23. $-\sin x\,dx$, $(1 - u^2)u$, 1, $\frac{1}{2}$, $-\frac{1}{2}$, u, $1 - u$, $1 + u$, $-\ln|u| + \frac{1}{2}\ln|1 - u| + \frac{1}{2}\ln|1 + u|$, $\left(1 - u^2\right)^{1/2}$, $|\cos x|$, $\ln|\tan x| + C'$

24. $\frac{C}{(x - 1)^2}$, Cx, $-2A - B + C$, $-2A - B + C$, A, 2, 0, x, $(x - 1)^2$, $2\ln|x| - \frac{5}{x - 1} + C'$

25. $\int \frac{(x^3 - x + 2)\,dx}{x^2 - x}$

Here the degree of the numerator fails to be less than the degree of the denominator so we must use long division and work with the remainder.

By long division, $\frac{x^3 - x + 2}{x^2 - x} = x + 1 +$ ________ . Therefore,

$$\int \frac{(x^3 - x + 2)\,dx}{x^2 - x} = \int (x + 1)\,dx + 2\int \frac{dx}{\_\_\_\_\_\_} .$$

By partial fractions, $\frac{1}{x(x - 1)} = \frac{A}{x} + \frac{B}{x - 1}$. Clearing fractions and equating numerators gives

$$1 = A(x - 1) + B \_\_\_\_\_\_ = (\_\_\_\_\_\_)x - A .$$

This will be an identity in x if and only if

$$A + B = \_\_\_\_\_\_ \quad \text{and} \quad \_\_\_\_\_\_ = 1 .$$

These equations determine the values of A and B to be $A =$ ________ and $B =$ ________. Therefore,

$$\int \frac{(x^3 - x + 2)\,dx}{x^2 - x} = \_\_\_\_\_\_ + 2\int \frac{dx}{\_\_\_\_\_\_} - 2\int \frac{dx}{\_\_\_\_\_\_}$$

$$= \frac{x^2}{2} + x + 2\,(\_\_\_\_\_\_\_\_\_\_) + C$$

$$= \_\_\_\_\_\_\_\_\_\_\_\_\_\_\_\_ .$$

7.4 TRIGONOMETRIC SUBSTITUTIONS

OBJECTIVE A: Find indefinite integrals of integrands involving the radicals $\sqrt{a^2 - u^2}$, $\sqrt{a^2 + u^2}$, $\sqrt{u^2 - a^2}$, and the binomials $a^2 + u^2$, $a^2 - u^2$.

26. $\int \frac{x^3\,dx}{\sqrt{4 - x^2}}$

Let $x = 2\sin u$ so that $dx =$ ________, and the integral becomes

$$\int \frac{x^3\,dx}{\sqrt{4 - x^2}} = \int \frac{\_\_\_\_ \cdot 2\cos u\,du}{\sqrt{4 - 4\sin^2 u}} = \int \frac{\_\_\_\_\_\_\,du}{\sqrt{4\cos^2 u}}$$

$$= \pm 8 \int \sin^2 u\,du \quad (\pm \text{ depends on sign of } \_\_\_\_\_\_)$$

$$= 8 \int \sin^3 u\,du, \quad \text{using only the principal value of } u = \sin^{-1}\frac{x}{2}$$

$$= 8 \int (1 - \cos^2 u)\ \_\_\_\_\_\_\_\_$$

$$= 8\,(-\cos u + \_\_\_\_\_\_\_\_) + C$$

---

25. $\frac{2}{x^2 - x}$, $x(x - 1)$, x, $A + B$, 0, $-A$, -1, 1, $\frac{1}{2}x^2 + x$, $x - 1$, x, $\ln|x - 1| - \ln|x|$, $\frac{x^2}{2} + x + 2\ln\left|\frac{x - 1}{x}\right| + C$

For the substitution $x = 2 \sin u$, finish labeling the diagram at the right. Then, from the diagram,

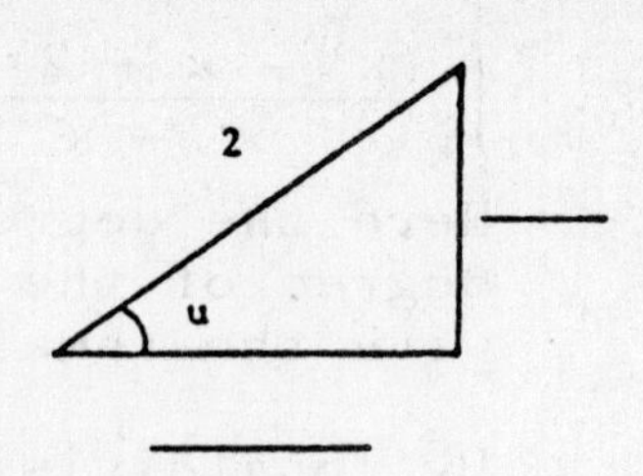

$\cos u =$ ______________, and

substitution gives

$$\int \frac{x^3\,dx}{\sqrt{4 - x^2}} = -4\sqrt{4 - x^2} + \underline{\hspace{4cm}} + C.$$

27. $\displaystyle\int \frac{dx}{\left(x^2 - 9\right)^{3/2}}$

Let $x = 3 \sec u$, $dx =$ ______________ and the integral becomes

$$\int \frac{dx}{\left(x^2 - 9\right)^{3/2}} = \int \frac{\underline{\hspace{4cm}}}{\left(9 \sec^2 u - 9\right)^{3/2}} = \int \frac{3 \sec u \tan u\,du}{27\underline{\hspace{2cm}}}$$

$$= \frac{1}{9}\int \frac{\sec u\,du}{\underline{\hspace{2cm}}} = \frac{1}{9}\int \frac{\cos u\,du}{\underline{\hspace{2cm}}} = \underline{\hspace{4cm}} + C.$$

For the substitution $x = 3 \sec u$, finish labeling the diagram at the right. Then, from the diagram,

$\sin u =$ ______________,

and substitution gives

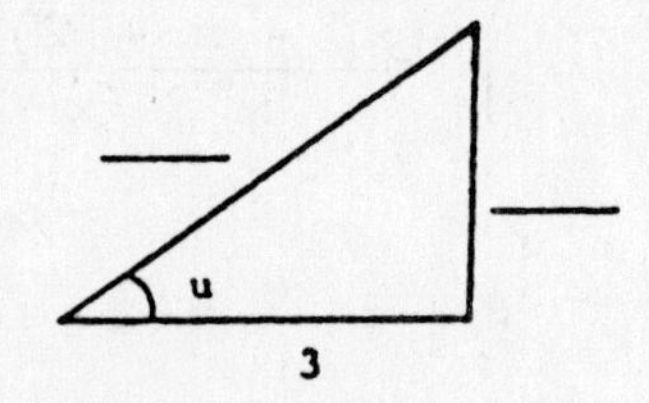

$$\int \frac{dx}{\left(x^2 - 9\right)^{3/2}} = \underline{\hspace{6cm}}.$$

28. $\displaystyle\int \frac{dx}{x^2\sqrt{x^2 + 16}}$

Let $x = 4 \tan u$, $dx =$ ______________ and the integral becomes

$$\int \frac{dx}{x^2\sqrt{x^2 + 16}} = \int \frac{4 \sec^2 u\,du}{\underline{\hspace{3cm}}} = \frac{1}{16}\int \frac{\sec u\,du}{\underline{\hspace{2cm}}}$$

$$= \frac{1}{16}\int \frac{(\cos^2 u) \sec u\,du}{\sin^2 u} = \frac{1}{16}\int \frac{\underline{\hspace{3cm}}}{\sin^2 u}\,du.$$

We substitute again: $z = \sin u$ so that $dz =$ ____________ and the integral becomes

---

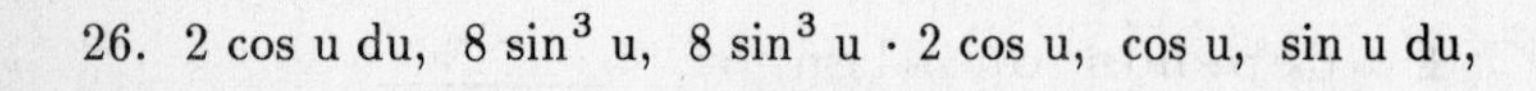

26. $2 \cos u\,du$, $8 \sin^3 u$, $8 \sin^3 u \cdot 2 \cos u$, $\cos u$, $\sin u\,du$,

$\frac{1}{3}\cos^3 u$, $\frac{1}{2}\sqrt{4 - x^2}$, $\frac{1}{3}\left(4 - x^2\right)^{3/2}$

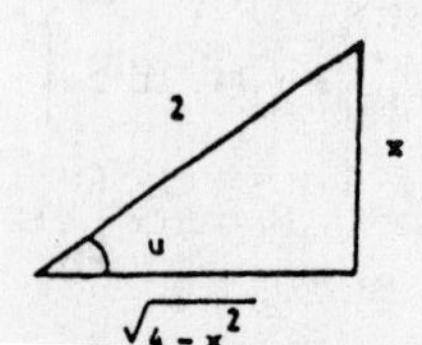

27. $3 \sec u \tan u\,du$, $3 \sec u \tan u\,du$, $\tan^3 u$, $\tan^2 u$,

$\sin^2 u$, $-\dfrac{1}{9 \sin u}$, $\dfrac{\sqrt{x^2 - 9}}{x}$, $\dfrac{-x}{9\sqrt{x^2 - 9}} + C$

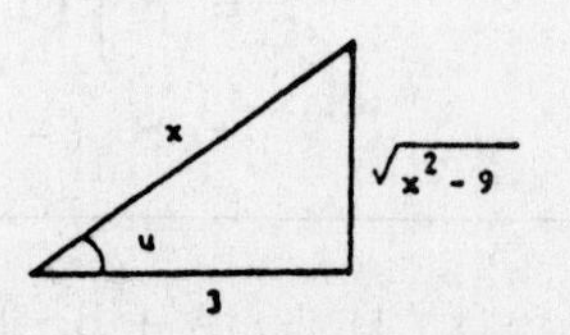

$$\frac{1}{16}\int \frac{\sec u\,du}{\tan^2 u} = \frac{1}{16}\int \frac{\cos u\,du}{\sin^2 u} = \frac{1}{16}\int \underline{\hspace{3cm}}\, dz$$

$$= \frac{1}{16}\left(\underline{\hspace{3cm}}\right) + C$$

$$= \underline{\hspace{5cm}}\ .$$

For the substitution $x = 4 \tan u$, finish labeling the diagram at the right. Then, from the diagram, $\csc u =$ ________________, and substitution gives

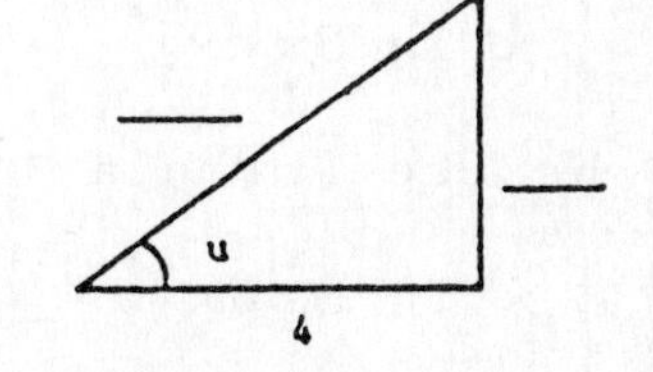

$$\int \frac{dx}{x^2\sqrt{x^2 + 16}} = \underline{\hspace{5cm}}\ .$$

OBJECTIVE B: Calculate definite integrals involving $\sqrt{a^2 - u^2}$, $\sqrt{a^2 + u^2}$, $\sqrt{u^2 - a^2}$, $a^2 + u^2$, and $a^2 - u^2$.

29. $\displaystyle\int_4^{4\sqrt{3}} \frac{dx}{x^2\sqrt{x^2 + 16}}$

As in the previous Problem 28, let $x = 4 \tan u$. When $x = 4$, $\tan u = 1$ or $u =$ ______; when $x = 4\sqrt{3}$, $\tan u = \sqrt{3}$ or $u =$ ________. Thus, upon substitution,

$$\int_4^{4\sqrt{3}} \frac{dx}{x^2\sqrt{x^2 + 16}} = \frac{1}{16}\int_{\_\_\_}^{\_\_\_} \frac{\sec u\,du}{\tan^2 u} \quad \text{(by Problem 28)}$$

$$= -\frac{1}{16} \csc u\Big]_{\_\_}^{\_\_}$$

$$= -\frac{1}{16}\left(\underline{\hspace{3cm}}\right) \approx 0.016.$$

7.5 INTEGRAL TABLES AND CAS

OBJECTIVE A: Find indefinite integrals with the aid of integral tables.

---

28. $4 \sec^2 u\, du$, $16 \tan^2 u \cdot 4 \sec u$, $\tan^2 u$,

$\cos u$, z^{-2}, $-\frac{1}{z}$, $-\frac{1}{16}\csc u + C$,

$\frac{\sqrt{x^2 + 16}}{x}$, $-\frac{\sqrt{x^2 + 16}}{16x} + C$

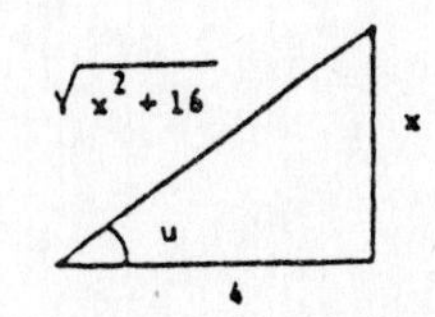

29. $\frac{\pi}{4}$, $\frac{\pi}{3}$, $\int_{\pi/4}^{\pi/3}$, $\Big]_{\pi/4}^{\pi/3}$, $\frac{2\sqrt{3}}{3} - \sqrt{2}$

30. $\int \frac{dx}{x^2\sqrt{4 + x^2}}$

We use Formula 27 with $a =$ ______ . Thus,

$\int \frac{dx}{x^2\sqrt{4 + x^2}} =$ ____________ $+ C$.

31. $\int x^2 \ln 3x \, dx$

We use Formula 110 with $n =$ ______ and $a =$ ______ . Thus,

$\int x^2 \ln 3x \, dx =$ ____________ $+ C$.

OBJECTIVE B: Use reduction formulas provided in integral tables to evaluate integrals.

32. $\int (x^2 - 16)^{3/2} dx$

We use Formula 38 with $n =$ ______, and $a =$ ______ . Thus,

$\int (x^2 - 16)^{3/2} dx = \frac{x\left(\sqrt{x^2 - 16}\right)^3}{4} -$ ____________ . We next use Formula 29 to evaluate the latter integral and obtain the result

$\int (x^2 - 16)^{3/2} dx =$ ____________ $+ C$.

7.6 IMPROPER INTEGRALS

33. A definite integral $\int_a^b f(x)\,dx$ is termed <u>improper</u> if,

(a) either limit of integration is ______ or ______, or
(b) $f(x)$ becomes ______ at some value $x = c$ satisfying ______ .

OBJECTIVE A: Determine whether a given improper integral converges or diverges. If convergent, evaluate it.

---

30. 2, $-\frac{\sqrt{4 + x^2}}{4x}$

31. 2, 3, $\frac{x^3}{3} \ln 3x - \frac{x^3}{9}$

32. 3, 4, $\frac{48}{4} \int \sqrt{x^2 - 16}\, dx$, $\frac{x}{4}(x^2 - 16)^{3/2} + 12\left[\frac{x}{2}\sqrt{x^2 - 16} + 8 \sin^{-1} \frac{x}{4}\right]$

33. $+\infty$, $-\infty$, infinite, $a \leq c \leq b$

34. $\int_1^{\infty} \frac{\ln x}{x^2}\,dx = \lim_{b\to\infty} \int_{\_\_\_}^{\_\_\_} \frac{\ln x}{x^2}\,dx$

To calculate the latter integral, we integrate by parts: let $u = \ln x$, $du =$ ________, $dv =$ ________, $v =$ ______ .

$\int_1^b \frac{\ln x\,dx}{x^2} = -\frac{\ln x}{x}\Big]_1^b + \int_1^b$ __________ $= \frac{-\ln b}{b} +$ ________; whence

$$\int_1^{\infty} \frac{\ln x\,dx}{x^2} = \lim_{b\to\infty}\left(\_\_\_\_\_\_\_\_\_\_\right) = \lim_{b\to\infty} \frac{-\ln b}{b} + \_\_\_\_\_\_\_\_\_\_$$

$$= \lim_{b\to\infty} \_\_\_\_\_\_\_\_\_\_ + 1 = \_\_\_\_\_\_\_\_ .$$

35. $\int_0^1 \ln x\,dx$

Here, $\ln x \to$ _______ as $x \to 0^+$ making the integral improper. Thus,

$$\int_0^1 \ln x\,dx = \lim_{b\to 0^+} \int_{\_\_\_}^{\_\_\_} \ln x\,dx.$$

To calculate the latter integral, we integrate by parts: let $u = \ln x$, $du =$ _______, $dv =$ _______, $v =$ _______ .

$\int_b^1 \ln x\,dx =$ ______________$\Big]_b^1 - \int_b^1 dx =$ ______________ . Therefore,

$$\int_0^1 \ln x\,dx = \lim_{b\to 0^+} (-b \ln b - 1 + b)$$

$$= \lim_{b\to 0^+} \left(\frac{-\ln b}{1/b}\right) - 1$$

$$= \lim_{b\to 0^+} (\_\_\_\_\_\_\_\_) - 1 = \_\_\_\_\_\_\_\_ .$$

36. $\int_0^{\infty} \frac{x\,dx}{\sqrt{1 + x^3}}$

Observe that $\frac{x}{\sqrt{1 + x^3}} > \frac{1}{x}$ is true whenever $x^2 >$ ____________

whenever $x^4 >$ ____________, and the last inequality certainly holds if $x \geq 2$. Thus, we consider the improper integral $\int_2^{\infty} \frac{dx}{x}$. Now, $\int_2^{\infty} \frac{dx}{x} = \lim_{b\to\infty} [\ln b - \ln 2] = +\infty$. Thus, since

---

34. $\int_1^b$, $\frac{1}{x}\,dx$, $\frac{1}{x^2}\,dx$, $-\frac{1}{x}$, $\frac{dx}{x^2}$, $-\frac{1}{x}\Big]_1^b$, $\frac{-\ln b}{b} - \frac{1}{b} + 1$, 1, $\frac{-1/b}{1}$ (l'Hôpital's rule), 1

35. $-\infty$, $\int_b^1$, $\frac{dx}{x}$, dx, x, x ln x, $-b \ln b - 1 + b$, $\frac{1/b}{1/b^2}$ (l'Hôpital's rule), -1

the original integral satisfies the inequalities,

$\int_0^{\infty} \frac{x\,dx}{\sqrt{1+x^3}} \geq \int_2^{\infty} \frac{x\,dx}{\sqrt{1+x^3}} \geq \int_2^{\infty} \frac{dx}{x}$, it ____________ to $+\infty$.

37. $\int_0^2 \frac{(x^2 - 3x + 1)\,dx}{x(x-1)^2}$

In this case the integrand becomes infinite when $x =$ ______ and $x =$ ______, so the integrand is improper. To simplify the notation momentarily, let $f(x)$ denote the integrand. Then,

$$\int_0^2 f(x)\,dx = \lim_{b \to 0^+} \int_b^{1/2} f(x)\,dx + \lim_{c \to 1^-} \int_{1/2}^{c} f(x)\,dx$$

$$+ \lim_{h \to 1^+} \int_{\_\_}^{\_\_} \_\_\_\_\_\_\_\_\_\_$$

To find the indefinite integral $\int \frac{(x^2 - 3x + 1)\,dx}{x(x-1)^2}$ we expand the integrand by partial fractions:

$$\frac{x^2 - 3x + 1}{x(x-1)^2} = \frac{A}{x} + \frac{B}{x-1} + \frac{C}{(x-1)^2}$$

Therefore,

$$x^2 - 3x + 1 = A(x-1)^2 + Bx(x-1) + \_\_\_\_\_\_$$
$$= (A + B)x^2 + (\_\_\_\_\_\_)x + A.$$

It follows that $A = 1$, $A + B =$ ______ and ____________ $= -3$ Solving, $B =$ ______ and $C =$ ______ . Hence,

$$\int \frac{(x^2 - 3x + 1)\,dx}{x(x-1)^2} = \int \frac{dx}{x} + \int \_\_\_\_\_\_ = \ln|x| + \_\_\_\_\_\_ + C'.$$

Therefore,

$$\int_0^2 \frac{(x^2 - 3x + 1)\,dx}{x(x-1)^2} = \lim_{b \to 0^+} (\_\_\_\_\_\_)\Big]_b^{1/2}$$

$$+ \lim_{c \to 1^-} (\_\_\_\_\_\_)\Big]_{1/2}^{c} + \lim_{h \to 1^+} (\_\_\_\_\_\_)\Big]_h^{2}$$

The first limit is ______, the second limit is ______, and the third is ______ . Therefore, the improper integral is ____________ .

---

36. $\sqrt{1+x^3}$ (since $x > 0$), $1 + x^3$, diverges

37. 0, 1, $\int_h^2 f(x)\,dx$, Cx, -2A - B + C, 1, -2A - B + C, 0, -1, $\frac{-dx}{(x-1)^2}$, $\frac{1}{x-1}$, $\ln x + \frac{1}{x-1}$, $\ln x + \frac{1}{x-1}$, $\ln x + \frac{1}{x-1}$, $+\infty$, $-\infty$, $-\infty$, divergent

OBJECTIVE B: Use the Comparison Test, if applicable, to determine the convergence or divergence of an improper integral $\int_a^{\infty} f(x)\,dx$.

38. The improper integrals in Problems 35 and 36 could be implicitly evaluated with the aid of indefinite integrals. Such is not the case, however, for the improper integral

$$\int_0^{\infty} \frac{\sin x\,dx}{1+x^2} = \lim_{b\to\infty} \int_{\_\_\_}^{\_\_\_} \frac{\sin x\,dx}{1+x^2}.$$

However, since $-1 \le \sin x \le 1$ we have that

$$\frac{\sin x}{1+x^2} \le \frac{1}{1+x^2} \quad \text{on} \quad [0,\infty).$$

Thus, evaluating the integral of the function on the right gives

$$\lim_{b\to\infty} \int_0^b \frac{dx}{1+x^2} = \lim_{b\to\infty} \_\_\_\_\_\_\_\_\Big]_0^b = \_\_\_\_\_\_\_\_.$$

Therefore, we know that the original integral $\int_0^{\infty} \frac{\sin x\,dx}{1+x^2}$ is

________.

OBJECTIVE C: Use the Limit Comparison Test, if applicable, to determine the convergence or divergence of an improper integral $\int_a^{\infty} f(x)\,dx$.

39. Consider the improper integral $\int_1^{\infty} \frac{1}{x(x^2-1)}\,dx$. With

$$f(x) = \frac{1}{x^3-x} \quad \text{and} \quad g(x) = \frac{1}{x^3}$$

we have

$$\lim_{x\to\infty} \frac{1}{x^3} \cdot \frac{x^3-x}{1} = \lim_{x\to\infty} \left(\_\_\_\_\_\_\_\_\right) = \_\_\_\_\_\_\_\_.$$

Since the limit is finite and positive we conclude that

$$\int_1^{\infty} \frac{dx}{x(x^2-1)} \_\_\_\_\_\_\_\_ \quad \text{because} \quad \int_1^{\infty} \frac{dx}{x^3} \_\_\_\_\_\_\_\_.$$

---

38. $\int_0^b$, $\tan^{-1} x$, $\frac{\pi}{2}$, convergent

39. $1 - \frac{1}{x^2}$, 1, converges, converges (see Example 11 in this section of the text)

CHAPTER 7 SELF-TEST

Evaluate the integrals in Problems 1-15.

1. $\int \frac{x\,dx}{x^4 + 1}$

2. $\int \frac{(4x + 1)\,dx}{x^2 - 3x - 4}$

3. $\int \frac{\ln x\,dx}{(x + 1)^2}$

4. $\int \frac{\sqrt{9 - 4x^2}\,dx}{x}$

5. $\int \frac{e^{3x/2}\,dx}{1 + e^{3x/4}}$

6. $\int \csc^3 x\,dx$ (Tables)

7. $\int \sin^3 x \cos^4 x\,dx$

8. $\int \cot^5 x\,dx$

9. $\int \frac{x\,dx}{x^2 + x + 1}$

10. $\int \cos^2 x \sin 2x\,dx$

11. $\int_{-\pi/3}^{\pi/4} x \sec^2 x\,dx$

12. $\int_0^1 \frac{(2x^3 + x + 3)\,dx}{x^2 + 1}$

13. $\int_5^{5\sqrt{3}} \frac{dx}{x\sqrt{25 + x^2}}$

14. $\int_0^a \ln(a^2 + x^2)\,dx, \quad a > 0$

15. $\int \frac{x - 2}{(3x - 1)^2}\,dx$

In Problems 16-19, determine the convergence or divergence of the integral. If the integral is convergent, find its value.

16. $\int_0^{\pi/2} \sec x \tan x\,dx$

17. $\int_{-1}^1 \frac{dx}{x^{2/3}}$

18. $\int_1^\infty \frac{\ln x\,dx}{x}$

19. $\int_0^\infty \frac{\sin x\,dx}{e^x}$

SOLUTIONS TO CHAPTER 7 SELF-TEST

1. Let $u = x^2$, $du = 2x\,dx$. Then,

$$\int \frac{x\,dx}{x^4 + 1} = \frac{1}{2}\int \frac{du}{u^2 + 1} = \frac{1}{2}\tan^{-1} u + C = \frac{1}{2}\tan^{-1} x^2 + C.$$

2. $$\frac{4x + 1}{x^2 + 3x - 4} = \frac{4x + 1}{(x - 1)(x + 4)} = \frac{A}{x - 1} + \frac{B}{x + 4}$$

Clearing fractions,

$$4x + 1 = A(x + 4) + B(x - 1) = (A + B)x + (4A - B) .$$

This last equation requires that

$$A + B = 4 \quad \text{and} \quad 4A - B = 1 .$$

The solution of these equations is $A = 1$, $B = 3$. Thus,

$$\int \frac{4x + 1}{x^2 + 3x - 4}\,dx = \int \frac{dx}{x - 1} + 3\int \frac{dx}{x + 4}$$
$$= \ln|x - 1| + 3\ln|x + 4| + C.$$

3. Let $u = \ln x$, $du = \frac{dx}{x}$, $dv = \frac{dx}{(x + 1)^2}$, $v = \frac{-1}{x + 1}$; $x > 0$

$$\int \frac{\ln x\,dx}{(x + 1)^2} = -\frac{\ln x}{x + 1} + \int \frac{dx}{x(x + 1)}$$
$$= -\frac{\ln x}{x + 1} + \int \frac{dx}{x} - \int \frac{dx}{x + 1}$$
$$= -\frac{\ln x}{x + 1} + \ln x - \ln(x + 1) + C$$
$$= -\frac{\ln x}{x + 1} + \ln\frac{x}{x + 1} + C, \quad x > 0 .$$

4. Let $x = \frac{3}{2}\sin u$, $dx = \frac{3}{2}\cos u\,du$. Then,

$$\int \frac{\sqrt{9 - 4x^2}\,dx}{x} = \int \frac{\sqrt{9 - 9\sin^2 u}\cdot\frac{3}{2}\cos u\,du}{\frac{3}{2}\sin u}$$
$$= 3\int \frac{\cos^2 u\,du}{\sin u} = 3\int \frac{(1 - \sin^2 u)\,du}{\sin u}$$
$$= 3\int (\csc u - \sin u)\,du$$
$$= 3\ln|\csc u - \cot u| + 3\cos u + C$$

From the substitution $x = \frac{3}{2}\sin u$, and the diagram at the right,

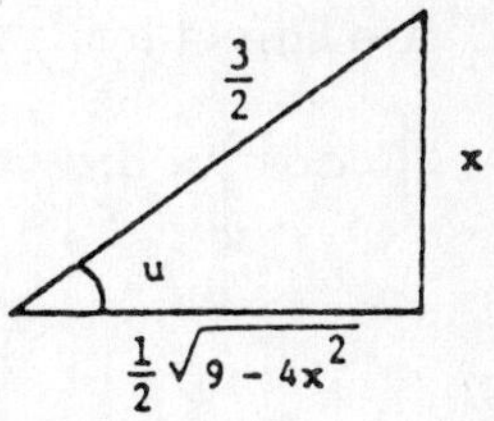

$\csc u = \frac{3}{2x}$ and $\cot u = \frac{\sqrt{9 - 4x^2}}{2x}$. Thus,

$$\int \frac{\sqrt{9 - 4x^2}\,dx}{x} = 3\ln\left|\frac{3}{2x} - \frac{\sqrt{9 - 4x^2}}{2x}\right| + \sqrt{9 - 4x^2} + C.$$

5. Let $u = e^x$, $du = e^x\,dx$. Then,

$$\int \frac{e^{3x/2}\,dx}{1 + e^{3x/4}} = \int \frac{e^{x/2}\cdot e^x\,dx}{1 + e^{3x/4}} = \int \frac{u^{1/2}\,du}{1 + u^{3/4}}$$

Next, let $z = u^{1/4}$, $dz = \frac{1}{4}u^{-3/4}\,du$. Hence, $z^2 = u^{1/2}$, $z^3 = u^{3/4}$, and $du = 4u^{3/4}\,dz = 4z^3\,dz$. Substitution into the last integral gives

$$\int \frac{e^{3x/2}\,dx}{1 + e^{3x/4}} = \int \frac{z^2(4z^3\,dz)}{1 + z^3} = 4\int \frac{z^5\,dz}{z^3 + 1}$$
$$= 4\int\left(z^2 - \frac{z^2}{z^3 + 1}\right)dz$$

$$= \frac{4}{3}z^3 - \frac{4}{3}\ln|z^3 + 1| + C$$
$$= \frac{4}{3}u^{3/4} - \frac{4}{3}\ln|u^{3/4} + 1| + C$$
$$= \frac{4}{3}e^{3x/4} - \frac{4}{3}\ln(e^{3x/4} + 1) + C .$$

6. $\int \csc^3 x\,dx$

We use the reduction Formula 93 in the integral tables:

$$\int \csc^3 x\,dx = \frac{-\csc x \cot x}{2} + \frac{1}{2}\int \csc x\,dx$$

The last integral is evaluated using Formula 89 to yield

$$\int \csc^3 x\,dx = -\frac{1}{2}\csc x \cot x - \frac{1}{2}\ln|\csc x + \cot x| + C.$$

7. Let $u = \cos x$, $du = -\sin x\,dx$. Then

$$\int \sin^3 x \cos^4 x\,dx = \int \sin^2 x \cos^4 x \sin x\,dx$$
$$= \int (1 - \cos^2 x)\cos^4 x \sin x\,dx$$
$$= \int -(1 - u^2)u^4\,du$$
$$= -\frac{1}{5}u^5 + \frac{1}{7}u^7 + C = -\frac{1}{5}\cos^5 x + \frac{1}{7}\cos^7 x + C.$$

8. Using the trigonometric identity $\cot^2 x = \csc^2 x - 1$

$$\int \cot^5 x\,dx = \int \cot^3 x(\csc^2 x - 1)\,dx$$
$$= \int \cot^3 x \csc^2 x\,dx - \int \cot^3 x\,dx$$
$$= \int \cot^3 x \csc^2 x\,dx - \int \cot x(\csc^2 x - 1)\,dx$$
$$= \int \cot^3 x \csc^2 x\,dx - \int \cot x \csc^2 x\,dx + \int \cot x\,dx$$

Let $u = \cot x$, $du = -\csc^2 x\,dx$ and we have

$$\int \cot^5 x\,dx = -\frac{1}{4}\cot^4 x + \frac{1}{2}\cot^2 x + \ln|\sin x| + C.$$

9. Completing the square, $x^2 + x + 1 = \left(x + \frac{1}{2}\right)^2 + \frac{3}{4}$. Let $u = x + \frac{1}{2}$, $du = dx$. Then,

$$\int \frac{x\,dx}{x^2 + x + 1} = \int \frac{\left(u - \frac{1}{2}\right)du}{u^2 + \frac{3}{4}} = \int \frac{u\,du}{u^2 + \frac{3}{4}} - \frac{1}{2}\int \frac{du}{u^2 + \frac{3}{4}}$$
$$= \frac{1}{2}\ln\left|u^2 + \frac{3}{4}\right| - \frac{1}{2}\cdot\frac{2}{\sqrt{3}}\tan^{-1}\frac{2u}{\sqrt{3}} + C$$

$$= \frac{1}{2} \ln |x^2 + x + 1| - \frac{1}{\sqrt{3}} \tan^{-1} \frac{2x + 1}{\sqrt{3}} + C.$$

10. $$\int \cos^2 x \sin 2x\,dx = \frac{1}{2}\int (1 + \cos 2x) \sin 2x\,dx$$
$$= \frac{1}{2}\int \sin 2x\,dx + \frac{1}{2}\int \sin 2x \cos 2x\,dx$$
$$= \frac{1}{2}\int \sin 2x\,dx + \frac{1}{4}\int \sin 4x\,dx$$
$$= -\frac{1}{4}\cos 2x - \frac{1}{16}\cos 4x + C.$$

11. Let $u = x$, $du = dx$, $dv = \sec^2 x\,dx$, $v = \tan x$. Then,

$$\int_{-\pi/3}^{\pi/4} x \sec^2 x\,dx = x \tan x\Big]_{-\pi/3}^{\pi/4} - \int_{-\pi/3}^{\pi/4} \tan x\,dx$$
$$= \frac{\pi}{4} \cdot 1 + \frac{\pi}{3} \cdot \sqrt{3} - \ln |\cos x|\Big]_{-\pi/3}^{\pi/4}$$
$$= \frac{\pi}{4} + \frac{\pi}{\sqrt{3}} - \ln \frac{\sqrt{2}}{2} + \ln \frac{1}{2}$$
$$= \frac{\pi}{4} + \frac{\pi}{\sqrt{3}} - \frac{1}{2}\ln 2 \approx 2.253.$$

12. $$\int_0^1 \frac{(2x^3 + x + 3)\,dx}{x^2 + 1} = \int_0^1 \left(2x + \frac{3 - x}{x^2 + 1}\right) dx$$
$$= \int_0^1 2x\,dx + 3\int_0^1 \frac{dx}{x^2 + 1} - \int_0^1 \frac{x\,dx}{x^2 + 1}$$
$$= x^2 + 3\tan^{-1} x - \frac{1}{2}\ln(x^2 + 1)\Big]_0^1$$
$$= \left(1 + \frac{3\pi}{4} - \frac{1}{2}\ln 2\right) - (0 + 0 - 0) \approx 3.01.$$

13. Let $x = 5 \tan u$, $dx = 5 \sec^2 u\,du$. When $x = 5$, $\tan u = 1$ or $u = \frac{\pi}{4}$; when $x = 5\sqrt{3}$, $\tan u = \sqrt{3}$ or $u = \frac{\pi}{3}$. Thus,

$$\int_5^{5\sqrt{3}} \frac{dx}{x\sqrt{25 + x^2}} = \int_{\pi/4}^{\pi/3} \frac{5 \sec^2 u\,du}{5 \tan u \sqrt{25 + 25 \tan^2 u}}$$
$$= \frac{1}{5}\int_{\pi/4}^{\pi/3} \frac{\sec u\,du}{\tan u} = \frac{1}{5}\int_{\pi/4}^{\pi/3} \csc u\,du$$
$$= \frac{1}{5}\ln|\csc u - \cot u|\Big]_{\pi/4}^{\pi/3}$$
$$= \frac{1}{5}\left(\ln\left|\frac{2\sqrt{3}}{3} - \frac{\sqrt{3}}{3}\right| - \ln|\sqrt{2} - 1|\right)$$
$$= \frac{1}{5}\ln\left|\frac{\sqrt{3}}{3(\sqrt{2} - 1)}\right| \approx 0.066.$$

14. Let $u = \ln\left(a^2 + x^2\right)$, $du = \frac{2x\,dx}{a^2 + x^2}$, $dv = dx$, and $v = x$. Then,

$$\int_0^a \ln\left(a^2 + x^2\right)dx = x\ln\left(a^2 + x^2\right)\Big]_0^a - \int_0^a \frac{2x^2\,dx}{a^2 + x^2}$$

$$= a\ln 2a^2 - 2\int_0^a \left(1 - \frac{a^2}{a^2 + x^2}\right)dx$$

$$= a\ln 2a^2 - (2x + 2a\tan^{-1}\frac{x}{a}\Big]_0^a$$

$$= a\ln 2a^2 - 2a + \frac{2\pi a}{4}.$$

15. $\dfrac{x - 2}{(3x - 1)^2} = \dfrac{A}{3x - 1} + \dfrac{B}{(3x - 1)^2}$

Clearing fractions,

$$x - 2 = A(3x - 1) + B = 3Ax + (B - A)\ .$$

Thus, $3A = 1$ and $B - A = -2$. The solution of these equations is $A = \frac{1}{3}$ and $B = -\frac{5}{3}$. Thus,

$$\int \frac{x - 2}{(3x - 1)^2}\,dx = \frac{1}{3}\int \frac{dx}{3x - 1} - \frac{5}{3}\int \frac{dx}{(3x - 1)^2}$$

$$= \frac{1}{9}\ln|3x - 1| + \frac{5}{9}(3x - 1)^{-1} + C\ .$$

16. $$\int_0^{\pi/2} \sec x \tan x\,dx = \lim_{b\to\frac{\pi}{2}^-} \int_0^b \sec x \tan x\,dx$$

$$= \lim_{b\to\frac{\pi}{2}^-} \sec x\Big]_0^b = \lim_{b\to\frac{\pi}{2}^-} (\sec b - 1\ ) = \infty.$$

Therefore, the improper integral diverges.

17. $$\int_{-1}^1 \frac{dx}{x^{2/3}} = \lim_{a\to 0^-} \int_{-1}^a \frac{dx}{x^{2/3}} + \lim_{a\to 0^+} \int_a^1 \frac{dx}{x^{2/3}}$$

$$= \lim_{a\to 0^-} 3x^{1/3}\Big]_{-1}^a + \lim_{a\to 0^+} 3x^{1/3}\Big]_a^1$$

$$= \lim_{a\to 0^-} \left(3a^{1/3} + 3\right) + \lim_{a\to 0^+} \left(3 - 3a^{1/3}\right) = 6.$$

18. For $x \geq e$, $\ln x \geq 1$. Hence,

$$\int_1^\infty \frac{\ln x\,dx}{x} \geq \int_e^\infty \frac{\ln x\,dx}{x} \geq \int_e^\infty \frac{dx}{x}.$$

Now, $\displaystyle\int_e^\infty \frac{dx}{x} = \lim_{b\to\infty} \ln x\Big]_e^b = \lim_{b\to\infty} \ln b - 1 = \infty.$

Therefore, the integral $\displaystyle\int_1^\infty \frac{\ln x\,dx}{x}$ diverges.

19. $\int_0^\infty \frac{\sin x\,dx}{e^x} = \lim_{b\to\infty} \int_0^b e^{-x} \sin x\,dx$

Let $u = e^{-x}$, $du = -e^{-x}dx$, $dv = \sin x\,dx$, $v = -\cos x$.

$\int_0^b e^{-x} \sin x\,dx = -e^{-x} \cos x]_0^b - \int_0^b e^{-x} \cos x\,dx$

Let $U = e^{-x}$, $dU = -e^{-x}dx$, $dV = \cos x\,dx$, $V = \sin x$, and

$\int_0^b e^{-x} \cos x\,dx = e^{-x} \sin x]_0^b + \int_0^b e^{-x} \sin x\,dx.$

Putting these results together,

$$2\int_0^b e^{-x} \sin x\,dx = -e^{-x} \cos x - e^{-x} \sin x]_0^b$$

$$= -e^{-b} \cos b + 1 - e^{-b} \sin b$$

Now, $\lim_{b\to\infty} \frac{\cos b}{e^b} = \lim_{b\to\infty} \frac{\sin b}{e^b} = 0$. Thus,

$\int_0^\infty \frac{\sin x\,dx}{e^x} = \lim_{b\to\infty} \int_0^b e^{-x} \sin x\,dx = \frac{1}{2}.$

NOTES.

CHAPTER 8 INFINITE SERIES

INTRODUCTION

An ancient Greek paradox, due to the mathematician Zeno, concerned the following problem. Suppose that a man wants to walk a certain distance, say two miles, along a straight line from A to B. First he must pass the half-way point, then the 3/4 point, then the 7/8 point, and so on as illustrated in the following figure.

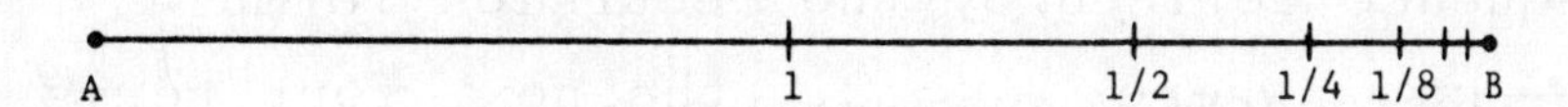

The fractional numbers in the figure indicate the distance in miles remaining to be covered. Therefore, on the assumption that a finite length contains an infinite number of points, the man must pass an infinite number of distance markers along the way. But that seems impossible. The paradox is that the man does get to B, and in a finite amount of time, assuming he walks at some steady pace.

An analysis of the problem is not difficult. The total distance s from A to B is an infinite sum expressible as,

$$s = 1 + \left(\frac{1}{2}\right) + \left(\frac{1}{2}\right)^2 + \left(\frac{1}{2}\right)^3 + \dots,$$

and the paradox is dispelled if this infinite sum equals the finite number of 2 miles. That turns out to be exactly the case, as you will see further on in this chapter when it is established that

$$\frac{1}{1-x} = 1 + x + x^2 + \dots + x^n + \dots, \quad \text{if} \quad |x| < 1.$$

For then,

$$2 = \frac{1}{1-\frac{1}{2}} = 1 + \frac{1}{2} + \left(\frac{1}{2}\right)^2 + \left(\frac{1}{2}\right)^3 + \dots + \left(\frac{1}{2}\right)^n + \dots.$$

8.1 LIMITS OF SEQUENCES OF NUMBERS

OBJECTIVE A: Given a defining rule for the sequence $\{a_n\}$, write the first few items of the sequence.

1. An infinite sequence is a __________ whose domain is the set of ______________________________ .

2. The numbers in the range of a sequence are called the __________ of the sequence. The number a_n is called the __________ of the sequence, or the term with __________ n.

---

1. function, positive integers 2. terms, nth term, index

3. For the sequence whose defining rule is $a_n = 2 + \frac{1}{n}$, the first four terms are
$a_1 = 3$, $a_2 =$ ________, $a_3 =$ ________, and $a_4 =$ ________ .

4. For the sequence whose defining rule is $a_n = \frac{4^n}{n!}$, the first four terms are
$a_1 = 4$, $a_2 =$ ________, $a_3 =$ ________, and $a_4 =$ ________ .

5. For the sequence whose defining rule is $a_n = (-1)^{n+1}\left(\frac{n + 1}{n^3}\right)$, the first four terms are
$a_1 = 2$, $a_2 =$ ________, $a_3 =$ ________, and $a_4 =$ ________ .

6. For the sequence defined by the recursion formula
$x_{n+1} = \left(\frac{2}{n + 1}\right)x_n$, where $x_1 = 2$, the next four terms are
$x_2 =$ ________, $x_3 =$ ________, $x_4 =$ ________, and $x_5 =$ ________ .

OBJECTIVE B: Given a sequence $\{a_n\}$ use the definition to determine if it converges or diverges.

7. The sequence $\{a_n\}$ converges to the number L if to every ________ ϵ there corresponds an ________ N such that if $n > N$, then ________ . If no such limit L exists, we say that $\{a_n\}$ ________ .

8. If $0 < b < 1$, then $\{b^n\}$ converges to 0. To see why, consider the inequality

$$|b^n - 0| = b^n < \epsilon.$$

Thus, we seek an integer N such that if $n > N$, then

________ .

Since the natural logarithm $y = \ln x$ is an increasing function for all x,

$$b^n < \epsilon \text{ is equivalent to } n \ln b < \ln \epsilon.$$

Also, because $\ln b$ is a negative number for $0 < b < 1$, this latter inequality is equivalent to

$$n > \text{________} .$$

Therefore, we need only choose an integer N satisfying

________,

and the criterion set forth in Problem 7 for convergence to 0 is satisfied.

---

3. $\frac{5}{2}, \frac{7}{3}, \frac{9}{4}$ 4. $8, \frac{32}{3}, \frac{32}{3}$ 5. $-\frac{3}{8}, \frac{4}{27}, -\frac{5}{64}$ 6. $2, \frac{4}{3}, \frac{2}{3}, \frac{4}{15}$

7. positive number, integer, $|a_n - L| < \epsilon$, diverges 8. $b^n < \epsilon$, $\frac{\ln \epsilon}{\ln b}$, $N > \frac{\ln \epsilon}{\ln b}$

OBJECTIVE C: Know the Nondecreasing Sequence Theorem and how it applies.

9. A nondecreasing sequence $\{a_n\}$ is a sequence with the property that ______________ for every n.

10. A sequence $\{a_n\}$ is said to be ______________ from above if there is a number M such that $a_n \le M$ for every n.

11. If $\{a_n\}$ is a nondecreasing sequence that is bounded from above, then it ______________.

12. If a nondecreasing sequence $\{a_n\}$ fails to be bounded, then it ______________.

8.2 THEOREMS FOR CALCULATING LIMITS OF SEQUENCES

OBJECTIVE: Given a sequence $\{a_n\}$ determine if it converges or diverges. If it converges use the theorems presented in this section of the text, or l'Hôpital's rule, or Table 8.1, to find the limit.

13. Consider the sequence defined by $a_n = \frac{4^n}{n!}$. Thus, if $n > 5$,

$$a_n = \frac{4\cdot4\cdot4\cdot4\cdots4}{1\cdot2\cdot3\cdot4\cdots n} = \_\_\_\_ \left(\frac{4\cdot4\cdots4}{5\cdot6\cdots n}\right)$$

$$\le \frac{32}{3}\left(\frac{4}{5}\right)^{\_\_\_\_} = \left(\frac{32}{3}\right)\left(\frac{5}{4}\right)^4\left(\frac{4}{5}\right)^n$$

Since $0 \le a_n$ for all n, the Sandwich Theorem 3 of the Thomas/Finney text gives $a_n \to$ ______ because $\left(\frac{4}{5}\right)^n \to 0$ from Problem 8.

14. For the sequence $\left\{\frac{n^3+5}{n^2-1}\right\}$, $\lim_{n\to\infty} \frac{n^3+5}{n^2-1} = \lim_{n\to\infty} \frac{n + (5/n^2)}{1 - (1/n^2)} =$ ______ .

Therefore, the sequence ______________.

15. For the sequence defined by $a_n = (-1)^{n+1}\left(\frac{n+1}{n^3}\right)$,

$0 \le |a_n| =$ __________ $= \frac{1}{n^2} +$ __________ .

Therefore,

$$(-1)^{n+1}\frac{n+1}{n^3} \to \_\_\_\_\_\_ \text{ as } n \to \infty,$$

by the Sandwich Theorem 3.

---

9. $a_n \le a_{n+1}$ 10. bounded 11. converges

12. diverges to plus infinity 13. $\frac{32}{3}$, $n-4$, 0 14. ∞, diverges

15. $\frac{n+1}{n^3}$, $\frac{1}{n^3}$, 0

16. Let $a_n = \left(\frac{3n - 1}{5n + 1}\right)^n$. Then,

$\frac{3n - 1}{5n + 1} < \frac{3n}{5n + 1} <$ ________ implies $0 \le \left(\frac{3n - 1}{5n + 1}\right)^n <$ __________ .

Therefore $a_n \to$ _______ because $\left(\frac{3}{5}\right)^n \to 0$.

17. Let $a_n = \left(\frac{3n + 1}{5n - 1}\right)^{1/n}$. Then, applying the natural logarithm to each side, $\ln a_n = \frac{1}{n}(3n + 1) -$ _________ . By l'Hôpital's rule,

$$\lim_{n\to\infty} \frac{\ln (3n + 1)}{n} = \lim_{n\to\infty} \frac{\_\_\_\_\_\_}{1} = \_\_\_\_\_\_\_\_, \text{ and}$$

$$\lim_{n\to\infty} \frac{\ln (5n - 1)}{n} = \lim_{n\to\infty} \frac{\_\_\_\_\_\_}{1} = \_\_\_\_\_\_\_\_ .$$

Then, $\ln a_n \to 0$ so that taking $f(x) = e^x$ and $L = 0$ in Theorem 4, $a_n = e^{\ln a_n} \to$ _________ .

18. Consider the sequence defined by $a_n = \left(1 + e^{-n}\right)^n$. Then,

$\ln a_n =$ ____________________,

and by l'Hôpital's rule,

$$\lim_{n\to\infty} \frac{\ln(1 + e^{-n})}{1/n} = \lim_{n\to\infty} \frac{\_\_\_\_\_\_\_\_}{-1/n^2} \quad (0/0)$$

$$= \lim_{n\to\infty} \frac{n^2}{e^n + 1} \quad (\infty/\infty)$$

$$= \lim_{n\to\infty} \_\_\_\_\_\_\_\_ \quad (\text{still } \infty/\infty)$$

$$= \lim_{n\to\infty} \frac{2}{e^n} = \_\_\_\_\_\_\_\_ .$$

Therefore,

$$a_n = e^{\ln a_n} \to \_\_\_\_\_\_\_\_ .$$

19. Let $a_n = \sqrt[n]{n^3}$. Then, $a_n = \left(n^3\right)^{1/n} = n^{\_\_\_} = (\sqrt[n]{n})^{\_\_\_}$.
Now, $\sqrt[n]{n} \to$ ________ by Rule 2 in Table 8.1 in the text.
Thus, if $f(x) = x^3$, then

$$a_n = f(\sqrt[n]{n}) \to \_\_\_\_\_\_\_\_ .$$

---

16. $\frac{3}{5}$, $\left(\frac{3}{5}\right)^n$, 0

17. $\frac{1}{n}\ln(5n - 1)$, $\frac{3}{3n + 1}$, 0, $\frac{5}{5n - 1}$, 0, $e^0 = 1$

18. $n \ln (1 + e^{-n})$, $\left(\frac{1}{1 + e^{-n}}\right)(-e^{-n})$, $\frac{2n}{e^n}$, 0, $e^0 = 1$

19. $\frac{3}{n}$, 3, 1, $1^3 = 1$

8.3 INFINITE SERIES

20. If $\{a_n\}$ is a sequence, and

$$s_n = a_1 + a_2 + \dots + a_n,$$

then the sequence $\{s_n\}$ is called an ____________ .

21. The number s_n is called the ____________ of the series.

22. Instead of $\{s_n\}$ we usually use the notation ________ for the series.

23. The series $\sum_{n=1}^{\infty} a_n$ is said to converge if the sequence ________ converges to a finite limit L. In that case we write ____________ or $a_1 + a_2 + \dots + a_n + \dots = L$. If no such limit exists, the series is said to __________ .

OBJECTIVE A: For a given geometric series $\sum_{n=1}^{\infty} ar^{n-1}$, determine if the series converges or diverges. If it does converge, then compute the sum of the series. The indexing of the series may be changed for a given problem.

24. Consider the series $\sum_{n=1}^{\infty} \frac{3}{5^{n-1}}$. This is a geometric series with $a =$ ________ and $r =$ ________ . Since $|r| < 1$, the geometric series __________, and its sum is given by

$$\sum_{n=1}^{\infty} \frac{3}{5^{n-1}} = \_\_\_\_\_\_\_\_\_\_ = \_\_\_\_\_\_\_\_ .$$

25. The series $\sum_{n=3}^{\infty} (-1)^{n-1} \frac{4}{3^{n-1}}$ is a geometric series with $a =$ ______ and $r =$ _______ . Since $|r| < 1$, the geometric series _________ . However, the index begins with $n = 3$ instead of $n = 1$. Now,

$$\sum_{n=3}^{\infty} (-1)^{n-1} \frac{4}{3^{n-1}} = \sum_{n=1}^{\infty} (-1)^{n-1} \frac{4}{3^{n}} - (\_\_\_\_\_\_\_\_)$$

$$= \_\_\_\_\_\_\_\_\_\_ - \frac{8}{3} = \_\_\_\_\_\_\_\_\_ .$$

---

20. infinite series

21. nth partial sum

22. $\sum_{n=1}^{\infty} a_n$

23. $\{s_n\}$, $\sum_{n=1}^{\infty} a_n = L$, diverge

24. 3, $\frac{1}{5}$, converges, $\frac{3}{1 - \frac{1}{5}}$, $\frac{15}{4}$

25. 4, $-\frac{1}{3}$, converges, $4 - \frac{4}{3}$, $\frac{4}{1 + \frac{1}{3}}$, $\frac{1}{3}$

26. The series $\sum_{n=3}^{\infty} \frac{2^n}{7}$ is a geometric series with a = ________ and r = ________ . Since ____________ the series diverges.

27. The repeating decimal 0.15 15 15 ... is a geometric series in disguise. It can be written as

$$0.15\ 15\ 15\ \ldots = \frac{15}{100} + \frac{15}{\_\_\_\_} + \frac{15}{\_\_\_\_} + \cdots$$

$$= \sum_{n=1}^{\infty} \_\_\_\_\_\_\_\_$$

$$= \frac{1}{100} \sum_{n=1}^{\infty} \_\_\_\_\_\_\_\_$$

$$= \frac{1}{100} \left(\_\_\_\_\_\_\_\_ \right) = \frac{15}{\_\_\_\_\_\_} = \frac{15}{99} .$$

28. If you write in the first four terms of the following geometric series you have

$$\sum_{n=0}^{\infty} ar^n = \_\_\_\_\_ + \_\_\_\_\_ + \_\_\_\_\_ + \_\_\_\_\_ + \cdots .$$

Thus, $\sum_{n=0}^{\infty} ar^n$ is simply another way of writing the geometric series $\sum_{n=1}^{\infty} ar^{n-1}$. Therefore, if $|r| < 1$ we conclude that

$$\sum_{n=0}^{\infty} ar^n = \_\_\_\_\_\_\_\_\_\_ .$$

29. Sometimes the terms of a given series are a sum or difference of terms, each of which beongs to a geometric series. For example,

$$\sum_{n=0}^{\infty} \left(\frac{7}{3^n} - \frac{1}{2^n} \right) = \sum_{n=0}^{\infty} \frac{7}{3^n} - \sum_{n=0}^{\infty} \frac{1}{2^n}$$

$$= \_\_\_\_\_\_ - \_\_\_\_\_\_ = \_\_\_\_\_\_ - \frac{4}{2} = \_\_\_\_\_\_ .$$

OBJECTIVE B: Use the nth-term test for divergence to test a given series $\sum_{n=1}^{\infty} a_n$ for divergence.

---

26. $\frac{1}{7}$, 2, $|r| > 1$

27. 100^2, 100^3, $\frac{15}{100^n}$, $\frac{15}{100^{n-1}}$, $\frac{15}{1 - \frac{1}{100}}$, 100 - 1

28. a, ar, ar^2, ar^3, $\frac{a}{1 - r}$

29. $\frac{7}{1 - \frac{1}{3}}$, $\frac{1}{1 - \frac{1}{2}}$, $\frac{21}{2}$, $\frac{17}{2}$

30. For the series $\sum_{n=1}^{\infty} \frac{n^n}{n!}$, we have for every index n,

$$a_n = \frac{n^n}{n!} = \frac{n \cdot n \cdot n \cdots n}{1 \cdot 2 \cdot 3 \cdots n} = \left(\frac{n}{1}\right)\left(\frac{n}{2}\right)\left(\rule{2em}{0.4pt}\right) \cdots \left(\frac{n}{n}\right) > \rule{4em}{0.4pt}\ .$$

Therefore, $\lim_{n\to\infty} a_n \neq 0$. We conclude that the series ________ .

31. Consider the series $\sum_{n=1}^{\infty} (-1)^{n+1}\left(1 + \frac{1}{3^n}\right)$. For large values of the index n, the absolute value of the nth term,

$$|a_n| = 1 + 3^{-n}$$

is close to ________ . Therefore, the limit

$\lim_{n\to\infty} (-1)^{n+1}\left(1 + \frac{1}{3^n}\right)$ ________ exist. We conclude that the series ________ .

32. For the series $\sum_{n=1}^{\infty} \left(\frac{n+2}{n}\right)^n$, we have

$\lim_{n\to\infty} \left(\frac{n+2}{n}\right)^n = \lim_{n\to\infty} \left(1 + \frac{2}{n}\right)^n =$ ________ . Thus, the series ________ because the limit of the nth term is not zero.

33. For the series $\sum_{n=1}^{\infty} \frac{2n^2 + 1}{n^3 - 1}$ the limit of the nth-term is

$$\lim_{n\to\infty} \left(\frac{2n^2 + 1}{n^3 - 1}\right) = \lim_{n\to\infty} \left(\frac{\frac{2}{n} + \frac{1}{n^3}}{\rule{3em}{0.4pt}}\right) = \frac{0 + 0}{\rule{3em}{0.4pt}} = 0\ .$$

Since the limit is equal to zero, the nth-term test gives no information concerning convergence or divergence.

OBJECTIVE C: Use partial fractions to find the sum of a telescoping series.

34. Consider the series $\sum_{n=1}^{\infty} \frac{1}{(2n - 1)(2n + 1)}$. This is not a geometric series. However, we can use partial fractions to re-write the kth term:

$$\frac{1}{(2k - 1)(2k + 1)} = \frac{1}{2}\left[\rule{3em}{0.4pt} - \frac{1}{2k + 1}\right].$$

This permits us to write the partial sum

$$\sum_{n=1}^{k} \frac{1}{(2n - 1)(2n + 1)} = \frac{1}{1 \cdot 3} + \frac{1}{3 \cdot 5} + \cdots + \frac{1}{(2k - 1)(2k + 1)}$$

---

30. $\frac{n}{3}$, 1, diverges

31. 1, does not, diverges

32. e^2, diverges

33. $1 - \frac{1}{n^3}$, $1 + 0$

as

$$s_k = \frac{1}{2}\left(\frac{1}{1} - \frac{1}{3}\right) + \frac{1}{2}\left(\underline{\qquad}\right) + \frac{1}{2}\left(\underline{\qquad}\right) + \dots + \frac{1}{2}\left(\frac{1}{2k-1} - \frac{1}{2k+1}\right).$$

By removing parentheses on the right, and combining terms, we find that

$$s_k = \underline{\qquad\qquad} .$$

Therefore, $s_k \to$ ______ and the series ______ converge. Hence,

$$\sum_{n=1}^{\infty} \frac{1}{(2n-1)(2n+1)} = \underline{\qquad} .$$

8.4 SERIES OF NONNEGATIVE TERMS. THE INTEGRAL TEST.

35. If Σa_n is a series of nonnegative terms, then its sequence $\{s_n\}$ of partial sums is ______ .

36. If Σa_n is a series of nonnegative terms, then it converges if and only if its sequence $\{s_n\}$ of partial sums is ______ .

OBJECTIVE: Use the integral test to determine whether a given series with nonnegative terms converges or diverges.

37. $\displaystyle\sum_{n=2}^{\infty} \frac{1}{n(\ln n)^2}$

We apply the integral test $\displaystyle\int_2^{\infty} \frac{dx}{x(\ln x)^2} = \lim_{b\to\infty}$ ______$\Big]_b^2 =$ ______ .

Therefore, the integral ______ and hence the series ______ .

38. The series $\displaystyle\sum_{n=1}^{\infty} \frac{1}{\sqrt[3]{n}}$ ______ because it is a p-series with $p =$ ______ .

---

34. $\frac{1}{2k-1}$, $\frac{1}{3} - \frac{1}{5}$, $\frac{1}{5} - \frac{1}{7}$, $\frac{1}{2}\left(1 - \frac{1}{2k+1}\right)$, $\frac{1}{2}$, does, $\frac{1}{2}$

35. nondecreasing

36. bounded from above

37. $\frac{1}{\ln x}$, $\frac{1}{\ln 2}$, converges, converges

38. diverges, $\frac{1}{3}$

8.5 COMPARISON TESTS FOR SERIES OF NONNEGATIVE TERMS

OBJECTIVE A: Use the Direct Comparison Tests to determine if a given series with nonnegative terms converges or diverges.

39. $\sum_{n=1}^{\infty} \frac{n+5}{n^2 - 3n + 5}$

For every index n, $5n > -3n + 5$ because n is a positive integer. Then, $n^2 + 5n >$ __________, and since $n^2 - 3n + 5$ is positive for every index it follows that

$$\frac{n+2}{n^2 - 3n + 5} > \text{\_\_\_\_\_\_\_\_} .$$

We conclude that the given series __________ by comparison with the series $\sum \frac{1}{n}$.

40. $\sum_{n=1}^{\infty} \frac{(\ln n)^2}{n^3}$

The following argument shows that $\ln n < \sqrt{n}$ for every index n. First, define the function $g(x) = \sqrt{x} - \ln x$. Now $g'(x) =$ __________ is nonnegative if $x \geq$ _______, and it follows that g is an increasing function of x for $x \geq 4$. Also, $g(4) = 2 - \ln 4 \approx 0.613$ is positive. Therefore, $g(x) > 0$ for $x \geq 4$. A simple verification using tables, or a calculator, shows that $g(1)$, $g(2)$, and $g(3)$ are positive. Hence, we have established that $\ln n < \sqrt{n}$ for every positive integer n. Using this fact,

$$\frac{(\ln n)^2}{n^3} < \frac{(\sqrt{n})^2}{n^3} = \text{\_\_\_\_\_\_\_\_} .$$

We conclude that the series $\sum_{n=1}^{\infty} \frac{(\ln n)^2}{n^3}$ __________ by the direct comparison test.

OBJECTIVE B: Use the Limit Comparison Test to determine if a given series with nonnegative terms converges or diverges.

41. $\sum_{n=1}^{\infty} \frac{3n^4 - n^3 + 2n}{5n^6 + n - 1}$

For n large, we expect $a_n = (3n^4 - n^3 + 2n)/(5n^6 + n - 1)$ to behave like ______________, so we let $b_n =$ ________ . Then,

$$\lim_{n\to\infty} \frac{a_n}{b_n} = \lim_{n\to\infty} \frac{\text{\_\_\_\_\_\_\_\_\_\_\_\_}}{5n^6 + n - 1} = \text{\_\_\_\_\_\_\_\_} .$$

---

39. $n^2 - 3n + 5$, $\frac{1}{n}$, diverges 40. $\frac{1}{2\sqrt{x}} - \frac{1}{x}$, 4, $\frac{1}{n^2}$, converges

Since $\sum_{n=1}^{\infty} \frac{1}{n^2}$ converges, we conclude that $\sum a_n$ __________ by part 2 of the Limit Comparison Test.

8.6 THE RATIO AND ROOT TESTS FOR SERIES OF NONNEGATIVE TERMS

OBJECTIVE: Given a series of nonnegative terms, investigate its convergence or divergence using the ratio and root tests.

42. $\sum_{n=1}^{\infty} \frac{n!}{3^n}$

We try the ratio test. Thus, $\frac{a_{n+1}}{a_n} = \frac{(n+1)!/3^{n+1}}{n!/3^n} =$ __________ .

Hence, $\rho = \lim_{n\to\infty} \frac{a_{n+1}}{a_n} =$ __________, and the series __________ .

43. $\sum_{n=1}^{\infty} (\sqrt[n]{n} - 1)^n$

We try the nth-root test. Thus, for $a_n = (\sqrt[n]{n} - 1)^n$,

$$\sqrt[n]{a_n} = \_\_\_\_\_\_\_\_ \to \_\_\_\_\_\_\_\_ .$$

Because $\rho =$ ______ we conclude that the series __________ according to the root test.

44. $\sum_{n=1}^{\infty} \frac{n \ln n}{(n+1)^2}$

We try the ratio test. Thus,

$$\frac{a_{n+1}}{a_n} = \frac{(n+1)\ln(n+1)}{(n+2)^2} \cdot \_\_\_\_\_\_\_\_ .$$

Hence,

$$\rho = \lim_{n\to\infty} \frac{a_{n+1}}{a_n} = \lim_{n\to\infty} \left[\frac{(n+1)^3}{n(n+1)^2} \cdot \_\_\_\_\_\_\_\_\right] = \_\_\_\_ \cdot 1 = \_\_\_\_ .$$

In this case the root test says the series ____________________ .

---

41. $\frac{3n^4}{5n^6}$, $\frac{1}{n^2}$, $3n^2 - n + \frac{2}{n}$, 0, converges

42. $\frac{1}{3} \cdot (n+1)$, $+\infty$, diverges

43. $\sqrt[n]{n} - 1$, 0, 0, converges

44. $\frac{(n+1)^2}{n \ln n}$, $\frac{\ln(n+1)}{\ln n}$, 1, 1, may converge or may diverge

8.7 ALTERNATING SERIES, ABSOLUTE AND CONDITIONAL CONVERGENCE

OBJECTIVE A: Use the Alternating Series Test (Leibniz's Theorem) to investigate the convergence of an alternating series.

45. $\sum_{n=1}^{\infty} (-1)^{n+1} \frac{n}{n^2 + 1}$

First we see that $a_n = \frac{n}{n^2 + 1}$ is positive for every n. Also, $\lim_{n\to\infty} a_n =$ ______ . Next, we compare a_{n+1} with a_n for arbitrary n. Now, $\frac{n + 1}{(n + 1)^2 + 1} \le \frac{n}{n^2 + 1}$ if and only if $(n + 1)(n^2 + 1) \le n[(n + 1)^2 + 1]$. This last inequality is equivalent to $n^3 + n^2 + n + 1 \le$ ______ or, $1 \le n^2 + n$ which is true. We conclude that the alternating series ______ .

46. $\sum_{n=1}^{\infty} (-1)^{n+1} \frac{1}{n^{1+1/n}}$

It is clear that $a_n = \frac{1}{n^{1+1/n}} = \frac{1}{n \cdot \sqrt[n]{n}}$ is positive for every n. Also, $\lim_{n\to\infty} a_n = \lim_{n\to\infty} \frac{1}{n} \cdot \lim_{n\to\infty} \frac{1}{\sqrt[n]{n}} =$ ______ . We would like to show that $\{a_n\}$ is a ______ sequence. One way is to replace n by the continuous variable x and show that the resultant function

$$y = f(x) = x^{1+1/x}, \quad x > 0,$$

which is the reciprocal of a_n for $x = n$, is an increasing function of x for every x. If we take the logarithm of both sides of this last equation, and differentiate implicitly with respect to x, we obtain

$$\ln y = \left(1 + \frac{1}{x}\right) \ln x, \quad \text{and} \quad \frac{y'}{y} = \frac{1}{x^2} (\_\_\_\_\_\_) + \frac{1}{x}.$$

Simplifying algebraically, $y' = \frac{y}{x^2}(1 +$ ______ $- \ln x)$.

Thus, since $x > \ln x$ for all $x > 0$, we find that y' is positive, and $y = f(x)$ is increasing for all x (so the reciprocal is ______). Therefore we have established that

$$\frac{1}{(n + 1) \cdot \sqrt[n+1]{n + 1}} < \frac{1}{n \cdot \sqrt[n]{n}} \quad \text{for every index } n.$$

We conclude that the alternating series ______ .

---

45. 0, $n^3 + 2n^2 + 2n$, converges

46. $0 \cdot 1 = 0$, decreasing, $1 - \ln x$, x, decreasing, converges

OBJECTIVE B: Use the Alternating Series Estimation Theorem to estimate the magnitude of the error if the first k terms, for some specified number k, are used to approximate a given alternating series.

47. It can be shown, with a little work, that the alternating harmonic series

$$\sum_{n=1}^{\infty} \frac{(-1)^{n+1}}{n}$$

converges to $\ln 2$. If we wish to approximate $\ln 2$ correct to four decimal places using this series, the alternating series error estimation gives

_______ $< 0.5 \times 10^{-5}$, or $n >$ ____________.

Therefore, we would need to sum the first 200,000 terms of the alternating harmonic series to ensure four decimal place accuracy in approximating $\ln 2$. This does not mean that fewer terms would not provide that accuracy. A more efficient approximation for $\ln 2$, accurate to four decimal places, uses Simpson's rule with $n = 6$ to estimate

$$\int_1^2 \frac{dx}{x} .$$

OBJECTIVE C: Given an infinite series, use the Flowchart 8.1 of the text to determine if the series is absolutely convergent, conditionally convergent, or divergent.

48. A series $\sum_{n=1}^{\infty} a_n$ is said to converge absolutely if

_______________________.

49. True or False:
(a) If a series converges absolutely, then it converges.
(b) If a series converges, then it converges absolutely.

50. If a series $\sum a_n$ converges, but the series of absolute values $\sum |a_n|$ diverges, we say that the original series $\sum a_n$ is

_______________________.

51. $\sum_{n=1}^{\infty} \frac{(-1)^{n+1} n \ln n}{3^n}$

The absolute value of the nth term of the series is $|a_n| = \frac{n \ln n}{3^n}$. Applying the ratio test,

---

47. $\frac{1}{n}$, 2×10^5

48. $\sum_{n=1}^{\infty} |a_n|$ converges

49. (a) True (b) False

50. conditionally convergent

$\dfrac{|a_{n+1}|}{|a_n|} = \dfrac{(n + 1) \ln (n + 1)/3^{n+1}}{n \ln n/3^n} =$ __________ . By l'Hôpital's rule, $\lim_{n\to\infty} \dfrac{\ln (n + 1)}{\ln n} = \lim_{n\to\infty}$ __________ = ________ .

Therefore, $\lim_{n\to\infty} \dfrac{|a_{n+1}|}{|a_n|} =$ ________ , so we conclude that $\sum_{n=1}^{\infty} \dfrac{(-1)^{n+1} n \ln n}{3^n}$ __________ converge absolutely.

52. $\sum_{n=1}^{\infty} \dfrac{2n - n^2}{n^3}$

The nth term of the series is $a_n = \dfrac{2n - n^2}{n^3} = \dfrac{n(2 - n)}{n^3}$ which is negative if $n >$ ________ . Thus, $-a_n = \dfrac{n - 2}{n^2}$ is positive for $n > 2$. Now, $\dfrac{n - 2}{n^2} > \dfrac{1}{2n}$ whenever $n >$ ________ . Since the series $\sum_{n=1}^{\infty} \dfrac{1}{2n}$ __________, we conclude that the series $\sum_{n=1}^{\infty} |a_n| = \sum_{n=1}^{\infty} \dfrac{n - 2}{n^2}$ __________ by the comparison test.

Therefore, the original series ________ converge absolutely.

53. In Problem 46 above, the series $\sum_{n=1}^{\infty} (-1)^{n+1} \dfrac{1}{n^{1+1/n}}$ was shown to be convergent. We want to know if the series converges absolutely. Using the same technique as in Problem 46, it is easy to establish that

$$\sqrt[n+1]{n + 1} < \sqrt[n]{n} \quad \text{if} \quad n \geq 3:$$

we define the function $y = x^{1/n}$, $x \geq 3$, and show that y' is always negative; whence we conclude that y is a ________ function of x. In particular, $\sqrt[n]{n} < \sqrt[3]{3} \approx 1.44 < \frac{3}{2}$ if $n \geq$ ________ . It follows that

$$\frac{1}{n \cdot \sqrt[n]{n}} > \text{\_\_\_\_\_\_\_\_} \quad \text{if} \quad n \geq 4.$$

Since the harmonic series $\sum \frac{1}{n}$ diverges, we find that the series $\sum_{n=1}^{\infty} \dfrac{1}{n^{1+1/n}}$ __________ by the comparison test.

---

51. $\frac{1}{3}\left(\frac{n+1}{n}\right) \dfrac{\ln (n + 1)}{\ln n}$, $\dfrac{1/(n + 1)}{1/n}$, 1, $\frac{1}{3}$, does

52. 2, 4, diverges, diverges, does not

Therefore, the original series $\sum_{n=1}^{\infty} (-1)^{n+1} \frac{1}{n^{1+1/n}}$ is

____________________________ .

54. $\sum_{n=1}^{\infty} (-1)^{n+1} \left(\frac{n+1}{n}\right)^n$

In this case, $a_n = \left(\frac{n+1}{n}\right)^n = \left(1 + \frac{1}{n}\right)^n \to$ __________ . Therefore the series _____________ .

55. $\sum_{n=2}^{\infty} \frac{\cos n\pi}{n\sqrt{\ln n}}$

Since $\cos n\pi$ is 1 when n is even, and -1 when n is odd, the series is alternating in sign. Let us see if the series converges absolutely. Now,

$|a_n| = \left|\frac{\cos n\pi}{n\sqrt{\ln n}}\right| =$ ___________ . Applying the integral test,

$$\int_1^{\infty} \frac{dx}{x\sqrt{\ln x}} = \lim_{b\to\infty} \underline{\hspace{3cm}}\Big]_1^b = \underline{\hspace{3cm}} .$$

Therefore, the series $\sum |a_n|$ ___________ . To see if the original series converges we check the three conditions of the Alternating Series Test (remember that the numerator $\cos n\pi$ simply determines the sign of the nth term of the series):

$\frac{1}{n\sqrt{\ln n}}$ is positive for all n, and converges to _______ .

It is clear that $\frac{1}{(n+1)\sqrt{\ln (n+1)}} < \frac{1}{n\sqrt{\ln n}}$ because $y = \ln x$ is an ___________ function of x. Therefore, the series $\sum_{n=1}^{\infty} \frac{\cos n\pi}{n\sqrt{\ln n}}$ is _______________________ .

8.8 POWER SERIES

56. A series of the form

$$\sum_{n=0}^{\infty} c_n x^n = c_0 + c_1 x + c_2 x^2 + c_3 x^3 + \dots + c_n x^n + \dots$$

is called a ___________________________________ .

57. For the power series

$$\sum_{n=0}^{\infty} c_n (x-a)^n = c_0 + c_1(x-a) + c_2(x-a)^2 + \dots + c_n(x-a)^n + \dots$$

---

53. decreasing, 4, $\frac{2}{3n}$, diverges, conditionally convergent

54. e, diverges

55. $\frac{1}{n\sqrt{\ln n}}$, $2\sqrt{\ln x}$, $+\infty$, diverges, 0, increasing, conditionally convergent

56. power series about $x = 0$

the constant a is called the __________ . The coefficients $c_0, c_1, c_2, \ldots, c_n, \ldots$ are all ______________ .

OBJECTIVE A: Given a power series $\sum_{n=0}^{\infty} c_n(x-a)^n$, find its interval of convergence. If the interval is finite, determine whether the series converges at each endpoint.

58. $\sum_{n=1}^{\infty} \frac{1}{\sqrt{n}\ 3^n} x^n$

We apply the ratio test to the series of absolute values, and find

$$\rho = \lim_{n\to\infty} \left| \frac{x^{n+1}}{\sqrt{n+1}\ 3^{n+1}} \cdot \frac{\quad}{\quad} \right| = \lim_{n\to\infty} \frac{\sqrt{n}}{\quad} |x| = \_\_\_\_\_ .$$

Therefore the original series converges absolutely if $|x| <$ ______ and diverges if ________ . When $x = 3$, the series becomes

$\sum_{n=1}^{\infty}$ __________ , the p-series with $p =$ ________ ;

this series __________ . When $x = -3$, the series becomes

$$\sum_{n=1}^{\infty} \_\_\_\_\_\_\_ ,$$

and this series __________ , by the Alternating Series Test. Therefore, the interval of convergence of the original power series is ______________ .

59. $\sum_{n=1}^{\infty} \frac{2^n}{n(3^{n+2})} x^{n+1}$

The power series converges for $x = 0$. For $x \neq 0$, we apply the root test to the series of absolute values, and find

$$\rho = \lim_{n\to\infty} \sqrt[n]{\frac{2^n\ |x|^n\ |x|}{n\cdot 3^n\cdot 3^2}} = \lim_{n\to\infty} \_\_\_\_\_\_\_\_\_\_\_\_$$

$$= \frac{2|x| \cdot 1}{\quad} < 1, \text{ if } |x| < \_\_\_\_\_ .$$

Therefore, the original series converges absolutely if $|x| < \frac{3}{2}$ and diverges if $|x| > \frac{3}{2}$. When $x = \frac{3}{2}$, the series becomes

$$\sum_{n=1}^{\infty} \frac{2^n}{n\left(3^{n+2}\right)} \left(\frac{3}{2}\right)^{n+1} = \sum_{n=1}^{\infty} \_\_\_\_\_\_\_ ,$$

---

57. center, constants

58. $\frac{\sqrt{n}\ 3^n}{x^n}$, $3\sqrt{n+1}$, $\frac{1}{3}|x|$, 3, $|x| > 3$, $\frac{1}{\sqrt{n}}$, $\frac{1}{2}$, diverges, $\frac{(-1)^n}{\sqrt{n}}$, converges, $-3 \le x < 3$

and this series ________ . When $x = -\frac{3}{2}$, the series becomes

$$\sum_{n=1}^{\infty} \frac{(-1)^{n+1}}{6n},$$

and this series ________, by the Alternating Series Test. Therefore, the interval of convergence of the original power series is ________ .

60. $\sum_{n=1}^{\infty} [\sin(5n)](x - \pi)^n$

For every value of x, $|[\sin(5n)](x - \pi)^n| \le |x - \pi|^n$. The geometric series $\sum_{n=1}^{\infty} |x - \pi|^n$ converges if ________ and diverges if ________ . Therefore, by the comparison test, the original series converges absolutely if ________ . Suppose $|x - \pi| = 1$. Then the series becomes,

$$\sum_{n=1}^{\infty} \sin(5n) \quad \text{or} \quad \sum_{n=1}^{\infty} \text{\_\_\_\_\_\_\_\_\_\_} .$$

However, $\lim_{n\to\infty} \sin(5n)$ fails to exist, so neither of these series can converge. We conclude that the interval of convergence of the original series is ________ .

OBJECTIVE B: Given a power series $f(x) = \sum a_n x^n$, find the power series for $f'(x)$.

61. Replacing x by $-x^2$ in Example 1 of the text we deduce that

$$\frac{1}{1 + x^2} = 1 - x^2 + x^4 - x^6 + \dots, \quad \text{for} \quad -1 < x < 1.$$

Therefore, using the Term-by-Term Differentiation Theorem,

$$\frac{-2x}{\left(1 + x^2\right)^2} = \text{\_\_\_\_\_\_\_\_\_\_}, \quad \text{for} \quad \text{\_\_\_\_\_\_\_\_\_\_} .$$

---

59. $\dfrac{2|x| \cdot \sqrt[n]{|x|}}{\sqrt[n]{n} \cdot 3 \cdot \sqrt[n]{9}}$, $1 \cdot 3 \cdot 1$, $\frac{3}{2}$, $\frac{1}{6n}$, diverges, converges, $-\frac{3}{2} \le x < \frac{3}{2}$

60. $|x - \pi| < 1$, $|x - \pi| \ge 1$, $|x - \pi| < 1$, $(-1)^n \sin(5n)$, $\pi - 1 < x < \pi + 1$

61. $-2x + 4x^3 - 6x^5 + \dots$, $-1 < x < 1$

8.9 TAYLOR AND MACLAURIN SERIES

OBJECTIVE: Find the Taylor Series at $x = a$, or the Maclaurin series, for a given function $y = f(x)$. Assume that $x = a$ is specified and that f has finite derivatives of all orders at $x = a$.

62. If $y = f(x)$ has finite derivatives of all orders at $x = a$, the particular power series

$$f(a) + f'(a)(x - a) + \frac{f''(a)}{2!}(x - a)^2 + \ldots + \frac{f^{(n)}(a)}{n!}(x - a)^n + \ldots$$

is called the ______________________________.
If $a = 0$, the series is known as the ______________ for f. The Taylor series for a function may or may not converge to the function.

63. If $y = f(x)$ has finite derivatives of order up to and including n, then the polynomial

$$P_n(x) = f(a) + f'(a)(x - a) + \frac{f''(a)}{2!}(x - a)^2 + \ldots + \frac{f^{(n)}(a)}{n!}(x - a)^n$$

is called the ______________________________ ______________. The graph of this polynomial passes through the point ______, and its first n derivatives match the first n derivatives of ______ at ______. Each nonnegative integer n corresponds to a Taylor polynomial for f at $x = a$, provided the first n derivatives of f exist at $x = a$.

64. Let us find the Maclaurin series for the function $f(x) = x^5 + 4x^4 + 3x^3 + 2x + 1$. We need to find the derivatives of f of all orders, and evalute them at $x = 0$:

$f'(x)$ = ______________, $f'(0) = 2$
$f^{(2)}(x)$ = ______________, $f^{(2)}(0)$ = ______
$f^{(3)}(x)$ = ______________, $f^{(3)}(0)$ = ______
$f^{(4)}(x) = 120x + 96$, $f^{(4)}(0) = 96$
$f^{(5)}(x)$ = ____ , $f^{(5)}(0)$ = ______
$f^{(6)}(x) = 0$, $f^{(6)}(0) = 0$

In general, $f^{(k)}(0)$ = ____ if $k \geq 6$. Thus, the Maclaurin series is

______________________________,

---

62. Taylor series generated by f at $x = a$, Maclaurin series

63. Taylor polynomial of order n generated by f at $x = a$, $(a, f(a))$, $y = f(x)$, $x = a$

64. $5x^4 + 16x^3 + 9x^2 + 2$, $20x^3 + 48x^2 + 18x$, 0, $60x^2 + 96x + 18$, 18, 120, 120, 0,

$1 + 2x + 0x^2 + \frac{18}{3!}x^3 + \frac{96}{4!}x^4 + \frac{120}{5!}x^5$

which simplifies to $1 + 2x + 3x^3 + 4x^4 + x^5$. Therefore, the Maclaurin series for a polynomial expressed in powers of x is the polynomial itself.

65. Suppose we want to express the polynomial in Problem 64 in powers of $(x + 1)$ instead of powers of x. We find the Taylor series of f at $x =$ ______ . From our previous calculations of the derivatives, we find that $f(-1) = -1$, $f'(-1) = 0$, $f^{(2)}(-1) =$ ______, $f^{(3)}(-1) =$ ______, $f^{(4)}(-1) =$ ______, $f^{(5)}(-1) =$ ______, and $f^{(k)}(-1) =$ ______ if $k \geq 6$. Thus the Taylor series of f at $x = -1$ is

______________________________,

which simplifies to

$$-1 + 5(x + 1)^2 - 3(x + 1)^3 - (x + 1)^4 + (x + 1)^5.$$

66. Let us find the Taylor polynomials $P_3(x)$ and $P_4(x)$ for the function $f(x) = a^x$, $a > 0$, at $x = 1$. To do this we complete the following table:

n	$f^{(n)}(x)$	$f^{(n)}(1)$
0	a^x	a
1	$a^x \ln a$	$a \ln a$
2	______	______
3	______	______
4	______	______

Then,

$P_3(x) = a + a(\ln a)(x - 1) + \frac{a(\ln a)^2}{2!}(x - 1)^2 +$ ______________ ,

$P_4(x) =$ ______________________________ .

67. For the function $f(x) = a^x$ in Problem 66, the Taylor series at $x = 1$ is

$$\sum_{n=0}^{\infty} \text{\_\_\_\_\_\_\_\_\_\_\_\_} .$$

---

65. -1, 10, -18, -24, 120, 0, $-1 + 0(x + 1) + \frac{10}{2!}(x + 1)^2 - \frac{18}{3!}(x + 1)^3 - \frac{24}{4!}(x + 1)^4 + \frac{120}{5!}(x + 1)^5$

66. For $n = k$, $f^{(k)}(1) = a(\ln a)^k$; $\frac{a(\ln a)^3}{3!}(x - 1)^3$,

$a + a(\ln a)(x - 1) + \frac{a(\ln a)^2}{2!}(x - 1)^2 + \frac{a(\ln a)^3}{3!}(x - 1)^3 + \frac{a(\ln a)^4}{4!}(x - 1)^4$

67. $\frac{a(\ln a)^n}{n!}(x - 1)^n$

8.10 CONVERGENCE OF TAYLOR SERIES; ERROR ESTIMATES

OBJECTIVE A: Know the statement of Taylor's Theorem and a formula for the remainder of order n.

68. The statement of Taylor's Theorem in the text gives the remainder term as

$R_n(x) =$ ______________________________,

where the number c lies between ____________________ . This remainder term measures the error in the approximation of $y = f(x)$ by the nth-degree Taylor polynomial at __________ . Thus, the Taylor series expansion for $f(x)$ will converge to $f(x)$ provided that

______________________ .

69. This remainder form is very useful because often we can bound the derivative $f^{(n+1)}(c)$ by some constant M: $|f^{(n+1)}(c)| \le M$. This ensures that $R_n(x)$ converges to ______ as $n \to \infty$.

OBJECTIVE B: Use the Remainder Estimation Theorem to estimate the truncation error when a Taylor polynomial is used to approximate a given function. Assume that the function has derivatives of all orders.

70. We will calculate $\cos \sqrt{2}$ with an error less than 10^{-6}. By Taylor's Theorem, $\cos \sqrt{2} =$ ____________________________ $+ R_{2k}(x)$. The Remainder Estimation Theorem, with $M =$ _______, $x =$ ______ and $r = 1$ gives $|R_{2k}| \le 1 \cdot$ _______________ . By trial we find that $\frac{(\sqrt{2})^{11}}{11!} = 0.0000011337 > 10^{-6}$ and $\frac{(\sqrt{2})^{13}}{13!} = 0.0000000145 < 10^{-6}$. Thus, we should take $(2k + 1)$ to be at least ______, or k to be at least 6. With an error less than 10^{-6},

$$\cos \sqrt{2} = 1 - \frac{2}{2!} + \frac{4}{4!} - \frac{8}{6!} + \ldots + \text{\_\_\_\_\_\_}$$

↑ last term

$$\approx 0.155944.$$

71. Let us determine for what values of $x > 0$ we can replace e^x by $1 + x + \left(\frac{x^2}{2}\right) + \left(\frac{x^3}{3!}\right)$ with an error of magnitude

---

68. $f^{(n+1)}(c) \frac{(x - a)^{n+1}}{(n + 1)!}$, a and x, $x = a$, $\lim_{n\to\infty} R_n(x) = 0$

69. 0

70. $1 - \frac{2}{2!} + \frac{4}{4!} - \frac{8}{6!} + \ldots + (-1)^k \frac{2^k}{(2k)!}$, 1, $\sqrt{2}$, $\frac{(\sqrt{2})^{2k+1}}{(2k + 1)!}$, 13, 6, $\frac{64}{12!}$

less than 5×10^{-5}. In Example 1 in this section of the text, if $x > 0$,

$$|R_3(x)| < \underline{\hspace{4cm}} .$$

We desire $|R_3(x)| < 5 \times 10^{-5}$. This is the case if $e^x|x|^4 < \underline{\hspace{2cm}}$ or, since $x > 0$, $x + 4 \ln x < -5.116$. By calculator experimentation this inequality holds if $0 < x < 0.26$. Thus, for instance,

$$e^{0.1} = 1 + (0.1) + \frac{(0.1)^2}{2} + \frac{(0.1)^3}{6} \approx 1.10517$$

is correct to five decimal places.

OBJECTIVE C: Using the Maclaurin series for the functions e^x, $\sin x$, and $\cos x$, write the Maclaurin series for functions which are combinations of sines, cosines, exponentials, or powers of x.

72. The Maclaurin series for e^x, $\sin x$, and $\cos x$ are $e^x = \underline{\hspace{2cm}}$, $\sin x = \underline{\hspace{2cm}}$, and $\cos x = \underline{\hspace{2cm}}$.

73. Let us find the Maclaurin series for $\sin^3 x$. A trigonometric identity gives

$$\sin^3 x = \frac{1}{4}(3 \sin x - \sin 3x) .$$

We use Maclaurin series for the terms on the right side:

$$3 \sin x = 3x - \frac{3x^3}{3!} + \frac{3x^5}{5!} - \frac{3x^7}{7!} + \dots,$$

$$\sin 3x = \underline{\hspace{4cm}},$$

$$3 \sin x - \sin 3x = 4x^3 - 2x^5 + \frac{52}{5!} x^7 - \dots$$

Therefore, $\sin^3 x = \underline{\hspace{4cm}}$

$$= \sum_{n=0}^{\infty} \frac{(-1)^n(3 - 3^{2n+1})}{4(2n + 1)!} x^{2n+1} .$$

74. Euler's formula asserts that

$$e^{i\theta} = \underline{\hspace{4cm}} .$$

---

71. $e^x \cdot \frac{x^4}{4!}$, 6×10^{-3}

72. $\sum_{n=0}^{\infty} \frac{x^n}{n!}$, $\sum_{n=0}^{\infty} \frac{(-1)^n x^{2n+1}}{(2n + 1)!}$, $\sum_{n=0}^{\infty} \frac{(-1)^n x^{2n}}{(2n)!}$

73. $3x - \frac{(3x)^3}{3!} + \frac{(3x)^5}{5!} - \frac{(3x)^7}{7!} + \dots$, $x^3 - \frac{1}{2}x^5 + \frac{13}{5!}x^7 - \dots$

74. $\cos \theta + i \sin \theta$

8.11 APPLICATIONS OF POWER SERIES

OBJECTIVE A: Use a suitable series to calculate a given quantity to three decimal places. Show that the remainder term does not exceed 5×10^{-4}. (Assume the quantity is the value of a function whose series expansion has been studied in this chapter of the text.)

75. Replacing x by $-x$ in the Taylor series expansion for $\ln(1 + x)$ gives the expansion

$$\ln(1 - x) = \underline{\qquad\qquad\qquad}$$

which is also valid for $|x| < 1$. Subtracting this result from the expansion for $\ln(1 + x)$ gives

$$\ln(1 + x) - \ln(1 - x) = \ln \frac{1 + x}{1 - x} = \underline{\qquad\qquad\qquad} .$$

76. Let N be a positive integer. Then

$$\ln(N + 1) = \ln N + \ln \frac{N + 1}{N} .$$

Now, solve the equation

$$\frac{1 + x}{1 - x} = \frac{N + 1}{N}$$

for x to obtain $x = \underline{\qquad\qquad}$. Substitution into the result from Problem 75 yields

$$\ln \frac{N + 1}{N} = \underline{\qquad\qquad\qquad} .$$

77. Let's use the result of Problem 76 to calculate $\ln 2$ by setting $N = 1$. Thus,

$$\ln 2 = 2\left(\frac{1}{3} + \frac{1}{3(3)^3} + \frac{1}{3(3)^5} + \frac{1}{7(3)^7} + \frac{1}{9(3)^9} + \frac{1}{11(3)^{11}} + \ldots\right)$$

$$\approx 2(0.3333333 + 0.0123457 + 0.0008230 + 0.0000653 + 0.0000056 + 0.0000005)$$

$$\approx \underline{\qquad\qquad\qquad} .$$

The error satisfies

$$|R_{11}(x)| \le \frac{1}{12} \cdot \frac{|x|^{12}}{1 - |x|} \quad \text{where} \quad x = \frac{1}{2N + 1} = \underline{\qquad\qquad} .$$

Thus,

$$|R_{11}(x)| \le \frac{1}{12} \cdot \frac{3}{2}\left|\frac{1}{3}\right|^{12} = 0.00000023252 < 5 \times 10^{-6}.$$

It follows that $\ln 2 \approx 0.69315$, accurate to five decimal places.

---

75. $-x - \frac{x^2}{2} - \frac{x^3}{3} - \ldots - \frac{x^n}{n} - \ldots$, $2\left(x + \frac{x^3}{3} + \frac{x^5}{5} + \ldots + \frac{x^{2k-1}}{2k - 1} + \ldots\right)$

76. $\frac{1}{2N + 1}$, $2\left(\frac{1}{2N + 1} + \frac{1}{3(2N + 1)^3} + \frac{1}{5(2N + 1)^5} + \ldots\right)$

77. 0.6931468, $\frac{1}{3}$

78. Now let's calculate $\ln 0.75$ using the result in Problem 76. First use the identity

$$\ln 0.75 = \ln \frac{3}{4} = \ln \frac{3}{2} - \underline{\qquad\qquad} .$$

We can use the calculation for $\ln 2$ obtained in Problem 77: $\ln 2 \approx 0.69315$. To obtain the first term on the right side of the previous equation we use the series

$$\ln \frac{N+1}{N} = 2\left(\underline{\qquad\qquad\qquad}\right) \text{ with } N = 2.$$

Then,

$$\ln \frac{3}{2} = 2\left(\frac{1}{5} + \frac{1}{3(5)^3} + \frac{1}{5(5)^5} + \frac{1}{7(5)^7} + \ldots\right)$$

$$= 2(0.2 + 0.00266667 + 0.000064 + 0.00000183 + \ldots)$$

$$\approx 0.40547.$$

The error satisfies

$$|R_7(x)| \le \frac{1}{8} \cdot \frac{|x|^8}{1 - |x|}, \text{ where } x = \frac{1}{2N+1} = \underline{\qquad} .$$

Thus,

$$|R_7(x)| \le \frac{5}{32} \left|\frac{1}{5}\right|^8 < 5 \times 10^{-6}.$$

It follows that

$$\ln \frac{3}{4} \approx 0.40547 - 0.69315 = -0.28768,$$

accurate to five decimal places.

OBJECTIVE B: If f is a function having a known power series $f(x) = \sum a_n x^n$, use the series and a calculator to estimate the integral $\int_0^b f(x)\,dx$, assuming that b lies within the interval of convergence.

79. Let us find $\int_0^{0.2} \cos \sqrt{x}\,dx$ accurate to five decimal places. Now,

$$\cos x = 1 - \frac{x^2}{2!} + \frac{x^4}{4!} - \frac{x^6}{6!} + \frac{x^8}{8!} - \ldots ,$$

so the power series for $\cos \sqrt{x}$ is given by

$$\cos \sqrt{x} = \underline{\qquad\qquad\qquad}, \quad x \ge 0.$$

Thus, using term-by-term integration,

$$\int_0^{0.2} \cos \sqrt{x}\,dx = \underline{\qquad\qquad\qquad}\Big]_0^{0.2}$$

$$= 0.2 - \frac{0.04}{4} + \frac{0.008}{72} - \frac{0.0016}{2880} + \frac{0.00032}{201600} - \ldots$$

$$\approx 0.2 - 0.01 + 0.00011 - 0.00000056 + \ldots$$

---

78. $\ln 2$, $\frac{1}{2N+1} + \frac{1}{3(2N+1)^3} + \frac{1}{5(2N+1)^5} + \ldots$, $\frac{1}{5}$

Hence,

$$\int_0^{0.2} \cos \sqrt{x}\, dx \approx \underline{\qquad\qquad}$$

with an error less than 5×10^{-6}.

OBJECTIVE C: Find power series solutions for differential equations or initial value problems.

80. Solve the initial value problem

$$y' - y = 0, \quad y(0) = 1 .$$

Solution.
We assume there is a solution of the form $y = \sum_{n=0}^{\infty} a_0 x^n$. Then $y' = \underline{\qquad\qquad\qquad\qquad\qquad}$. The series for $y' - y$ is given by

$$y' - y = (a_1 - a_0) + (\underline{\qquad\qquad})x^2 + (\underline{\qquad\qquad})x^3 + \ldots$$
$$+ (na_n - a_{n-1})x^{n-1} + \ldots$$

Equating the coefficients of $y' - y$ to the coefficients of 0 on the right side of the differential equation gives $a_1 - a_0 = 0$, $2a_2 - a_1 = \underline{\qquad}$, $3a_3 - a_2 = \underline{\qquad}$, and, in general, $\underline{\qquad\qquad} = 0$ for $n \geq 1$. Solving these equations we see that

$a_1 = a_0$, $a_2 = \frac{1}{2}a_1 = \frac{1}{2}a_0$, $a_3 = \frac{1}{3}a_2 = \frac{1}{2 \cdot 3}a_0$. In general,

$a_n = \frac{1}{n}a_{n-1} = \underline{\qquad}\ a_0$. The solution of the differential equation is

$$y = \sum_{n=0}^{\infty} a_n x^n = \sum_{n=0}^{\infty} \left(\frac{1}{n!}a_0\right)x^n = a_0\,(\underline{\qquad\qquad}) .$$

Since $y(0) = 1$ from the initial condition, we see that $a_0 = \underline{\qquad}$. Thus the solution of the initial value problem is $y = \underline{\qquad}$.

81. Find a power series solution for

$$y'' + xy' + y = 0 .$$

---

79. $1 - \frac{x}{2!} + \frac{x^2}{4!} - \frac{x^3}{6!} + \frac{x^4}{8!} - \ldots$, $x - \frac{x^2}{2 \cdot 2!} + \frac{x^3}{3 \cdot 4!} - \frac{x^4}{4 \cdot 6!} + \frac{x^5}{5 \cdot 8!} - \ldots$, 0.19011

80. $a_1 + 2a_2x + 3a_3x^2 + \ldots + na_nx^{n-1} + \ldots$, $2a_2 - a_1$, $3a_3 - a_2$, 0, 0, $na_n - a_{n-1}$, $\frac{1}{n!}$, e^x, 1, $e^x = \sum_{n=0}^{\infty} \frac{1}{n!}x^n$

Solution. Assume there is a solution of the form

$$y = a_0 + a_1x + a_2x^2 + \dots + a_nx^n + \dots = \sum_{n=0}^{\infty} a_nx^n .$$

Then $y' = \sum_{n=1}^{\infty} na_nx^{n-1}$ and $y'' =$ ______________. Substitution into the differential equation gives

$$\sum_{n=2}^{\infty} n(n-1)a_nx^{n-2} + x\sum_{n=1}^{\infty} na_nx^{n-1} + \sum_{n=0}^{\infty} a_nx^n = 0 .$$

We next collect like powers of x on the left-hand side and set each coefficient equal to 0:

Power of x	Associated coefficient
x^0	$2(1)a_2 + a_0 = 0$ so $a_2 =$ ________
x^1	$3(2)a_3 + a_1 + a_1 = 0$ so $a_3 =$ ________
x^2	______________ $= 0$ so $a_4 = -\frac{1}{4}a_2$
x^3	______________ $= 0$ so $a_5 = -\frac{1}{5}a_3$
x^4	______________ $= 0$ so $a_6 = -\frac{1}{6}a_4$

Thus, a_0 and a_1 are arbitrary with

$a_2 = -\frac{1}{2}a_0$, $a_3 = -\frac{1}{3}a_1$, $a_4 = -\frac{1}{4}a_2 = \frac{1}{2\cdot 4}a_0$, $a_5 =$ ______ a_1,

$a_6 =$ ______ a_0. Therefore, the series solution is given by

$$y = a_0(1 - \tfrac{1}{2}x^2 + \tfrac{1}{8}x^4 - \tfrac{1}{48}x^6 + \dots) + a_1(\text{\_\_\_\_\_\_\_\_\_\_\_\_\_\_\_\_})$$

Remark: If you calculate the coefficient of the power x^{2k-2} you will find $a_{2k} = -\frac{1}{2k}a_{2k-2}$; likewise, the power x^{2k-1} produces the equation $a_{2k+1} = -\frac{1}{2k+1}a_{2k-1}$. Hence the even coefficients will be expressed (ultimately) in terms of a_0 and the odd coefficients in terms of a_1. In fact,

$$a_{2k} = \frac{(-1)^k}{2\cdot 4\cdot 6\cdots 2k}a_0 , \quad k = 1,2,3, \dots$$

$$a_{2k+1} = \text{\_\_\_\_\_\_\_\_\_\_\_\_}\ a_1 , \quad k = 1,2,3, \dots$$

Then the solution becomes

$$y = a_0\sum_{k=0}^{\infty}(-1)^k\frac{x^{2k}}{2^k\cdot k!} + a_1\sum_{k=0}^{\infty}(-1)^k\frac{x^{2k+1}}{1\cdot 3\cdot 5\cdots(2k+1)} .$$

81. $\sum_{n=2}^{\infty} n(n-1)a_nx^{n-2}$, $-\frac{1}{2}a_0$, $-\frac{1}{3}a_1$, $4(3)a_4 + 2a_2 + a_2$, $5(4)a_5 + 3a_3 + a_3$, $6(5)a_6 + 4a_4 + a_4$, $\frac{1}{3\cdot 5}$, $\frac{-1}{2\cdot 4\cdot 6}$, $x - \frac{1}{3}x^3 + \frac{1}{15}x^5 - \dots$, $\frac{(-1)^k}{3\cdot 5\cdot 7\cdots(2k+1)}$

CHAPTER 8 SELF-TEST

1. Determine if each sequence $\{a_n\}$ converges or diverges. Find the limit of the sequence if it does converge.

   (a) $a_n = \sqrt{n+1} - \sqrt{n}$ (b) $a_n = \dfrac{1 + (-1)^n}{\sqrt[n]{n}}$

   (c) $a_n = \left(\dfrac{n - 0.05}{n}\right)^n$ (d) $a_n = \dfrac{2^n}{5^{3+1/n}}$

2. Find the sum of each series.

   (a) $\sum_{n=0}^{\infty} (-1)^n \dfrac{3}{5^n}$ (b) $\sum_{n=4}^{\infty} \dfrac{2}{(4n-3)(4n+1)}$

   (c) $\sum_{n=0}^{\infty} \left(\dfrac{5}{3^n} - \dfrac{2}{7^n}\right)$

   (d) $\dfrac{127}{1000} + \dfrac{127}{1000^2} + \dfrac{127}{1000^3} + \ldots + \dfrac{127}{1000^n} + \ldots$

In Problems 3-8, determine whether the given series converges or diverges. In each case, give a reason for your answer.

3. $\sum_{n=1}^{\infty} \dfrac{\sqrt{n}}{n^2 + 3}$ 4. $\sum_{n=1}^{\infty} \dfrac{n!\,3^n}{10^n}$ 5. $\sum_{n=1}^{\infty} \sin\left(\dfrac{n\pi - 2}{3n}\right)$

6. $\sum_{n=1}^{\infty} \left(\dfrac{n}{2n+5}\right)^n$ 7. $\sum_{n=1}^{\infty} \dfrac{1}{n + \sqrt{n}}$ 8. $\sum_{n=1}^{\infty} \dfrac{\tan^{-1} n}{n^2 + 1}$

In Problems 9-12, determine whether the series are absolutely convergent, conditionally convergent, or divergent.

9. $\sum_{n=1}^{\infty} (-1)^{n+1} \dfrac{\sin n}{n^2 + 1}$ 10. $\sum_{n=1}^{\infty} (-1)^{n+1} \dfrac{1}{(n+1)^{1/n}}$

11. $\sum_{n=2}^{\infty} (-1)^n \dfrac{1}{(\ln n)^2}$ 12. $\sum_{n=1}^{\infty} (-1)^{n+1} \dfrac{n+1}{7n-2}$

13. Estimate the magnitude of the error if the first five terms are used to approximate the series,

$$\sum_{n=1}^{\infty} (-1)^{n+1} \frac{2^n}{3^n}.$$

Sum the first five terms, and state whether your approximation underestimates or overestimates the sum of the series.

14. Find the interval of convergence for the power series

$$\sum_{n=2}^{\infty} \frac{\ln n}{n} x^n$$

15. Find the Taylor series of $f(x) = \sqrt{x}$ at $a = 9$. Do not be concerned with whether the series converges to the given function f.

16. Find the Maclaurin series for the function $f(x) = x \ln(1 + x^2)$ using series that have already been obtained in the Thomas/Finney text.

17. Use series to estimate the number $e^{-1/3}$ with an error of magnitude less than 0.001.

18. Find the first three nonzero terms in the Maclaurin series for the function $f(x) = \sec^2 x$ using the Maclaurin series for $\tan x$.

19. (Calculator) Use series and a calculator to estimate the integral

$$\int_0^{0.5} \cos x^2\, dx$$

with an error of magnitude less than 0.0001.

20. Find a power series solution to

$$2y'' + xy' - 4y = 0 .$$

SOLUTIONS TO CHAPTER 8 SELF-TEST

1. (a) $a_n = \sqrt{n+1} - \sqrt{n} = \dfrac{(\sqrt{n+1} - \sqrt{n})(\sqrt{n+1} + \sqrt{n})}{(\sqrt{n+1} + \sqrt{n})} = \dfrac{(n+1) - n}{\sqrt{n+1} + \sqrt{n}}$

$$= \frac{1}{\sqrt{n+1} + \sqrt{n}} \to 0 \quad \text{as} \quad n \to \infty$$

(b) $\sqrt[n]{n} \to 1$, but $1 + (-1)^n$ alternates back and forth between 0 and 2. Thus, for n large, a_n alternates between numbers very close to 2 and 0; hence the sequence diverges.

(c) $a_n = \left(\dfrac{n - 0.05}{n}\right)^n = \left(1 + \dfrac{-0.05}{n}\right)^n \to e^{-0.05} \approx 0.951.$

(d) $5^{3+1/n} = 125\sqrt[n]{5} \to 125$, but $2^n \to +\infty$. Therefore, the sequence $\{a_n\}$ is unbounded and diverges.

2. (a) $\displaystyle\sum_{n=0}^{\infty} (-1)^n \frac{3}{5^n} = \sum_{n=0}^{\infty} 3\left(-\frac{1}{5}\right)^n = \frac{3}{1 + \frac{1}{5}} = \frac{5}{2}.$

(b) Using the partial fraction decomposition,

$\dfrac{2}{(4k-3)(4k+1)} = \dfrac{1}{2}\left(\dfrac{1}{4k-3}\right) - \dfrac{1}{2}\left(\dfrac{1}{4k+1}\right)$, we write the partial sum

$$s_k = \sum_{n=4}^{k} \frac{2}{(4n-3)(4n+1)}$$

as

$$s_k = \frac{1}{2}\left(\frac{1}{13} - \frac{1}{17}\right) + \frac{1}{2}\left(\frac{1}{17} - \frac{1}{21}\right) + \frac{1}{2}\left(\frac{1}{21} - \frac{1}{25}\right) + \cdots + \frac{1}{2}\left(\frac{1}{4k - 3} - \frac{1}{4k + 1}\right).$$

Thus,

$$s_k = \frac{1}{2}\left(\frac{1}{13} - \frac{1}{4k + 1}\right) \to \frac{1}{26} \quad \text{as} \quad k \to \infty$$

so that

$$\sum_{n=4}^{\infty} \frac{2}{(4n - 3)(4n + 1)} = \frac{1}{26}.$$

(c) $\displaystyle\sum_{n=0}^{\infty}\left(\frac{5}{3^n} - \frac{2}{7^n}\right) = \sum_{n=0}^{\infty}\frac{5}{3^n} - \sum_{n=0}^{\infty}\frac{2}{7^n} = \frac{5}{1 - \frac{1}{3}} - \frac{2}{1 - \frac{1}{7}} = \frac{31}{6}.$

(d) $\displaystyle\sum_{n=1}^{\infty} 127\left(\frac{1}{1000}\right)^n = \sum_{n=0}^{\infty} 127\left(\frac{1}{1000}\right)^n - 127 = \frac{127}{1 - \frac{1}{1000}} - 127$

$$= \frac{127{,}000 - 126{,}873}{999} = \frac{127}{999}.$$

3. $\displaystyle\frac{\sqrt{n}}{n^2 + 3} < \frac{\sqrt{n}}{n^2} = \frac{1}{n^{3/2}}$ so that $\displaystyle\sum_{n=1}^{\infty}\frac{\sqrt{n}}{n^2 + 3}$ converges by comparison with the convergent p-series for $p = \frac{3}{2}$.

4. Using the ratio test, $\displaystyle\lim_{n\to\infty}\frac{(n + 1)!\,3^{n+1}}{10^{n+1}} \cdot \frac{10^n}{n!\,3^n} = \lim_{n\to\infty}\frac{(n + 1)3}{10} = \infty$. Thus, $\displaystyle\sum_{n=1}^{\infty}\frac{n!\,3^n}{10^n}$ diverges by the ratio test.

5. $\displaystyle\lim_{n\to\infty}\sin\left(\frac{n\pi - 2}{3n}\right) = \lim_{n\to\infty}\sin\left(\frac{\pi}{3} - \frac{2}{3n}\right) = \sin\frac{\pi}{3} = \frac{\sqrt{3}}{2} \neq 0$, so the series $\displaystyle\sum_{n=1}^{\infty}\sin\left(\frac{n\pi - 2}{3n}\right)$ diverges by the nth-term test for divergence.

6. If $a_n = \left(\dfrac{n}{2n + 5}\right)^n$, then $\sqrt[n]{a_n} = \dfrac{n}{2n + 5} \to \dfrac{1}{2}$. Thus, the series $\displaystyle\sum_{n=1}^{\infty}\left(\frac{n}{2n + 5}\right)^n$ converges by the root test.

7. $\displaystyle\frac{1}{n + \sqrt{n}} > \frac{1}{n + n} = \frac{1}{2n}$ so that $\displaystyle\sum_{n=1}^{\infty}\frac{1}{n + \sqrt{n}}$ diverges by comparison to the divergent series $\displaystyle\sum_{n=1}^{\infty}\frac{1}{2n}$.

8. $\int_1^{\infty} \frac{\tan^{-1} x\,dx}{x^2+1} = \lim_{b\to\infty} \frac{1}{2}\left(\tan^{-1} x\right)^2\Big]_1^b = \lim_{b\to\infty} \frac{1}{2}\left(\tan^{-1} b\right)^2 - \frac{1}{2}\tan^{-1} 1$

$= \frac{1}{2}\left(\frac{\pi}{2}\right)^2 - \frac{1}{2}\left(\frac{\pi}{4}\right)$

Therefore, the improper integral converges, so the original series $\sum_{n=1}^{\infty} \frac{\tan^{-1} n}{n^2+1}$ converges by the integral test.

9. $\left|(-1)^{n+1} \frac{\sin n}{n^2+1}\right| \le \frac{1}{n^2+1}$, so the original series $\sum_{n=1}^{\infty} (-1)^{n+1} \frac{\sin n}{n^2+1}$ converges absolutely by the comparison test.

10. Since $\frac{1}{n^2} < \frac{1}{n+1} < \frac{1}{n}$, it follows that $\left(\frac{1}{\sqrt[n]{n}}\right)\left(\frac{1}{\sqrt[n]{n}}\right) < \frac{1}{\sqrt[n]{n+1}} < \frac{1}{\sqrt[n]{n}}$. Thus, $\lim_{n\to\infty} \frac{1}{\sqrt[n]{n+1}} = 1$ so the original series $\sum_{n=1}^{\infty} (-1)^{n+1} \frac{1}{(n+1)^{1/n}}$ diverges by the nth-term test.

11. $\lim_{n\to\infty} \frac{1}{(\ln n)^2} = 0$, and $\frac{1}{[\ln (n+1)]^2} < \frac{1}{(\ln n)^2}$ because $y = \ln x$ is an increasing function of x. Therefore the alternating series $\sum_{n=2}^{\infty} (-1)^n \frac{1}{(\ln n)^2}$ converges by the Alternating Series Test. However, since $\ln n < \sqrt{n}$ implies $\frac{1}{(\ln n)^2} > \frac{1}{n}$ if $n \ge 2$, the series of absolute values $\sum_{n=2}^{\infty} \frac{1}{(\ln n)^2}$ diverges by comparison with the divergent harmonic series. Therefore, $\sum_{n=2}^{\infty} (-1)^n \frac{1}{(\ln n)^2}$ is conditionally convergent.

12. $\lim_{n\to\infty} \frac{n+1}{7n-2} = \frac{1}{7}$ so that $\sum_{n=1}^{\infty} (-1)^{n+1} \frac{n+1}{7n-2}$ diverges by the nth-term test.

13. $\sum_{n=1}^{\infty} (-1)^{n+1} \frac{2^n}{3^n} \approx \frac{2}{3} - \frac{4}{9} + \frac{8}{27} - \frac{16}{81} + \frac{32}{243} \approx 0.4527$ with an error of magnitude less than $2^6/3^6 < 0.0878$. Since the sign of the first unused term is negative, the sum 0.4527 overestimates the value of the series. In fact, the given geometric series sums to 0.4.

14. Using the ratio test,

$$\lim_{n\to\infty} \frac{\ln (n + 1) |x|^{n+1}}{(n + 1)} \cdot \frac{n}{\ln n \; |x|^n}$$

$$= \lim_{n\to\infty} \frac{\ln (n + 1)}{\ln n} \cdot \frac{n + 1}{n} |x| = |x|.$$

Thus, the given power series converges absolutely for $|x| < 1$ and diverges for $|x| > 1$. Test the endpoints of the interval:

For $x = 1$, the power series is $\sum_{n=2}^{\infty} \frac{\ln n}{n}$. Now

$\int_2^{\infty} \frac{\ln x}{x} dx = \lim_{b\to\infty} \frac{1}{2}(\ln x)^2]_2^b = +\infty$ diverges, so the series

$\sum_{n=2}^{\infty} \frac{\ln n}{n}$ is divergent by the integral test.

For $x = -1$, the power series is $\sum_{n=2}^{\infty} \frac{(-1)^n \ln n}{n}$. Since

$0 \le \lim_{n\to\infty} \frac{\ln n}{n} \le \lim_{n\to\infty} \frac{\sqrt{n}}{n} = 0$, and $\frac{d}{dx}\left(\frac{\ln x}{x}\right) = \frac{1 - \ln x}{x^2} < 0$ for

$x \ge 3$ implies that $\frac{\ln (n + 1)}{n + 1} < \frac{\ln n}{n}$, the alternating series

$\sum_{n=2}^{\infty} \frac{(-1)^n \ln n}{n}$ converges by the Alternating Series Test.

Therefore, the power series $\sum_{n=2}^{\infty} \frac{\ln n}{n} x^n$ converges for all x

satisfying $-1 \le x < 1$.

15. We calculate the derivatives of $f(x) = \sqrt{x}$, and evaluate f and these derivatives at $a = 9$:

$f(x) = \sqrt{x}$ $\qquad f(9) = 3$

$f'(x) = \frac{1}{2}x^{-1/2}$ $\qquad f'(9) = \frac{1}{6}$

$f^{(2)}(x) = (-1)\left(\frac{1}{2}\right)\left(\frac{1}{2}\right)x^{-3/2}$ $\qquad f^{(2)}(9) = -\frac{1}{108}$

$f^{(3)}(x) = (-1)^2\left(\frac{1}{2}\right)\left(\frac{1}{2}\right)\left(\frac{3}{2}\right)x^{-5/2}$ $\qquad f^{(3)}(9) = \frac{1}{648}$

$f^{(4)}(x) = (-1)^3 \frac{3 \cdot 5}{2^4} x^{-7/2}$ $\qquad f^{(4)}(9) = -\frac{5}{11664}$

$\vdots$ $\qquad \vdots$

$f^{(k)}(x) = (-1)^{k+1} \frac{3 \cdot 5 \cdots (2k-3)}{2^k} x^{-(2k-1)/2}$

$f^{(k)}(9) = (-1)^{k+1} \frac{3 \cdot 5 \cdots (2k-3)}{2^k \, 3^{2k-1}}$

Therefore, the Taylor series for $f(x) = \sqrt{x}$ at $a = 9$ is

$$3 + \frac{1}{6}(x-9) - \frac{1}{216}(x-9)^2 + \dots + (-1)^{k+1}\frac{3\cdot 5\cdots(2k-3)}{2^k\, 3^{2k-1}\cdot k!}(x-9)^k + \dots$$

16. $\ln(1 + x) = x - \frac{x^2}{2} + \frac{x^3}{3} - \frac{x^4}{4} + \dots, \quad -1 < x \le 1$

$\ln(1 + x^2) = x^2 - \frac{x^4}{2} + \frac{x^6}{3} - \frac{x^8}{4} + \dots, \quad -1 < x \le 1$

$x\ln(1 + x^2) = x^3 - \frac{x^5}{2} + \frac{x^7}{3} - \frac{x^9}{4} + \dots, \quad -1 < x \le 1$

or, in closed form, $x\ln(1 + x^2) = \sum_{n=0}^{\infty} (-1)^n \frac{1}{n+1} x^{2n+3}$, valid for all x satisfying $-1 < x \le 1$.

17. $e^{-1/3} = 1 - \frac{1}{3} + \frac{(-1/3)^2}{2!} + \frac{(-1/3)^3}{3!} + \frac{(-1/3)^4}{4!} + \dots$

By trial, $\frac{(-1/3)^4}{4!} < 0.00052$ and $\frac{(1/3)^3}{3!} > 0.001$.

Since the series is an alternating series,

$e^{-1/3} \approx 1 - \frac{1}{3} + \frac{1/9}{2!} - \frac{1/27}{3!} = 0.71605$ with an error in magnitude less than 0.00052.

18. The Maclaurin series for $\tan x$, through the first three nonzero terms, is

$$\tan x = x + \frac{x^3}{3} + \frac{2x^5}{15} + \dots$$

Hence, $\sec^2 x = \frac{d}{dx}\tan x = 1 + x^2 + \frac{2}{3}x^4 + \dots$.

19. The Maclaurin series for $\cos x^2$ is

$$\cos x^2 = 1 - \frac{x^4}{2!} + \frac{x^8}{4!} - \frac{x^{12}}{6!} + \dots + (-1)^k \frac{x^{4k}}{(2k)!} + \dots,$$

obtained by substituting x^2 for x in the Maclaurin series for $\cos x$. Hence,

$$\int_0^{0.5} \cos x^2\,dx = x - \frac{x^5}{5\cdot 2!} + \frac{x^9}{9\cdot 4!} - \frac{x^{13}}{13\cdot 6!} + \dots\Big]_0^{0.5}$$

$$\approx 0.5 - 0.00313 + 0.0000090 - \dots = 0.49687,$$

with an error in magnitude less than 0.000009 because the series is an alternating series.

20. Let $y = \sum_{n=0}^{\infty} a_n x^n$ and the differential equation becomes

$$2\sum_{n=2}^{\infty} n(n-1)a_n x^{n-2} + x\sum_{n=1}^{\infty} n a_n x^{n-1} - 4\sum_{n=0}^{\infty} a_n x^n = 0 ,$$

or

$$\sum_{n=2}^{\infty} 2n(n-1)a_n x^{n-2} + \sum_{n=1}^{\infty} n a_n x^n - 4\sum_{n=0}^{\infty} a_n x^n = 0 .$$

Equating the coefficients of each power of x to zero gives:

x^0: $2\cdot 2(1)a_2 - 4a_0 = 0$ or $a_2 = a_0$

x^1: $2\cdot 3(2)a_3 + a_1 - 4a_1 = 0$ or $a_3 = \frac{1}{4}a_1$

x^2: $2\cdot 4(3)a_4 + 2a_2 - 4a_2 = 0$ or $a_4 = \frac{1}{12}a_2$

x^3: $2\cdot 5(4)a_5 + 3a_3 - 4a_3 = 0$ or $a_5 = \frac{1}{40}a_3$

x^4: $2\cdot 6(5)a_6 + 4a_4 - 4a_4 = 0$ or $a_6 = 0$

x^5: $2\cdot 7(6)a_7 + 5a_5 - 4a_5 = 0$ or $a_7 = -\frac{1}{84}a_5$

x^6: $2\cdot 8(7)a_8 + 6a_6 - 4a_6 = 0$ or $a_8 = 0$

Note: $a_{2k} = 0$ for $k \geq 3$.

x^{2k-1}: $2\cdot(2k+1)(2k)a_{2k+1} + (2k-1)a_{2k-1} - 4a_{2k-1} = 0$

$$a_{2k+1} = \frac{5-2k}{4k(2k+1)}a_{2k-1} , \quad k \geq 1 .$$

We can write the first few terms of the series solution:

$$y = a_0(1 + x^2 + \frac{1}{12}x^4) + a_1(x + \frac{1}{4}x^3 + \frac{1}{160}x^5 - \frac{1}{13,440}x^7 + \ldots)$$

NOTES.

CHAPTER 9 CONIC SECTIONS, PARAMETRIZED CURVES AND POLAR COORDINATES

9.1 CONIC SECTIONS AND QUADRATIC EQUATIONS

OBJECTIVE A: Given the coordinates for the vertex V and the focus F of a parabola, both of which lie along a line parallel to a coordinate axis, find an equation of the parabola and of its directrix. Sketch the graph showing the focus, vertex, and directrix.

1. Vertex at V(1,2) and focus at F(3,2):
The axis of symmetry of the parabola is the line containing both V and F, or the line with equation ________ . Since V is to the left of F on the axis of symmetry, the parabola will open to the ____________ . Thus, an equation of the parabola has the form __________ .
To calculate the number p, note that it is the distance between the vertex and the focus: thus, p = ______ . Therefore, an equation of the parabola is given by ________________ . The directrix is given by __________ . Graph the equation showing the focus, vertex, and directrix.

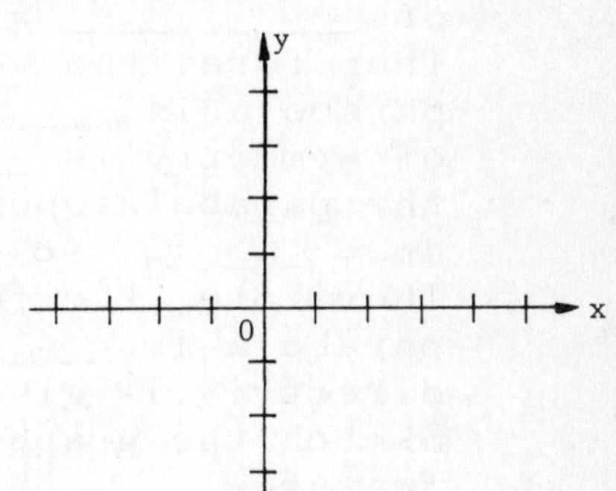

OBJECTIVE B: Given the coordinates for the vertex V, and the directrix L parallel to one of the coordinate axes, find an equation of the parabola so determined and give the coordinates of its focus. Sketch the graph showing the focus, vertex, and directrix.

---

1. $y = 2$, right, $(y - k)^2 = 4p(x - h)$,
   2, $(y - 2)^2 = 8(x - 1)$, $x = -1$

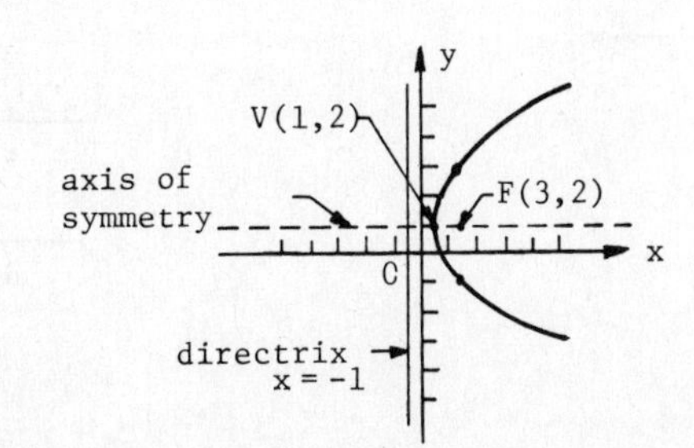

2. Vertex at $V(-1,2)$ and directrix L: $y = 3$
The axis of symmetry of the parabola is ________________ to the directrix. Since V is below the directrix, the parabola opens ________________ and so an equation of the parabola has the form ________________. The number p is the distance from the vertex to the directrix: thus, $p =$ ________. Therefore an equation of the parabola is given by ________________. The focus is located on the axis of symmetry p units from the vertex in the direction which the parabola opens. Thus the focus is ________. Graph the equation showing the focus, vertex, and directrix.

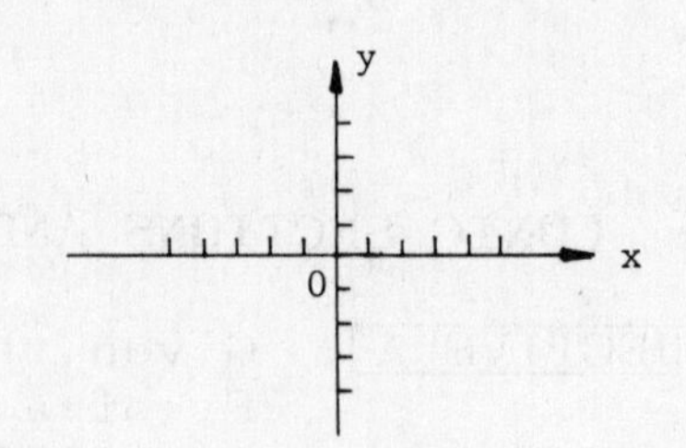

OBJECTIVE C: Given an equation of a parabola, find the vertex, axis of symmetry, focus, and directrix. Sketch the graph showing these features.

3. Consider the parabola $x^2 - 2x - 10y + 6 = 0$. Completing the square in the quadratic terms, ________________ $- 10y + 6 = 1$ or ________________ $= 10$ (________). Therefore, the vertex of the parabola is ________. The axis of symmetry is ________ because the parabola opens ________. $4p =$ ______, so that $p =$ ______. Therefore, the focus of the parabola is ________________, and the directrix is given by ________________. Sketch the graph showing these features.

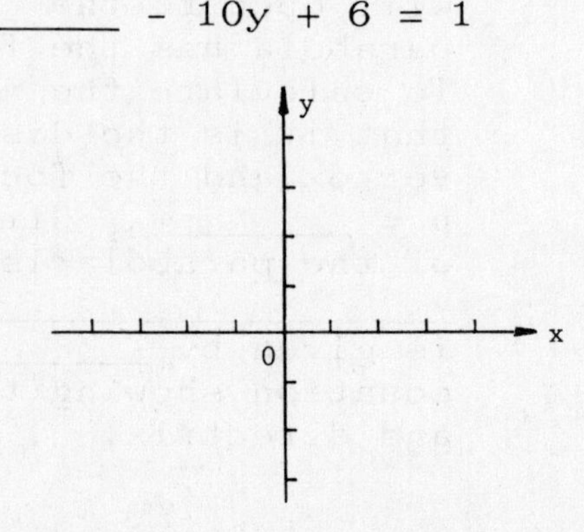

---

2. perpendicular, downward, $(x - h)^2 = -4p(y - k)$, 1, $(x + 1)^2 = -4(y - 2)$, $F(-1,1)$

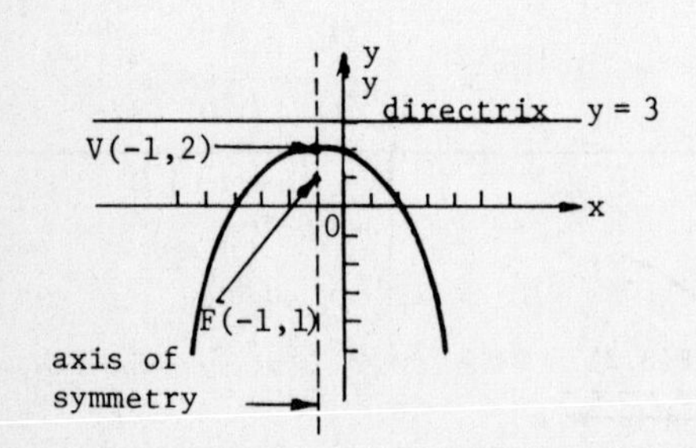

3. $x^2 - 2x + 1$, $(x - 1)^2$, $y - \frac{1}{2}$, $V\left(1, \frac{1}{2}\right)$, $x = 1$, upward, 10, $\frac{5}{2}$, $F(1,3)$, $y = -2$

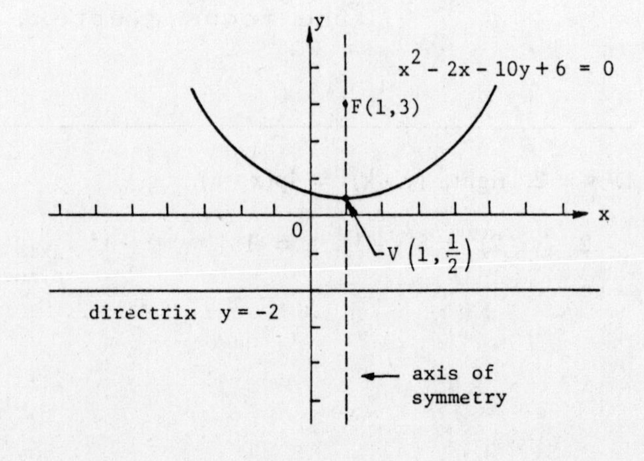

OBJECTIVE D: Given an equation of an ellipse, find its center, vertices, and foci. Identify the major and minor axes, and sketch the graph showing all these features.

4. In analyzing an ellipse, we find the intercepts are located on the axes of ______________ . The major axis is always in the direction of ______________ axis length, and the foci always lie on the ________ axis.

5. If we use the letters a, b, and c to represent the lengths of the semimajor axis, semiminor axis, and half-distance between the foci, then it is always true that $c^2 =$ ____________ .

6. Consider the ellipse given by $9x^2 + y^2 - 18x + 2y + 9 = 0$. Thus, $9(x^2 - 2x) + ($__________$) = -9$, so completing the squares in each set of parentheses gives $9(x - 1)^2 +$ __________ $= 1$, or $\frac{(x-1)^2}{1/9} +$ ____________ $= 1$. Hence, $c^2 = a^2 - b^2 =$ ____________, or $c =$ _________ . The center of the ellipse is ____________ and the foci are $F_1\left(1, -1 + \frac{2\sqrt{2}}{3}\right)$ and ______________ .

The length of the semimajor axis is $a = 1$ and the length of the semimajor axis is $b =$ _______ . The vertices are the four points $(1, 0)$, $\left(\frac{2}{3}, -1\right)$, _______, and _______ . The eccentricity of the ellipse is $e =$ _______ . Sketch the graph.

OBJECTIVE E: Find an equation of an ellipse having a given center C, focus F, and semimajor axis a.

7. Center at $C(6, -1)$, focus $F(6 + 3\sqrt{3}, -1)$, $a = 6$
The two foci are located along the major axis at a distance of c units from the center of the ellipse. Thus, $c =$ ________ .

---

4. symmetry, largest, major

5. $a^2 - b^2$

6. $y^2 + 2y$, $(y + 1)^2$, $\frac{(y + 1)^2}{1}$, $\frac{8}{9}$, $\frac{2\sqrt{2}}{3}$, $C(1,-1)$, $F_2\left(1, -1 - \frac{2\sqrt{2}}{3}\right)$, $\frac{1}{3}$, $(1,-2)$, $\left(\frac{4}{3}, -1\right)$, $\frac{2\sqrt{2}}{3}$

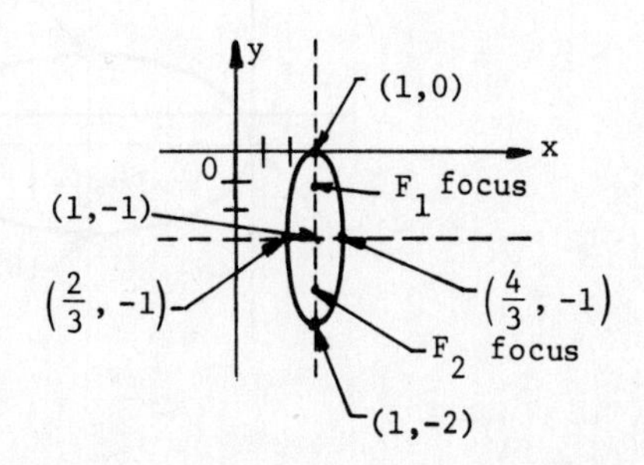

Since $b^2 = a^2 - c^2$, we find $b^2 = 36 -$ ______ so $b^2 =$ ______ . Thus, an equation of the ellipse is given by

______________________ .

The second focus of the ellipse is ______________ . The vertices are the points (6,2), (12,-1), ________, and ________ . Sketch the graph.

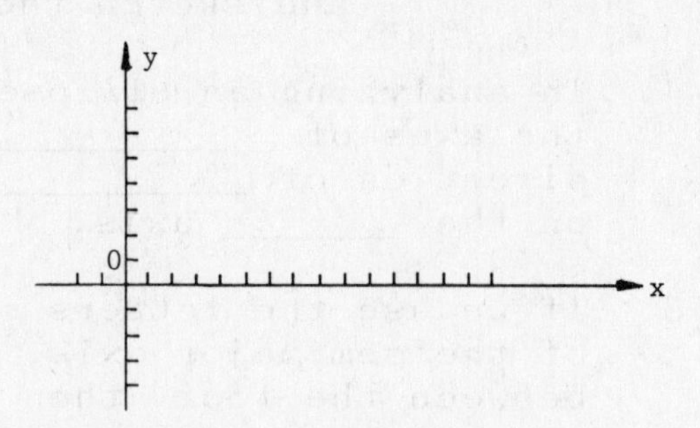

8. To find an equation of the ellipse with center C(-4,-2), $c = \sqrt{7}$, and one vertex of the major axis at (0,-2), we note first that the major axis is parallel to the ______ axis. The value of a is the distance between the points C(-4,-2) and ________, so a = ______ . Therefore, $b^2 = a^2 - c^2 =$ ______ - 7 = ______ . Thus, an equation of the ellipse is given by

______________________ .

OBJECTIVE F: Given an equation representing a hyperbola, find the center, vertices, foci, and asymptotes. Sketch a graph showing all these features.

9. The only differences between the equation of the ellipse and the equation of the hyperbola are the ____________ in the equation of the hyperbola, and the new relation among a, b, and c; namely, $a^2 - c^2 =$ ________ .

10. There is no restriction $a > b$ for the hyperbola as there is for the ellipse. The direction in which the hyperbola opens is controlled by the ________ rather than by the relative ________ of the coefficients of the quadratic terms.

---

7. $3\sqrt{3}$, 27, 9, $\frac{(x - 6)^2}{36} + \frac{(y + 1)^2}{9} = 1$,

$F_2(6 - 3\sqrt{3},-1)$, (6,-4), (0,-1)

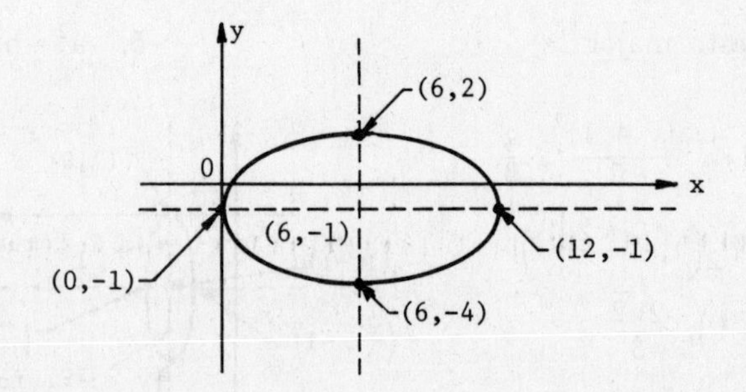

8. x, (0,-2), 4, 16, 9, $\frac{(x + 4)^2}{16} + \frac{(y + 2)^2}{9} = 1$

9. minus sign, $-b^2$

10. signs, sizes

11. To obtain the <u>asymptotes</u> for a hyperbola, set the expression ____________ or ____________ equal to zero (depending on how the hyperbola is defined), and then factor and solve the resulting quadratic equation.

12. Consider the hyperbola given by $\frac{(x - 2)^2}{4} - \frac{(y + 1)^2}{5} = 1$. Because of the minus sign associated with the y terms, the hyperbola opens __________ and __________ . The center of the hyperbola is $C(h,k) =$ __________ . Now, $a^2 =$ ______ and $b^2 =$ ______, so $c^2 = a^2 + b^2 =$ ________ . The coordinates of the foci are $F_1(h-c,k)$ and $F_2(h+c,k)$ or __________ and __________ . The coordinates of the vertices are $V_1(h-a,k)$ and $V_2(h+a,k)$ or __________ and __________ . Equations may be found for the asymptotes by setting $\frac{(x - 2)^2}{4} - \frac{(y + 1)^2}{5}$ equal to __________ . Thus, equations of the asymptotes are ____________________ . Sketch the graph of the hyperbola.

13. Consider the hyperbola $9x^2 - 4y^2 + 18x + 16y + 29 = 0$. Thus, $9(x^2 + 2x) - 4($____________$) = -29$. Completing the squares in each set of parentheses gives $9(x + 1)^2 - 4($________$) =$ $-29 + 9 -$ ________ $=$ ________, or $\frac{(y - 2)^2}{9} -$ __________ $= 1$.

---

11. $\frac{x^2}{a^2} - \frac{y^2}{b^2}$, $\frac{y^2}{a^2} - \frac{x^2}{b^2}$

12. right, left, C(2,-1), 4, 5, 9,

$F_1(-1,-1)$, $F_2(5,-1)$, $V_1(0,-1)$

$V_2(4,-1)$, zero, $y + 1 = \pm \frac{\sqrt{5}}{2}(x - 2)$,

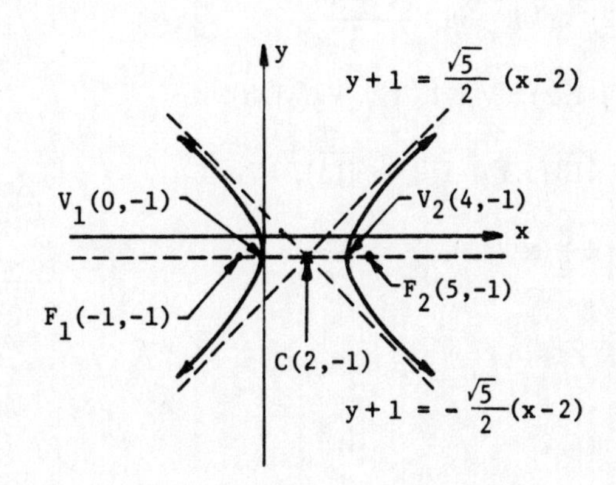

Because of the minus sign, the hyperbola opens _____ and _____ . The center of the hyperbola is $C(h,k) =$ _____ . The coordinates of the vertices are $V_1(h,k-a)$ and $V_2(h,k+a)$ or _____ and _____ . Now, $c^2 = a^2 + b^2 =$ _____, and the coordinates of the foci are $F_1(h,k-c)$ and $F_2(h,k+c)$ or _____ and _____ . Equations for the asymptotes may be found by setting the expression $\frac{(y-2)^2}{9} - \frac{(x+1)^2}{4}$ equal to _____ . Thus, equations of the asymptotes are _____ . Sketch the graph showing the features of the hyperbola.

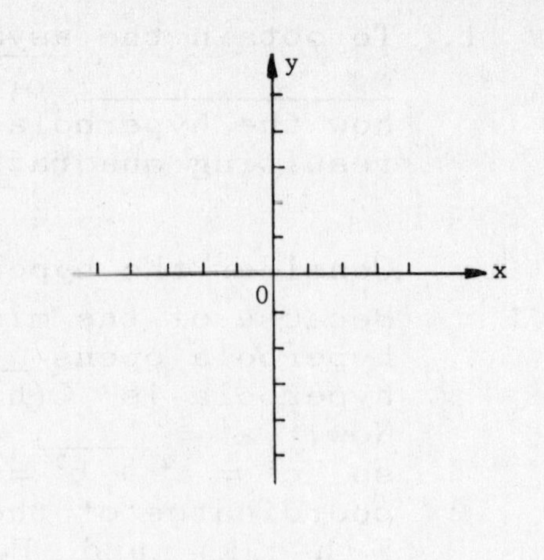

9.2 CLASSIFYING CONIC SECTIONS BY ECCENTRICITY

OBJECTIVE A: Given an equation of an ellipse $Ax^2 + Cy^2 = F$, where A, C and F are positive numbers, put the equation in standard form and find the ellipse's eccentricity. Sketch the ellipse and include the foci in your sketch.

14. The standard equation for an ellipse centered at the origin with foci on the x-axis is

_____ , $a > b$.

The foci are located at the points _____ where $c =$ _____ . The points $(\pm a, 0)$ are called the

---

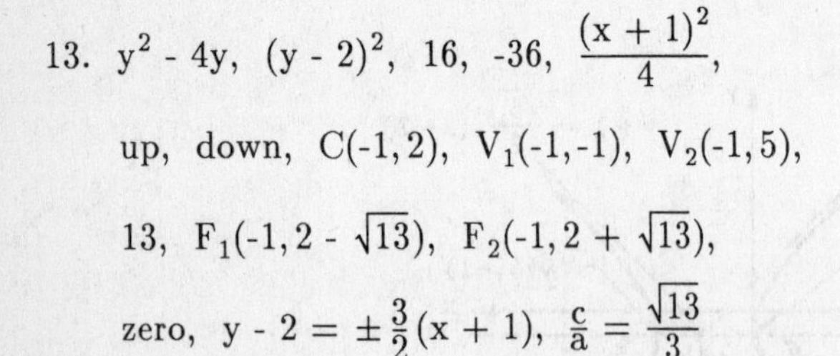

13. $y^2 - 4y$, $(y-2)^2$, 16, -36, $\frac{(x+1)^2}{4}$,

up, down, $C(-1,2)$, $V_1(-1,-1)$, $V_2(-1,5)$,

13, $F_1(-1,2-\sqrt{13})$, $F_2(-1,2+\sqrt{13})$,

zero, $y - 2 = \pm\frac{3}{2}(x+1)$, $\frac{c}{a} = \frac{\sqrt{13}}{3}$

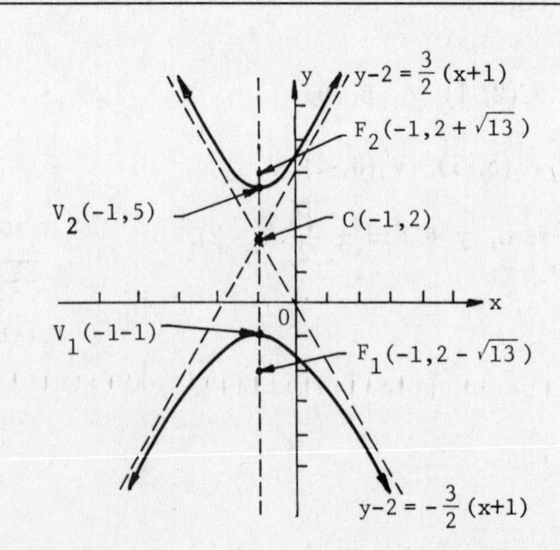

__________ of the ellipse, and a is the ______________ axis.

15. If the foci lie on the y-axis, the standard equation for an ellipse centered at the origin is

________________________, $a > b$.

The points $(0, \pm a)$ are the ___________ and the foci are ___________ where $c = \sqrt{a^2 - b^2}$.

16. The ratio

$$\frac{\text{Distance between foci}}{\text{Distance between vertices}} = \frac{c}{a}$$

is called the _______________________ of the ellipse. If $c = 0$ the ellipse takes the shape of a ___________ . As c increases, the ellipse tends to ______________, degenerating into a __________________ when $c = a$.

17. In an elliptical mirror, light is reflected from

________________________________ .

18. The equation $16x^2 + 5y^2 = 80$ represents an ellipse. In standard form it is written

____________________________ .

In this case $c =$ ______________ ≈ 3.32, and the foci are located at the points _____________ . The vertices are __________ . The eccentricity is

$e =$ __________ .

Sketch the ellipse in the coordinate system at the right, and include the foci in your sketch.

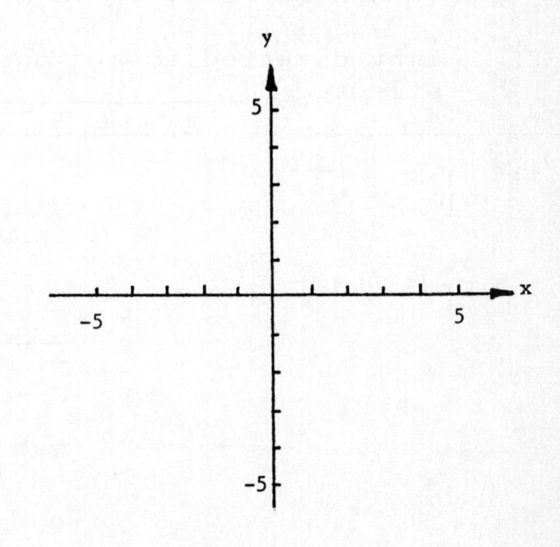

14. $\frac{x^2}{a^2} + \frac{y^2}{b^2} = 1$, $(\pm c, 0)$, $\sqrt{a^2 - b^2}$, vertices, semimajor

15. $\frac{x^2}{b^2} + \frac{y^2}{a^2} = 1$, vertices, $(0, \pm c)$

16. eccentricity, circle, flatten out, line segment

17. one focus to the other

OBJECTIVE B: Given an equation of a hyperbola $Ax^2 - Cy^2 = F$ or $Cy^2 - Ax^2 = F$, where A, C and F are positive numbers, put the equation in standard form and find the hyperbola's eccentricity and asymptotes. Sketch the hyperbola, including the asymptotes and foci in your sketch.

19. The standard equation for a hyperbola centered at the origin with foci on the x-axis is

_______________________ .

The foci are located at the points ________ where c = __________ . The points $(\pm a, 0)$ are called the ________ of the hyperbola. The asymptotes are straight lines passing through the origin with equations __________ . Note that in the case of the hyperbola it is not necessary for $a > b$ or $b > a$. It is also permissible for $a = b$.

20. If the foci lie on the y-axis, the standard equation for a hyperbola centered at the origin is

_______________________ .

The foci are located at the points $(0, \pm c)$ where c = ________ . The points ________ are the vertices, and the asymptotes are given by the equations __________ .

21. The eccentricity of a hyperbola is always given by

e = ________ .

For a hyperbola, it is always true that e ______ 1.

22. Light directed toward one focus of a hyperbolic mirror is reflected __________________ .

---

18. $\frac{x^2}{5} + \frac{y^2}{16} = 1$, $\sqrt{16 - 5}$,

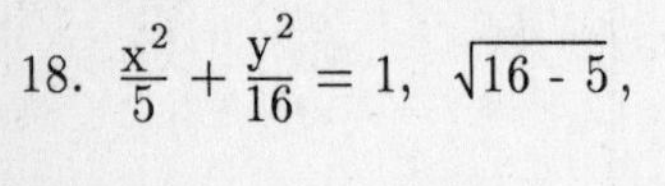

$(0, \pm\sqrt{11})$, $(0, \pm 4)$, $\frac{\sqrt{11}}{4}$

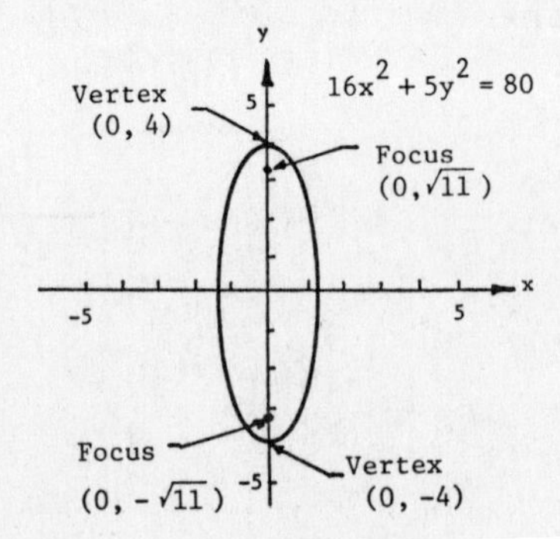

19. $\frac{x^2}{a^2} - \frac{y^2}{b^2} = 1$, $(\pm c, 0)$, $\sqrt{a^2 + b^2}$, vertices, $y = \pm \frac{b}{a}x$

20. $\frac{y^2}{a^2} - \frac{x^2}{b^2} = 1$, $\sqrt{a^2 + b^2}$, $(0, \pm a)$, $y = \pm \frac{a}{b}x$

21. $\frac{c}{a}$ or $\frac{\sqrt{a^2 + b^2}}{a}$, $e > 1$

22. toward the other focus

23. The equation $16y^2 - 5x^2 = 80$ represents a hyperbola. In standard form it is written

_______________________________ .

In this case $c =$ __________ ≈ 4.58, and the foci are located at the points __________ . The vertices are __________, and the asymptotes are given by __________ . The eccentricity is $e =$ ______ . Sketch the hyperbola in the coordinate system at the right, including the foci and asymptotes in your sketch.

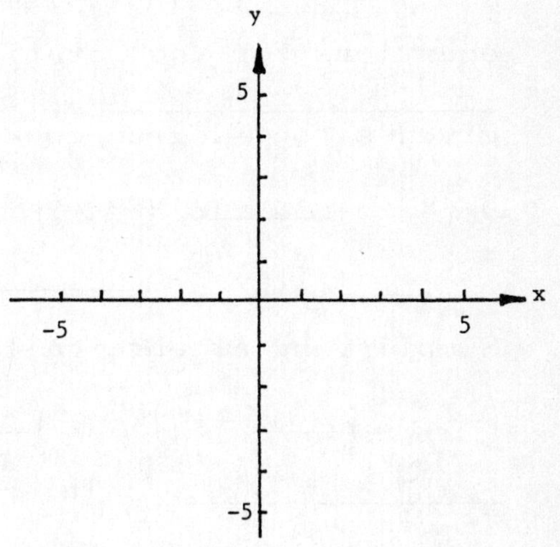

9.3 QUADRATIC EQUATIONS AND ROTATIONS

24. Any second degree equation in x and y represents a circle, parabola, ellipse, or hyperbola (although it may degenerate). To find the curve given its equation

$$Ax^2 + Bxy + Cy^2 + Dx + Ey + F = 0,$$

(1) First rotate axes, to force $B = 0$, through an angle α satisfying $\cot 2\alpha =$ ______________ .

(2) Next, ______________ axes by completing the squares (if necessary) to reduce the equation to a standard form.

---

23. $\frac{y^2}{5} - \frac{x^2}{16} = 1$, $\sqrt{5 + 16}$, $(0, \pm\sqrt{21})$,

$(0, \pm\sqrt{5})$, $y = \pm\frac{\sqrt{5}}{4}x$,

$\frac{\sqrt{21}}{\sqrt{5}} \approx 2.05$

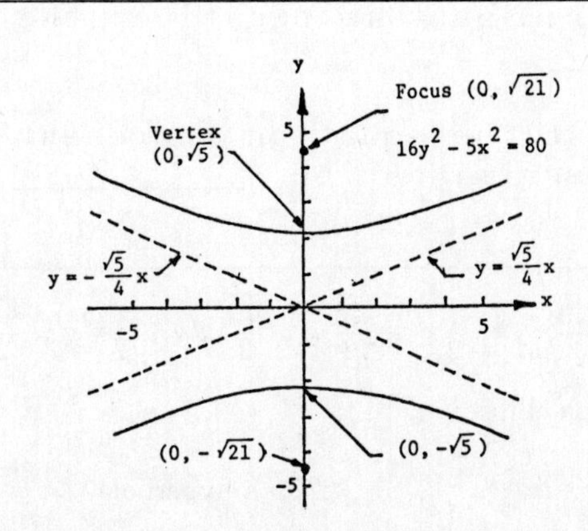

24. $\frac{A - C}{B}$, translate

OBJECTIVE A: Given an equation of the form $Ax^2 + Bxy + Cy^2 + F = 0$, transform the equation by a rotation of axes into an equation that has no cross-product term. Then identify the graph of the equation.

25. Consider the equation $5x^2 - 2\sqrt{3}xy + 7y^2 = 6$. The required angle of rotation to eliminate the xy term satisfies $\cot 2\alpha = \frac{5-7}{\_\_\_\_} = \_\_\_\_\_\_\_\_$; hence $2\alpha = \_\_\_\_$ radians, or $\alpha = \_\_\_\_\_\_\_\_$. Thus, $\sin \alpha = \_\_\_\_\_\_$ and $\cos \alpha = \_\_\_\_\_\_$. The equations for the rotation of axes are $x = \frac{\sqrt{3}}{2}x' - \frac{1}{2}y'$ and ________________ . Substitution for x and y into the given second-degree equation gives

$$5\left(\frac{3}{4}x'^2 - \_\_\_\_\_\_ + \frac{1}{4}y'^2\right) - 2\sqrt{3}\left(\frac{\sqrt{3}}{4}x'^2 + \_\_\_\_\_\_\_\_ - \frac{\sqrt{3}}{4}y'^2\right)$$

$$+ 7\left(\frac{1}{4}x'^2 + \_\_\_\_\_\_\_\_ + \frac{3}{4}y'^2\right) = 6 .$$

Simplifying algebraically,

$$\left(\frac{15}{4} - \_\_\_\_\_\_ + \frac{7}{4}\right)x'^2 + \left(\_\_\_ + \frac{3}{2} + \frac{21}{4}\right)y'^2 = 6, \quad \text{or}$$

____________ . This is an equation of ____________ .

OBJECTIVE B: Given a second degree equation of the form $Ax^2 + Bxy + Cy^2 + Dx + Ey + F = 0$, use the discriminant to classify it as representing a circle, an ellipse, a parabola, or a hyperbola.

26. The discriminant is the expression ________ . The discriminant is not changed by a rotation of axes.

27. If the discriminant is positive, the equation represents ____________ .

28. If the discriminant is zero, the equation represents ____________ .

29. If the discriminant is negative, the equation represents ____________ .

30. In order that the equation represent a circle, it is necessary that the discriminant be ________ and that __________ .

---

25. $-2\sqrt{3}$, $\frac{1}{\sqrt{3}}$, $\frac{\pi}{3}$, $\frac{\pi}{6}$, $\frac{1}{2}$, $\frac{\sqrt{3}}{2}$, $y = \frac{1}{2}x' + \frac{\sqrt{3}}{2}y'$, $\frac{\sqrt{3}}{2}x'y'$, $\frac{1}{2}x'y'$, $\frac{\sqrt{3}}{2}x'y'$, $\frac{3}{2}$, $\frac{5}{4}$, $4x'^2 + 8y'^2 = 6$, an ellipse

26. $B^2 - 4AC$, $A + C$

27. a hyperbola

28. a parabola

29. an ellipse

30. negative, $A = C$

31. Consider the equation given by

$$2x^2 - 4xy - y^2 + 20x - 2y + 17 = 0 .$$

The discriminant is $B^2 - 4AC = 16 - (\_\_\_\_\_\_) = \_\_\_\_\_\_$.
Thus, the equation represents ______________ .

32. For $4x^2 - 12xy + 9y^2 - 52x + 26y + 81 = 0$ the discriminant is $B^2 - 4AC = \_\_\_\_\_\_ - 4(36) = \_\_\_\_\_\_$. Thus, the equation represents ______________ .

9.4 PARAMETRIZATIONS OF CURVES

OBJECTIVE A: Given parametric equations $x = f(t)$ and $y = g(t)$ for the motion of a particle in the xy-plane, eliminate the parameter t to find a Cartesian equation for the particle's path. Graph the Cartesian equation.

33. Consider the curve given by the parametric equations $x = t - 2$, $y = 2t + 3$, $-\infty < t < \infty$. Complete the following table providing some of the points $P(x,y)$ on the curve:

t	-2	-1	0	1	2	3
x	-4					
y	-1					

To eliminate the parameter t, note that $t = x + 2$. Substitution for t in the parametric equation for y gives $y = \_\_\_\_\_\_$. This is a Cartesian equation for a ______ with slope $m = \_\_\_\_\_\_$ and y-intercept $b = \_\_\_\_\_\_$.

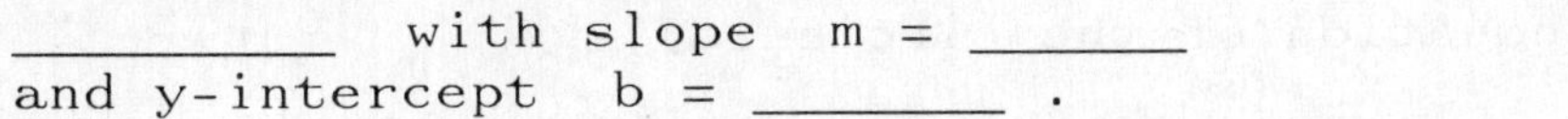

Sketch the curve in the coordinate system to the right.

---

31. -8, 24, a hyperbola

32. 144, 0, a parabola

33.

t	-2	-1	0	1	2	3
x	-4	-3	-2	-1	0	1
y	-1	1	3	5	7	9

2x + 7, line, 2, 7

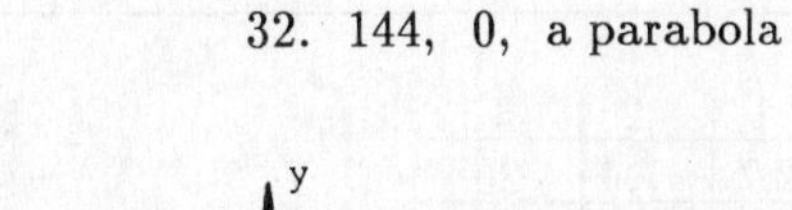

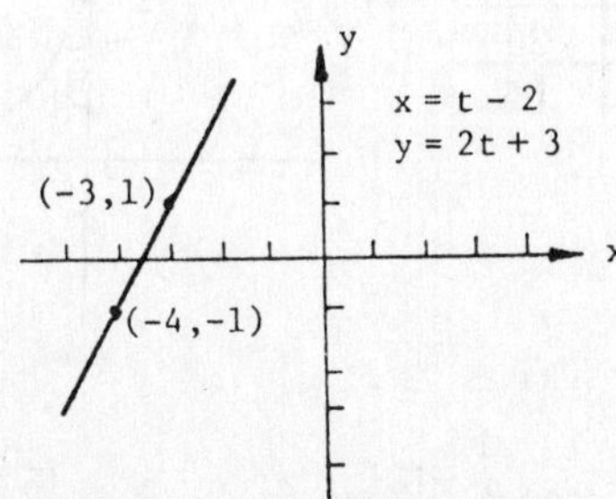

34. For the curve given by the parametric equations $x = e^t$ and $y = e^{-t}$, $-\infty < t < \infty$, complete the following table:

t	-2	-1	0	1	2	3
x						
y						

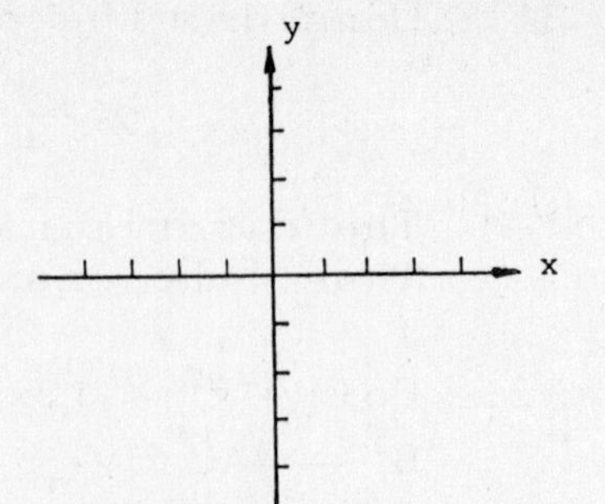

To eliminate the parameter t, notice that $xy =$ _______ . This equation describes a _____________ . Sketch the graph in the coordinate system at the right.

35. For the curve given parametrically in Problem 34, notice that x and y are always positive. Are the parametric equations and the cartesian equation coextensive? ______ , because x and y can both be ___________ in the cartesian equation $xy = 1$.

OBJECTIVE B: Find parametric equations for a curve described geometrically, or by an equation, in terms of some specified or arbitrary parameter.

36. Find parametric equations for the circle with center $C(-2,3)$ and radius $r = \sqrt{2}$.

Solution. An equation of the circle is $(x + 2)^2 +$ ________ = ______, or ______________________ = 1.

This suggests the substitutions $\frac{x + 2}{\sqrt{2}} = \sin\theta$ and $\frac{y - 3}{\sqrt{2}} =$ ________ . Hence, parametric equations for the circle are $x =$ _________ and $y =$ _________ , $0 \le \theta \le 2\pi$.

---

34. 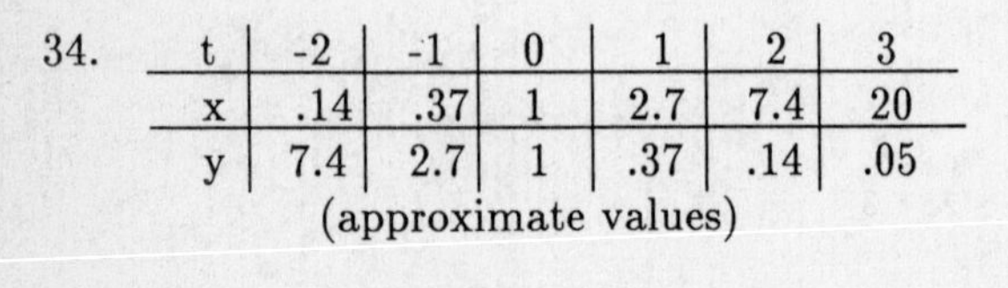

t	-2	-1	0	1	2	3
x	.14	.37	1	2.7	7.4	20
y	7.4	2.7	1	.37	.14	.05

(approximate values)

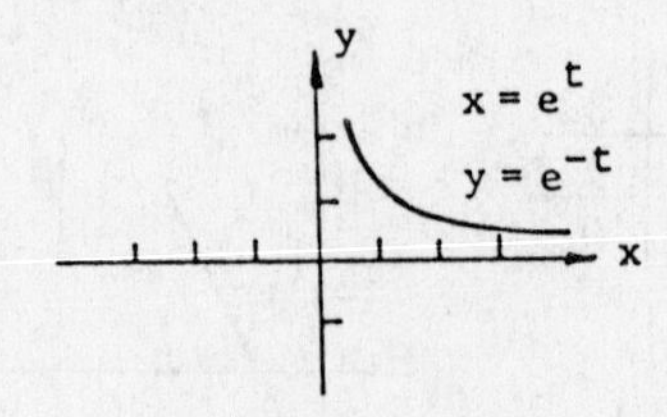

$xy = 1$, hyperbola

35. No, negative

36. $(y - 3)^2$, 2, $\left(\frac{x + 2}{\sqrt{2}}\right)^2 + \left(\frac{y - 3}{\sqrt{2}}\right)^2$, $\cos\theta$, $\sqrt{2}\sin\theta - 2$, $3 + \sqrt{2}\cos\theta$

37. Find parametric equations for the line in the plane through the point (a, b) with slope m, where the parameter t is the change $x - a$.

 Solution. For any point $P(x, y)$ on the line, $y - b = m(\_\_\_\_\_\_) = \_\_\_\_\_\_$. Thus, $x = \_\_\_\_\_$ and $y = \_\_\_\_\_$ give parametric equations of the line in terms of the specified parameter t.

9.5 CALCULUS WITH PARAMETRIZED CURVES

OBJECTIVE A: Given parametric equations $x = f(t)$ and $y = g(t)$, find $\frac{dy}{dx}$ in terms of $\frac{dy}{dt}$ and $\frac{dx}{dt}$. Find $\frac{d^2y}{dx^2}$ in terms of t.

38. The equations $x = f(t)$ and $y = g(t)$, which express x and y in terms of t, are called ________ equations. The variable t is called a ________. From the chain rule, the derivative $\frac{dy}{dx}$ is given by $\frac{dy}{dx} = \_\_\_\_\_\_\_\_$.

39. Let $x = t^2 - 1$ and $y = \frac{1}{t}$. Then $\frac{dx}{dt} = \_\_\_\_\_\_$ and $\frac{dy}{dt} = \_\_\_\_\_\_$. It follows that

$$y' = \frac{dy}{dx} = \_\_\_\_\_\_\_\_ .$$

 To calculate the second derivative $\frac{d^2y}{dx^2}$ we first find

$$\frac{dy'}{dt} = \_\_\_\_\_\_\_\_ .$$

 Then,

$$\frac{d^2y}{dx^2} = \frac{dy'/dt}{\_\_\_\_} = \_\_\_\_\_\_ .$$

40. If $x = t^2$ and $y = t^2 - 2t$, then $\frac{dx}{dt} = \_\_\_\_\_\_$ and $\frac{dy}{dt} = \_\_\_\_\_\_$. Thus,

$$y' = \frac{dy}{dx} = \_\_\_\_\_\_ .$$

 Next, $\frac{dy'}{dt} = \_\_\_\_\_\_$ so that

---

37. $x - a$, mt, $a + t$, $b + mt$

38. parametric, parameter, $\frac{dy/dt}{dx/dt}$

39. $2t$, $-\frac{1}{t^2}$, $-\frac{1}{2t^3}$, $\frac{3}{2}t^{-4}$, $\frac{dx}{dt}$, $\frac{3}{4t^5}$

$$\frac{d^2y}{dx^2} = \frac{\rule{3em}{0.4pt}}{dx/dt} = \frac{\rule{3em}{0.4pt}}{2t} = \rule{5em}{0.4pt} .$$

When $t = 2$, $x =$ ______, $y =$ ______ and $\frac{dy}{dx} =$ ______. Thus an equation of the line tangent to the curve at $(4,0)$ is

______________________________ .

OBJECTIVE B: Find the length of a smooth curve specified parametrically by continuously differentiable equations $x = f(t)$ and $y = g(t)$ over a given interval $a \le t \le b$.

41. To compute the length of the curve given by $x = t^2 \cos t$ and $y = t^2 \sin t$ for $0 \le t \le 1$, we first calculate $\frac{dx}{dt}$ and $\frac{dy}{dt}$.

$\frac{dx}{dt} =$ ______________________; $\frac{dy}{dt} =$ ______________________ so that

$$\left(\frac{dx}{dt}\right)^2 + \left(\frac{dy}{dt}\right)^2 = t^2[(2 \cos t - t \sin t)^2 + (\rule{6em}{0.4pt})^2]$$

$$= t^2[4 \cos^2 t - 4t \cos t \sin t + t^2 \sin^2 t$$

$$+ (\rule{10em}{0.4pt})]$$

$$= t^2[4 + \rule{5em}{0.4pt}].$$

Hence the arc length is given by,

$$s = \int_0^1 \sqrt{\left(\frac{dx}{dt}\right)^2 + \left(\frac{dy}{dt}\right)^2}\, dt = \int_0^1 \rule{6em}{0.4pt}\, dt$$

$$= \rule{6em}{0.4pt}\Big]_0^1 = \frac{1}{3}(\rule{5em}{0.4pt}) \approx 1.0601 \text{ units.}$$

OBJECTIVE C: Find the area of a surface generated by rotating the arc of a smooth curve specified parametrically by equations $x = f(t)$ and $y = g(t)$ over $a \le t \le b$ about an indicated axis. (Again, f and g are assumed to have continuous derivatives.)

42. Find the surface area generated when the arc $x = 2t$ and $y = \sqrt{2}t^2$ from $t = 0$ to $t = 2$ is rotated about the y-axis.
Solution. We write $dS = 2\pi$ ____ ds, since the rotation occurs about the y-axis. Next, we calculate the derivatives $\frac{dx}{dt} =$ ______ and $\frac{dy}{dt} =$ ______ so that the arc length differential is given by

---

40. $2t$, $2t - 2$, $1 - \frac{1}{t}$, $\frac{dy'}{dt}$, $\frac{1}{t^2}$, $\frac{1}{2t^3}$, 4, 0, $\frac{1}{2}$, $y = \frac{1}{2}(x - 4)$ or $2y - x + 4 = 0$

41. $2t \cos t - t^2 \sin t$, $2t \sin t + t^2 \cos t$, $2 \sin t + t \cos t$, $4 \sin^2 t + 4t \cos t \sin t + t^2 \cos^2 t$, t^2, $t\sqrt{4 + t^2}$, $\frac{1}{3}\left(4 + t^2\right)^{3/2}$, $5\sqrt{5} - 8$

$$ds = \sqrt{\_\_\_\_\_\_\_\_\_\_}\ dt = 2\sqrt{\_\_\_\_\_\_\_\_\_\_}\ dt.$$

Therefore, the surface area is

$$S = 2\pi \int_0^2 \_\_\_\_\_\_\_\_\_\_\ dt = \frac{4\pi}{3}\ (\_\_\_\_\_\_\_\_\_\_)\Big]_0^2$$

$$= \frac{4\pi}{3}\ (\_\_\_\_\_\_\_\_) = \_\_\_\_\_\_\_\_\_\_\_\_ \approx 108.90854 \text{ sq. units.}$$

9.6 POLAR COORDINATES

OBJECTIVE A: Given a point P in polar coordinates (r,θ), give the Cartesian coordinates (x,y) of P.

43. The polar and Cartesian coordinates are related by the equations $x =$ __________ and $y =$ __________ .

44. If P is the point $(-2, \frac{\pi}{6})$ in polar coordinates, then $x =$ __________ and $y =$ __________ so that P can be expressed in Cartesian coordinates by (____, ____).

45. For $P = (-2, -\frac{\pi}{6})$ in polar coordinates, $x =$ __________ and $y =$ __________ so that P = (____, ____) in Cartesian coordinates.

OBJECTIVE B: Graph the points $P(r,\theta)$ whose polar coordinates satisfy a given equation, inequality or inequalities.

46. $\theta = -\frac{\pi}{4}$, $-2 \le r$
Sketch the graph at the right.

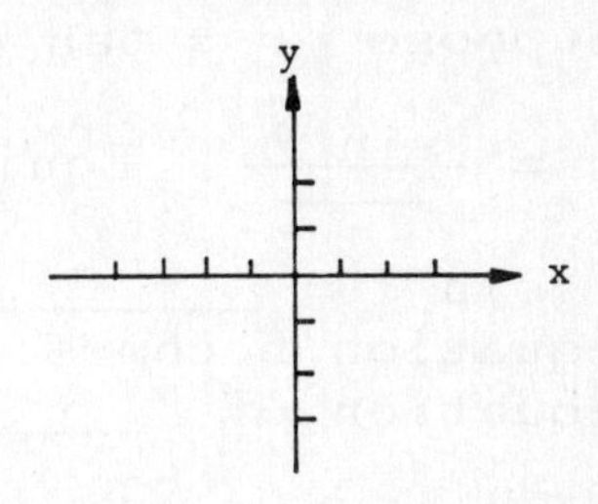

---

42. x, 2, $2\sqrt{2}\,t$, $\left(\frac{dx}{dt}\right)^2 + \left(\frac{dy}{dt}\right)^2$, $2t^2 + 1$, $4t\sqrt{2t^2+1}$, $\left(2t^2+1\right)^{3/2}$, 27 - 1, $\frac{104\pi}{3}$

43. $r\cos\theta$, $r\sin\theta$

44. $-2\cos\frac{\pi}{6}$, $-2\sin\frac{\pi}{6}$, $\left(-\sqrt{3}, -1\right)$

45. $-2\cos\left(-\frac{\pi}{6}\right)$, $-2\sin\left(-\frac{\pi}{6}\right)$, $(-\sqrt{3}, 1)$

46.

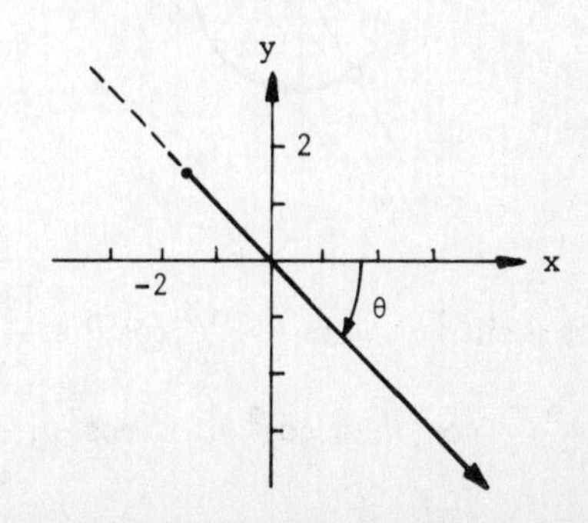

47. $r = 2, \quad -\frac{3\pi}{4} < \theta \le \frac{\pi}{6}$
Sketch the graph at the right.

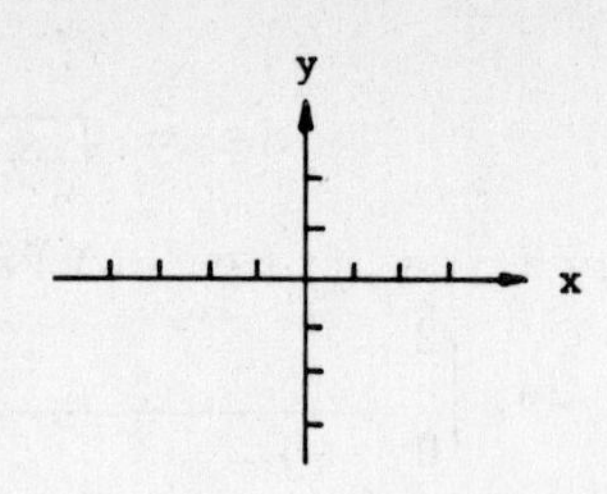

OBJECTIVE C: Given an equation in polar coordinates, replace it by an equivalent equation in Cartesian coordinates and identify the graph.

48. Consider the equation $r = -3 \sec \theta$. Then, $-3 = r$ ________, or since $x =$ ________ the equation is equivalent to ________ . This is an equation of a ________ line 3 units to the left of the ________ axis.

49. Consider the equation $r \sin\left(\theta - \frac{\pi}{3}\right) = \frac{1}{2}$. By the trigonometric summation identities, $\sin\left(\theta - \frac{\pi}{3}\right) = \sin\theta \cos\frac{\pi}{3} -$ ________ $=$ ________ .
Therefore, the polar equation can be written
$$\frac{1}{2} = \frac{1}{2} r \sin\theta - \text{\_\_\_\_\_\_\_\_} = \frac{1}{2} y - \text{\_\_\_\_\_\_\_\_} .$$
Simplifying algebraically, $y =$ ________ . This is an equation of a line with slope $m =$ ________ and y-intercept $b =$ ________ .

50. Suppose $r = \tan\theta \sec\theta$. Then, $r = \sin\theta \cdot$ ________ or $r = \frac{\sin\theta}{\_\_\_\_}$. Equivalently, $\sin\theta =$ ________ or $r \sin\theta =$ ________ . In terms of Cartesian coordinates the equation becomes ________, which is readily recognized as an equation of ________ .

47.

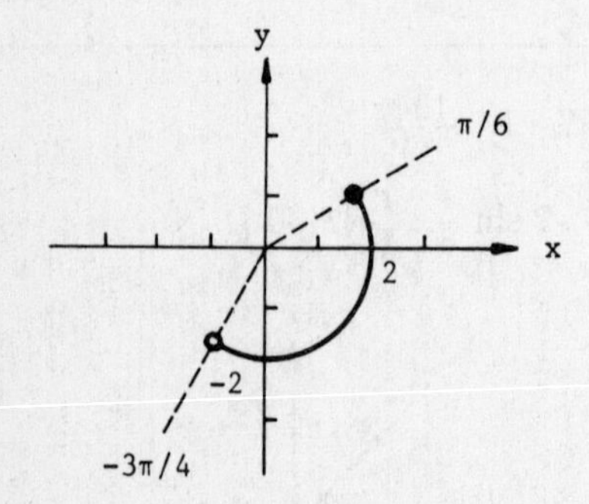

48. $\cos\theta$, $r\cos\theta$, $x = -3$, vertical, y

49. $\cos\theta \sin\frac{\pi}{3}$, $\frac{1}{2}\sin\theta - \frac{\sqrt{3}}{2}\cos\theta$, $\frac{\sqrt{3}}{2} r\cos\theta$, $\frac{\sqrt{3}}{2} x$, $\sqrt{3}x + 1$, $\sqrt{3}$, 1

50. $\sec^2\theta$, $\cos^2\theta$, $r\cos^2\theta$, $r^2\cos^2\theta$, $y = x^2$, a parabola

51. Given the equation $r = \dfrac{1}{3\cos\theta + 2\sin\theta}$, clear fractions and obtain $3r\cos\theta +$ ________ $= 1$. Next, substitute $x = r\cos\theta$ and $y =$ ________ to obtain $3x +$ ______ $= 1$. This is an equation of a straight line with slope $m =$ ______ and y-intercept $b =$ ______ .

52. The equation $r^2 - 5r + 4 = 0$ factors into $(r-4)($______$) = 0$. Thus, $r = 4$ or $r =$ ______ . The graph is two concentric circles, one of radius 4 and the other of radius 1, centered at the origin.

53. Consider $r^2 = 2\csc 2\theta$. Then, $r^2\sin 2\theta =$ ______ . Now, $\sin 2\theta = 2$ ________, so the equation becomes $2r\sin\theta \cdot$ ________ $= 2$ or ________ $= 2$. That is, $xy = 1$ which is an equation of ________ with center ________ .

54. Given the equation $r = \dfrac{3}{1 - 2\cos\theta}$, clear fractions to obtain $r -$ ______ $= 3$; or substituting $x = r\cos\theta$, $r =$ ________ . Hence, $r^2 =$ ________ $= 9 + 12x +$ ______ . Since $r^2 = x^2 + y^2$ this last equation simplifies to $y^2 = 9 + 12x +$ ______ or $y^2 = 3(x + 2)^2 + ($______$)$. Therefore, $(x + 2)^2 -$ ______ $= 1$. This is an equation of a hyperbola.

55. For $r = 2\sin(\theta + \frac{\pi}{4})$ we can expand the right side by the summation formula for the sine: $r = 2\sin\theta\cos\frac{\pi}{4} +$ ________ or $r =$ ________ . Hence, $r^2 = \sqrt{2}\,r\sin\theta +$ ________ . Since $x^2 + y^2 = r^2$, $x = r\cos\theta$, and $y = r\sin\theta$, substitution and algebraic simplification yields

$\left(x - \frac{\sqrt{2}}{2}\right)^2 +$ ________ $= 1$. This equation represents ________ with center ________ and radius $r = 1$.

9.7 GRAPHING IN POLAR COORDINATES

OBJECTIVE A: Given an equation $F(r, \theta) = 0$ in polar coordinates, analyze and sketch its graph.

---

51. $2r\sin\theta$, $r\sin\theta$, $2y$, $-\frac{3}{2}$, $\frac{1}{2}$

52. $r - 1$, 1

53. 2, $\sin\theta\cos\theta$, $r\cos\theta$, $2xy$, a hyperbola, $C(0,0)$

54. $2r\cos\theta$, $3 + 2x$, $(3 + 2x)^2$, $4x^2$, $3x^2$, -3, $\frac{y^2}{3}$

55. $2\cos\theta\sin\frac{\pi}{4}$, $\sqrt{2}\sin\theta + \sqrt{2}\cos\theta$, $\sqrt{2}\,r\cos\theta$, $\left(y - \frac{\sqrt{2}}{2}\right)^2$, a circle, $C\left(\frac{\sqrt{2}}{2}, \frac{\sqrt{2}}{2}\right)$

56. The graph of $F(r,\theta) = 0$ is symmetric about the x-axis if the equation is unchanged when θ is replaced by ______ or r is replaced by $-r$ and θ is replaced by ______ .

57. The graph of $F(r,\theta) = 0$ is symmetric about the origin if the equation is unchanged when r is replaced by ______ or θ is replaced by ______ .

58. The graph of $F(r,\theta) = 0$ is symmetric about the y-axis if the equation is unchanged when θ is replaced by ______, or r is replaced by ______ and θ is replaced by ______ .

59. Consider the curve given by $r = 1 - 2\cos\theta$. Since $\cos(-\theta) = \cos\theta$, the curve is symmetric about the ______ . Next, $\frac{dr}{d\theta}$ = ______ . Thus, as θ varies from 0 to $\frac{\pi}{3}$, r increases from r = ______ to r = ______; and as θ varies from $\frac{\pi}{2}$ to π, r increases from r = ______ to r = ______ . Complete the following table of values for the curve, and sketch its graph using its symmetries.

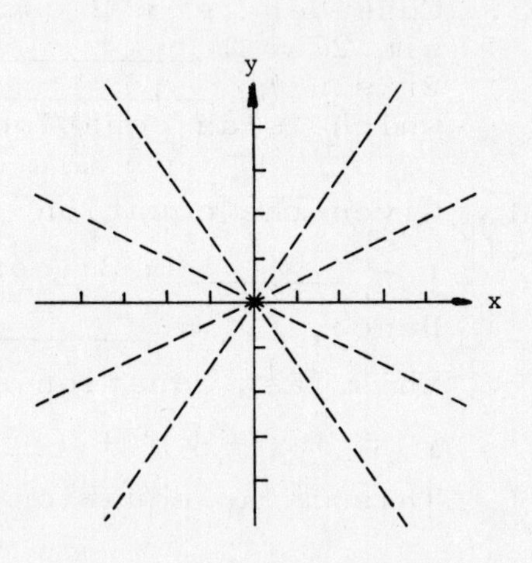

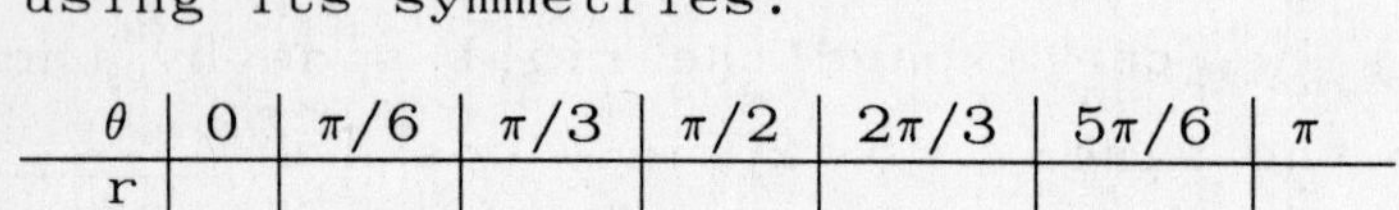

θ	0	$\pi/6$	$\pi/3$	$\pi/2$	$2\pi/3$	$5\pi/6$	π
r							

---

56. $-\theta$, $\pi - \theta$

57. $-r$, $\theta + \pi$

58. $\pi - \theta$, $-r$, $-\theta$

59. x-axis, $2\sin\theta$, -1, 0, 1, 3

θ	0	$\pi/6$	$\pi/3$	$\pi/2$	$2\pi/3$	$5\pi/6$	π
r	-1	$1 - \sqrt{3}$	0	1	2	$1 + \sqrt{3}$	3

The dashed portion of the curve is the rest of it due to its symmetry about the x-axis.

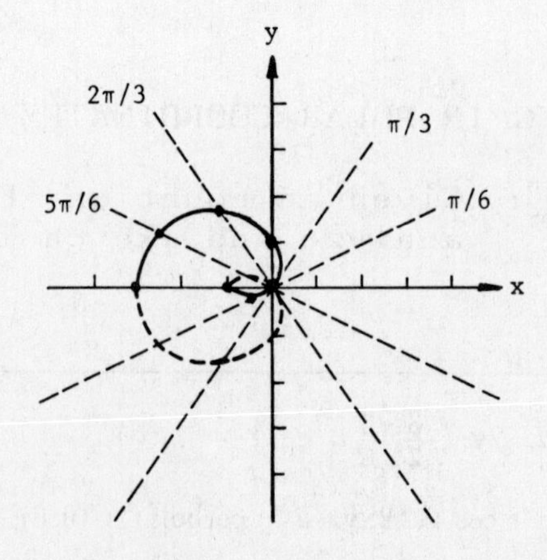

60. Consider the curve given by $r^2 = \sin\theta$. Since the $\sin\theta$ must be nonnegative in order to equal the square of a real number, we must restrict θ to the interval ____________ . Since the equation remains unchanged when r is replaced by $-r$, the curve is symmetric about the ____________ . Also, since $\sin(\pi - \theta) = \sin\theta$, the curve is symmetric about the ____________ . Complete the following table and sketch the graph using these symmetries:

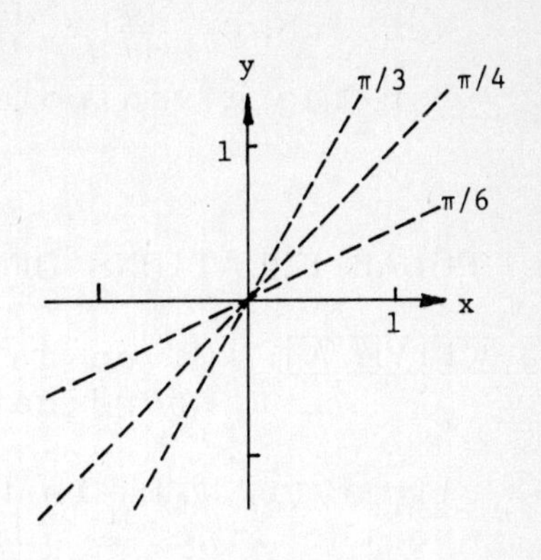

θ	0	$\pi/6$	$\pi/4$	$\pi/3$	$\pi/2$
r^2					
r					

OBJECTIVE B: If $r = f(\theta)$ is differentiable, find the slope $\frac{dy}{dx}$ at the point (r, θ) on the graph of f.

61. If $r = f(\theta)$ is differentiable and $\frac{dx}{d\theta} \neq 0$, then

slope at (r, θ) = ____________________ .

62. For the polar curve $r = 1 - 2\cos\theta$ in Problem 59,

r' = ________ . When $\theta = \frac{2\pi}{3}$ radians, r = ________ and r' = ________ . Thus,

$$\text{Slope at } \left(2, \frac{2\pi}{3}\right) = \frac{\sqrt{3}\sin\left(\frac{2\pi}{3}\right) + 2\cos\left(\frac{2\pi}{3}\right)}{\_\_\_\_\_\_\_\_\_\_}$$

$$= \frac{1/2}{\_\_\_\_} = \_\_\_\_\_\_\_\_$$

$$\approx -0.192 .$$

---

60. $0 \le \theta \le \pi$, origin, y-axis

θ	0	$\pi/6$	$\pi/4$	$\pi/3$	$\pi/2$
r^2	0	1/2	$1/\sqrt{2}$	$\sqrt{3}/2$	1
r	0	$\pm.71$	$\pm.84$	$\pm.93$	±1

The portions of the graph in QII, QIII, and QIV are obtained by the symmetries.

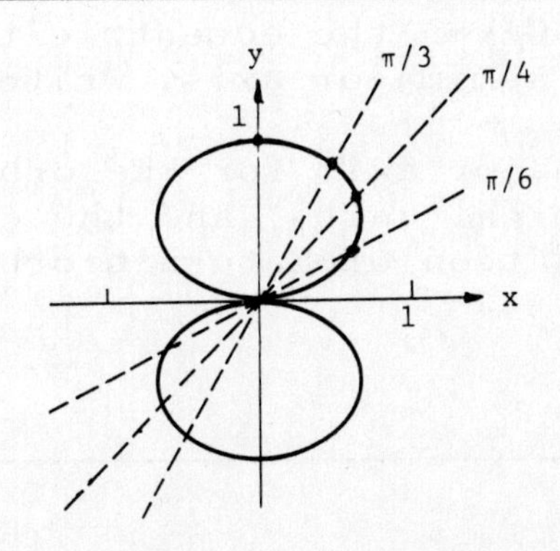

61. $\frac{r'\sin\theta + r\cos\theta}{r'\cos\theta - r\sin\theta}$ where $r' = \frac{dr}{d\theta} = f'(\theta)$

62. $2\sin\theta$, 2, $\sqrt{3}$, $\sqrt{3}\cos\left(\frac{2\pi}{3}\right) - 2\sin\left(\frac{2\pi}{3}\right)$, $\frac{-3\sqrt{3}}{2}$, $\frac{-1}{3\sqrt{3}}$

Note that $\frac{dx}{d\theta} = \frac{dr}{d\theta}\cos\theta - r\sin\theta = \frac{-3\sqrt{3}}{2} \neq 0$ when $\theta = \frac{2\pi}{3}$ for the given polar curve.

9.8 POLAR EQUATIONS OF CONIC SECTIONS

OBJECTIVE A: Given an equation of a straight line in polar coordinates, find a corresponding Cartesian equation.

63. If $P_0(r_0, \theta_0)$ is the foot of the perpendicular from the origin to the line L with $r_0 \geq 0$, then an equation for L in polar coordinates is

______________________________ .

64. A polar equation for the line displayed in the figure at the right is

____________________ .

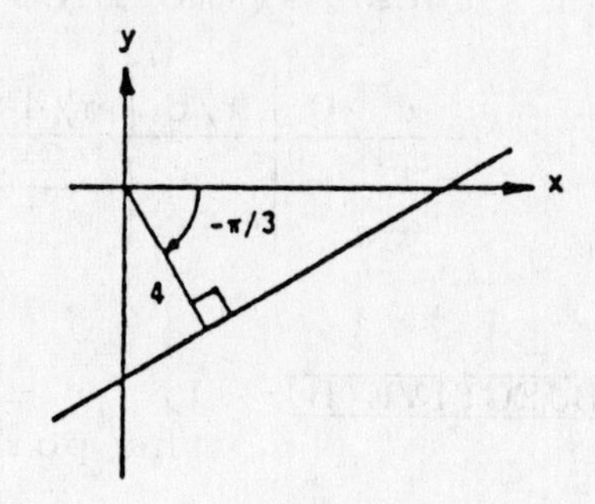

65. Using the identity for $\cos(A + B)$, we find the following Cartesian equation for the line in Problem 64:

$$r\cos\left(\theta + \frac{\pi}{3}\right) = 4$$

$$r[\_\_\_\_\_\_\_\_\_\_] = 4$$

$$\frac{1}{2}r\cos\theta - \_\_\_\_\_\_\_\_ = 4$$

$$\_\_\_\_\_\_\_\_\_\_ = 4$$

or

$$y = \_\_\_\_\_\_\_\_\_\_ .$$

OBJECTIVE B: Given the eccentricity of an ellipse together with its semimajor axis, write a polar equation for the ellipse.

66. The semimajor axis for the orbit of Saturn is 9.539 AU (astronomical units) and the eccentricity is 0.0543. Thus a polar equation of Saturn's orbit around the sun is

---

63. $r\cos(\theta - \theta_0) = r_0$

64. $r\cos\left(\theta + \frac{\pi}{3}\right) = 4$

65. $\cos\theta\cos\frac{\pi}{3} - \sin\theta\sin\frac{\pi}{3}$, $\frac{\sqrt{3}}{2}r\sin\theta$, $\frac{1}{2}x - \frac{\sqrt{3}}{2}y$, $\frac{1}{\sqrt{3}}(x - 8)$

$$r = \frac{\_\_\_\_\_\_[1 - (0.0543)^2]}{\_\_\_\_\_\_\_\_\_\_\_\_}$$

$$= \frac{9.511}{1 + 0.0543 \cos\theta}$$

$$= \frac{\_\_\_\_\_\_}{18.42 + \cos\theta} .$$

At its most distant point (aphelion), Saturn is ____________ (astronomical units) from the sun.

9.9 INTEGRATION IN POLAR COORDINATES

OBJECTIVE A: Find the total plane area enclosed by a polar graph $r = f(\theta)$ and the rays $\theta = \alpha$, $\theta = \beta$.

67. The area bounded by the polar curve $r = f(\theta)$ and the rays $\theta = \alpha$, $\theta = \beta$ is given by the integral

$$A = \_\_\_\_\_\_\_\_\_\_\_\_ .$$

68. Find the area inside the larger loop and outside the smaller loop of the polar graph $r = 1 - 2\cos\theta$ given in Problem 59.

Solution. The graph of the curve is sketched in the figure below. That part of the curve traced out as θ varies from $\theta = 0$ to $\theta = \pi$ is drawn in with a broader ink stroke. Now, as θ varies from $\frac{\pi}{3}$ to π, the radius vector r sweeps out the larger loop of the curve including that portion of the smaller loop lying above the x-axis. By symmetry, the area of that smaller half-loop is the same as the area of the half-loop swept out by r as θ varies from 0 to $\frac{\pi}{3}$. Thus, the total area inside the larger loop and outside the smaller loop is

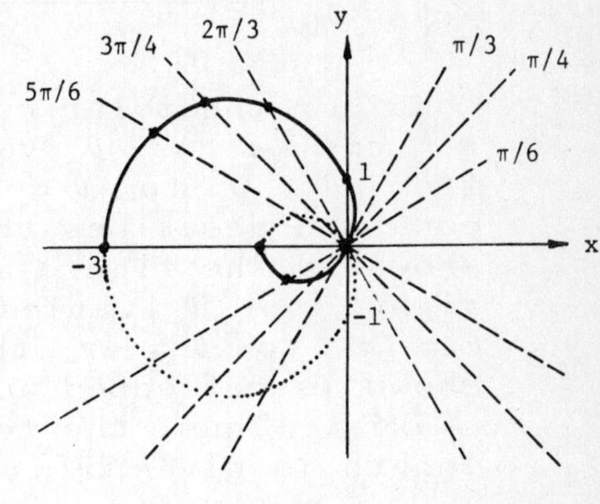

$$A = 2\left[\int_{\pi/3}^{\pi} \tfrac{1}{2}f^2(\theta)\, d\theta - \_\_\_\_\_\_\_\_\_\_\_\_\right]$$

$$= \int_{\pi/3}^{\pi} (1 - 4\cos\theta + 4\cos^2\theta)\, d\theta - \_\_\_\_\_\_\_\_\_\_\_\_$$

$$= \left[\theta - 4\sin\theta + 4\left(\frac{\theta}{2} + \frac{1}{4}\sin 2\theta\right)\right]_{\pi/3}^{\pi} - \_\_\_\_\_\_\_\_\_\_\_\_$$

---

66. $\dfrac{9.539\left[1 - (0.0543)^2\right]}{1 + 0.0543\cos\theta}$, 175.2, 10 AU

67. $\int_{\alpha}^{\beta} \frac{1}{2}\left[f(\theta)\right]^2 d\theta$

$= (\pi + 2\pi) - [\frac{\pi}{3} - 4\left(\frac{\sqrt{3}}{2}\right) + 4\left(\frac{\pi}{6} + \frac{1}{4} \cdot \frac{\sqrt{3}}{2}\right)] - $ ________________

$=$ ____________ $\approx 8.338.$

OBJECTIVE B: Given a polar curve $r = f(\theta)$, calculate its length as θ varies from $\theta = a$ to $\theta = b$.

69. The differential element of length ds for the polar curve $r = f(\theta)$ satisfies the equation

$$ds^2 = \_\_\_\_\_\_\_\_\_\_\_\_ .$$

Thus the length of arc traced out by the curve as θ varies from $\theta = \alpha$ to $\theta = \beta$ is given by

$$s = \_\_\_\_\_\_\_\_\_\_\_\_\_\_\_\_ .$$

70. To determine the length of the curve $r = 3 \sec \theta$ as θ varies from $\theta = 0$ to $\theta = \frac{\pi}{4}$, we find $\frac{dr}{d\theta} =$ __________ . Then the arc length is,

$$s = \int_0^{\pi/4} \sqrt{\_\_\_\_\_\_\_\_\_\_\_\_}\, d\theta = 3 \int_0^{\pi/4} \sec \theta \sqrt{\_\_\_\_\_\_\_\_}\, d\theta$$

$$= 3 \int_0^{\pi/4} \_\_\_\_\_\_\_\_\_\_\_\_\, d\theta = \_\_\_\_\_\_\_\_\Big]_0^{\pi/4} = \_\_\_\_\_\_ .$$

71. Consider the polar curve $r = \cos^4 \frac{\theta}{4}$. As θ varies from $\theta = 0$ to $\theta = 2\pi$, the equation describes the path shown in the figure at the right. As θ varies from $\theta = 2\pi$ to $\theta = 4\pi$ the curve shown is reflected across the x-axis. Thus, the total arc length is given by

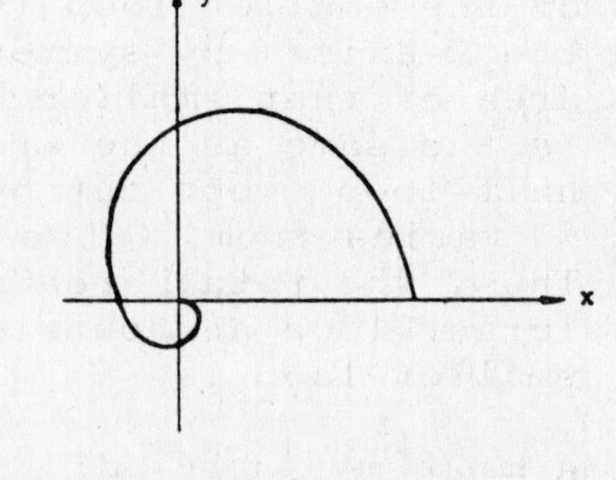

$$s = 2 \int_0^{2\pi} \sqrt{r^2 + \_\_\_\_\_\_\_\_}\, d\theta.$$

---

68. $\int_0^{\pi/3} \frac{1}{2} f^2(\theta)\, d\theta$, $\int_0^{\pi/3} (1 - 4 \cos \theta + 4 \cos^2 \theta)\, d\theta$, $[\theta - 4 \sin \theta + 4\left(\frac{\theta}{2} + \frac{1}{4} \sin 2\theta\right]_0^{\pi/3}$,

$[\frac{\pi}{3} - 4\left(\frac{\sqrt{3}}{2}\right) + 4\left(\frac{\pi}{6} + \frac{1}{4} \cdot \frac{\sqrt{3}}{2}\right)]$, $3\sqrt{3} + \pi$

69. $r^2\, d\theta^2 + dr^2$, $\int_\alpha^\beta \sqrt{r^2 + \left(\frac{dr}{d\theta}\right)^2}\, d\theta$

70. $3 \sec \theta \tan \theta$, $9 \sec^2 \theta + 9 \sec^2 \theta \tan^2 \theta$, $1 + \tan^2 \theta$, $\sec^2 \theta$, $3 \tan \theta$, 3

Now, $\frac{dr}{d\theta}$ = __________ so that

$r^2 + \left(\frac{dr}{d\theta}\right)^2 = \cos^8 \frac{\theta}{4} +$ __________ $= \cos^6 \frac{\theta}{4}$.

Thus, $s = 2\int_0^{2\pi}$ __________ $d\theta$.

Since $\cos \frac{\theta}{4} \geq 0$ for $0 \leq \theta \leq 2\pi$, the integral becomes

$s = 2\int_0^{2\pi} (1 - \sin^2 \frac{\theta}{4})$ __________ $d\theta$

$= 2\int_{\_\_}^{\_\_}$ __________ du, where $u = \sin \frac{\theta}{4}$

= _____ (________)]$_{\_\_}^{\_\_}$ = _________ .

OBJECTIVE C: Find the area of the surface generated when a polar graph is revolved about the x-axis or the y-axis.

72. The graph of the polar equation $r = 5 \cos \theta$ is the circle shown at the right. If the graph is rotated about the x-axis, the total surface area is generated by that portion of the graph as θ varies from $\theta = 0$ to θ = ________ because of the symmetry of the graph across the x-axis.

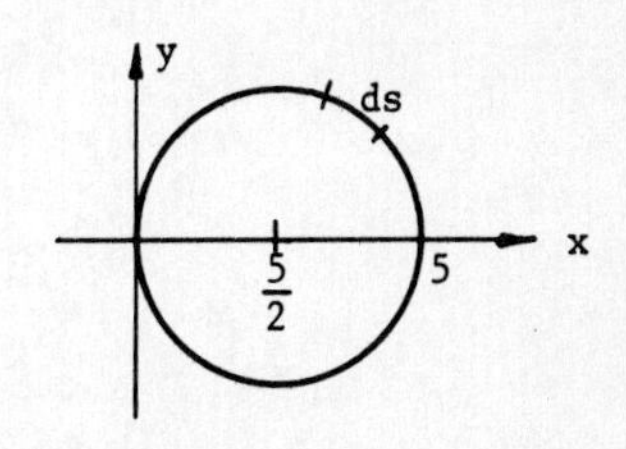

An element of arc length ds (see the figure) generates a portion of surface area dS = _________ ds, where $y = r \sin \theta$ and ds = ___________ . Thus,

$dS = 2\pi \cdot$ ________________________ $d\theta = 2\pi(5 \cos \theta)$ _________ $d\theta$.

Hence, the total surface area is given by

$S = 50\pi \int_{\_\_}^{\_\_}$ ___________ $d\theta$ = ______________]$_{\_\_}^{\_\_}$ = ________

≈ 78.54.

---

71. $\left(\frac{dr}{d\theta}\right)^2$, $-\cos^3 \frac{\theta}{4} \sin \frac{\theta}{4}$, $\cos^6 \frac{\theta}{4} \sin^2 \frac{\theta}{4}$, $\left|\cos^3 \frac{\theta}{4}\right|$, $\cos \frac{\theta}{4}$, $\int_0^1 4(1 - u^2)\, du$, $8(u - \frac{1}{3} u^3)]_0^1$, $\frac{16}{3}$

72. $\frac{\pi}{2}$, $2\pi y$, $\sqrt{dr^2 + r^2\, d\theta^2}$, $r \sin \theta \sqrt{25 \sin^2 \theta + 25 \cos^2 \theta}$, $5 \sin \theta$,

$\int_0^{\pi/2} \sin \theta \cos \theta\, d\theta$, $50\pi \cdot \frac{1}{2} \sin^2 \theta]_0^{\pi/2}$, 25π

73. If the graph in Problem 72 is rotated about the axis $\theta = \frac{\pi}{2}$, the total surface area is generated by the graph as θ varies from $\theta = 0$ to $\theta =$ ________ . An element of arc length ds now generates a portion of the surface area

$$dS = \_\_\_\_\_\_\_\_ = 2\pi \left(\_\_\_\_\_\_\_\_ \right) \cdot 5\ d\theta .$$

Thus, the total surface area is given by

$$S = 50\pi \int_{\_\_}^{\_\_} \_\_\_\_\_\_\_\_ \ d\theta = 50\pi \ [\_\_\_\_\_\_\_\_\_\_\_\_]_{\_\_}^{\_\_}$$

$$= \_\_\_\_\_\_\_\_ \approx 246.74.$$

---

73. π, $2\pi x$ ds, $r \cos \theta$, $\int_0^{\pi} \cos^2 \theta \, d\theta$, $\frac{1}{2}\theta + \frac{1}{4}\sin 2\theta \,]_0^{\pi}$, $25\pi^2$

CHAPTER 9 SELF-TEST

1. Find an equation of the parabola whose directrix is the line $y = -5$ with vertex $V(4,-3)$. Sketch.

2. Find the vertex, the focus, and the directrix of the parabola $x^2 - 4x + 6y + 34 = 0$, and sketch the graph.

3. Write an equation for the circle with center $C(4,6)$ and radius $r = 3$.

4. Find the eccentricity of the ellipse $81x^2 + 25y^2 = 2025$.

5. Find the center, vertices, foci, and eccentricity of the ellipse $9x^2 + 16y^2 - 54x + 128y + 193 = 0$, and sketch the graph.

6. Write an equation of the ellipse with center $C(2,5)$, focus $F(-1,5)$, and semiminor axis $= 4$.

7. Find the center, vertices, foci, and asymptotes of the hyperbola $9x^2 - 4y^2 - 18x + 32y - 91 = 0$, and sketch the graph.

8. Consider the equation $x^2 - 2xy - y^2 = 12$.
   (a) Use the discriminant to classify it.
   (b) Transform the equation by a rotation of axes into an equation with no cross-product term.

9. Use the invariants $B^2 - 4AC$ and $A + C$ to determine an equation to which $x^2 + xy + y^2 = 10$ reduces when the axes are rotated to eliminate the cross-product term.

10. Sketch the graph of the curve described by $x = t - 2$ and $y = t^2 - t + 1$ for $-\infty < t < \infty$. Also find a Cartesian equation of the curve.

11. Find parametric equations for the circle $x^2 + y^2 = 2x$, using as parameter the length s measured counterclockwise from the point $(2,0)$ to the point (x,y).

12. Given the parametric equations $x = t^2 - 1$ and $y = t + 1$
    (a) Express dx and dy in terms of t and dt,
    (b) Find $\frac{d^2y}{dx^2}$ in terms of t,
    (c) Find an equation of the tangent line to the curve at the point for which $t = 1$.

13. Consider the curve given by the parametric equations $x = 3t - 1$ and $y = t^2 - t$. Eliminate the parameter t to find an equation of the form $y = F(x)$ and find $\frac{dy}{dx}$ in terms of $\frac{dy}{dt}$ and $\frac{dx}{dt}$. For what value of t is $\frac{dy}{dx} = 0$? Sketch the graph of the curve over the interval $-2 \leq t \leq 3$.

14. Find the length of the curve given by $x = t^3 + 3t^2$ and $y = t^3 - 3t^2$ for $0 \leq t \leq 2$.

15. Find the area of the surface generated by rotating the cardiod $x = 2\cos\theta - \cos 2\theta$, $y = 2\sin\theta - \sin 2\theta$, for $0 \le \theta \le \pi$, about the x-axis. First sketch the curve.

16. Convert the following from polar coordinates to Cartesian coordinates.

(a) $\left(-6, \frac{\pi}{4}\right)$ (b) $\left(1, -\frac{5\pi}{6}\right)$ (c) $\left(-2, -\frac{7\pi}{12}\right)$

17. Write the following simple polar equations in Cartesian form.

(a) $r = 5$ (b) $\theta = \frac{3\pi}{4}$ (c) $r = -5\csc\theta$

In Problems 18 and 19, graph the polar equation.

18. $r = 2\cos 4\theta$ 19. $r^2 = -\sin 2\theta$

20. Determine a Cartesian equation, and sketch the curve, for $r = \cos\theta + 5\sin\theta$.

21. Find a polar equation of the line with slope $m = -2$ passing through the Cartesian point $(1, -3)$.

22. Find a polar equation of the circle centered at the Cartesian point $(\frac{1}{3}, 0)$ and passing through the origin.

23. Find the length of the polar curve $r = a\cos(\theta + b)$ from $\theta = 0$ to $\theta = \pi$, where a and b are constants.

24. Find the area of the region bounded on the outside by the graph of $r = 2 + 2\sin\theta$ for $\theta = 0$ to $\theta = \pi$, and on the inside by the graph of $r = 2\sin\theta$.

25. Write an integral expressing the surface area generated by rotating the portion of the polar curve $r = 1 + \cos\theta$ in the first quadrant about $\theta = \frac{\pi}{2}$.

SOLUTIONS TO CHAPTER 9 SELF-TEST

1. Since the directrix is parallel to the x-axis and the vertex lies above it (i.e., $-3 > -5$), the parabola opens upward and hence has an equation of the form $(x - h)^2 = 4p(y - k)$. The value of p is the distance from the vertex to the directrix, so $p = 2$. Thus, an equation of the parabola is $(x - 4)^2 = 8(y + 3)$. The focus is $F(4, -1)$.

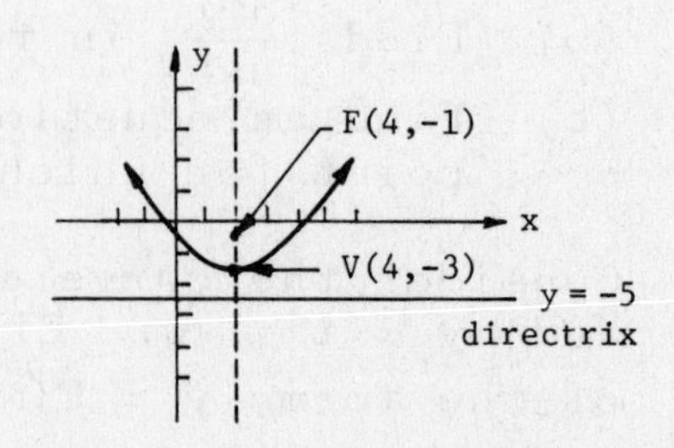

2. Completing the square in x, $(x - 2)^2 = -6(y + 5)$. The parabola opens downward with $4p = 6$ or $p = 3/2$. The vertex is located at $V(2,-5)$ and the focus is $F(2,-5-p) = F(2,-13/2)$. The directrix is $y = -5 + p$ or $y = -7/2$. The parabola is sketched at the right.

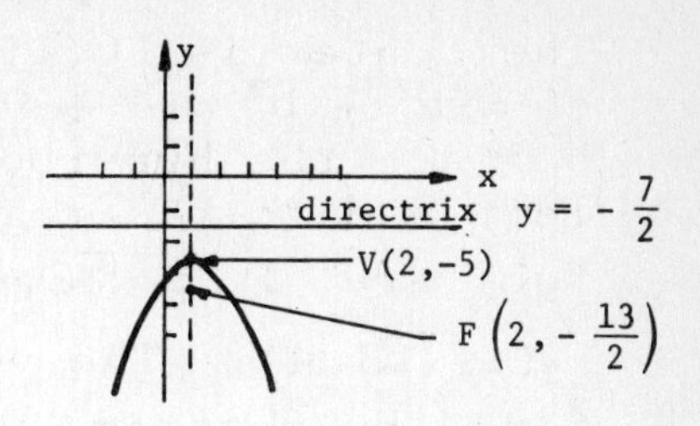

3. $(x - 4)^2 + (y - 6)^2 = 9$, or $x^2 + y^2 - 8x - 12y + 43 = 0$.

4. The standard form of the ellipse is

$$\frac{x^2}{25} + \frac{y^2}{81} - 1 \ ,$$

Thus, since $81 > 25$, the foci lie along the y-axis at $(0, \pm c)$ where $c = \sqrt{81 - 25} = 2\sqrt{14}$. The eccentricity is $e = \frac{c}{a} = \frac{2\sqrt{14}}{9}$ ≈ 0.83.

5. Completing the squares in the x and y terms,

$$9(x^2 - 6x + 9) + 16(y^2 + 8y + 16)$$
$$= 81 + 256 - 193,$$

which may be written as

$$9(x - 3)^2 + 16(y + 4)^2 = 144,$$

or

$$\frac{(x - 3)^2}{16} + \frac{(y + 4)^2}{9} = 1.$$

The center is $C(3,-4)$. The semimajor axis is $a = 4$, and the semiminor axis is $b = 3$. The foci lie on the line $y = -4$. Now $c^2 = a^2 - b^2 = 16 - 9 = 7$. Thus, the foci are located at $F_1(3-\sqrt{7},-4)$ and $F_2(3+\sqrt{7},-4)$. The vertices are $V_1(-1,-4)$, $V_2(7,-4)$, $V_3(3,-1)$, and $V_4(3,-7)$. The graph is shown at the right. The eccentricity is $e = c/a = \sqrt{7}/4 \approx 0.661$.

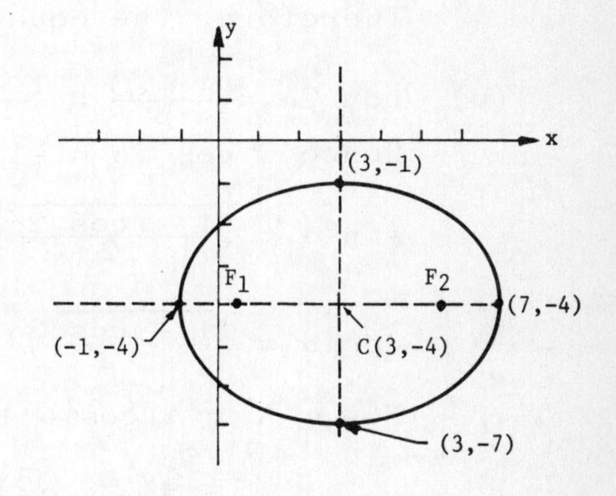

6. The focus lies on the major axis at a distance $c = 2 - (-1) = 3$ units from the center $C(2,5)$. The major axis is along the line $y = 5$ which contains the center and focus. Now, $a^2 = b^2 + c^2 = 16 + 9 = 25$. Hence, an equation of the ellipse is

$$\frac{(x - 2)^2}{25} + \frac{(y - 5)^2}{16} = 1 \quad \text{or} \quad 16(x - 2)^2 + 25(y - 5)^2 = 400.$$

7. Completing the squares in the x and y terms,

$$9(x^2 - 2x + 1) - 4(y^2 - 8y + 16) = 36,$$

or

$$\frac{(x - 1)^2}{4} - \frac{(y - 4)^2}{9} = 1.$$

The center is $C(1,4)$. Now, $c^2 = a^2 + b^2 = 4 + 9 = 13$. Since $b > a$, the hyperbola opens to the left and to the right. Thus, the foci are $F_1(1-\sqrt{13},4)$ and $F_2(1+\sqrt{13},4)$. The vertices are found by setting $y = 4$; whence $x - 1 = \pm 2$. We find $V_1(-1,4)$ and $V_2(3,4)$. Finally, the asymptotes are obtained by setting

$$\frac{(x-1)^2}{4} + \frac{(y-4)^2}{9} \text{ equal to zero,}$$

so $y - 4 = \pm\frac{3}{2}(x - 1)$. The graph is sketched at the right.

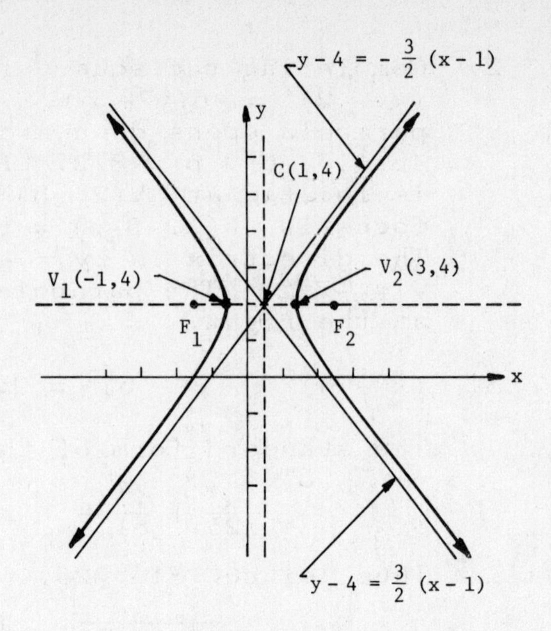

8. (a) The discriminant is $B^2 - 4AC = (-2)^2 - 4(1)(-1) = 8 > 0$. Therefore, the equation represents a hyperbola.

(b) $\cot 2\alpha = \dfrac{A - C}{B} = \dfrac{1 + 1}{-2} = -1$; thus $2\alpha = \dfrac{3\pi}{4}$ radians. Hence, $\cos 2\alpha = -1/\sqrt{2}$. Using the half-angle formulas,

$$\sin\alpha = \sqrt{\frac{1 - \cos 2\alpha}{2}} = \sqrt{\frac{1 + \sqrt{2}}{2\sqrt{2}}} = \frac{1}{2}\sqrt{2 + \sqrt{2}}, \text{ and}$$

$$\cos\alpha = \sqrt{\frac{1 + \cos 2\alpha}{2}} = \sqrt{\frac{1 - \sqrt{2}}{2\sqrt{2}}} = \frac{1}{2}\sqrt{2 - \sqrt{2}}.$$

Now, $A' = A\cos^2\alpha + B\cos\alpha\sin\alpha + C\sin^2\alpha$

$$= \frac{1}{4}(2 - \sqrt{2}) - \frac{2}{4}(\sqrt{-2 + 4}) - \frac{1}{4}(2 + \sqrt{2}) = -\sqrt{2}$$

$C' = A\sin^2\alpha - B\cos\alpha\sin\alpha + C\cos^2\alpha$

$$= \frac{1}{4}(2 + \sqrt{2}) + \frac{2}{4}(\sqrt{-2 + 4}) - \frac{1}{4}(2 - \sqrt{2}) = \sqrt{2}.$$

Thus, since $B' = 0$ because of the choice $\alpha = \dfrac{3\pi}{8}$, the original equation is reduced to $A'x'^2 + C'y'^2 = F$, or $-\sqrt{2}x'^2 + \sqrt{2}y'^2 = 12$, or $y'^2 - x'^2 = 6\sqrt{2}$.

9. Now, $B^2 - 4AC = 1 - 4(1)(1) = -3 = -4A'C'$; thus $A'C' = \dfrac{3}{4}$. Also, $A + C = 2 = A' + C'$. Therefore, $\dfrac{3}{4} = A'(2 - A') = 2A' - A'^2$, or $4A'^2 - 8A' + 3 = 0$. Solving this quadratic equation,

$$A' = \frac{8 \pm \sqrt{64 - 48}}{8}. \text{ Thus, } A' = \frac{3}{2} \text{ or } A' = \frac{1}{2}.$$

For $A' = \frac{3}{2}$, $C' = \frac{1}{2}$ and the original equation reduces to

$$\frac{3}{2}x'^2 + \frac{1}{2}y'^2 = 10 \text{ or } 3x'^2 + y'^2 = 20.$$

On the other hand, if $A' = \frac{1}{2}$, $C' = \frac{3}{2}$ and the original equation reduces to

$$x'^2 + 3y'^2 = 20.$$

We recognize either of these equations as representing an ellipse.

10. We have the following table giving some of the points on the curve:

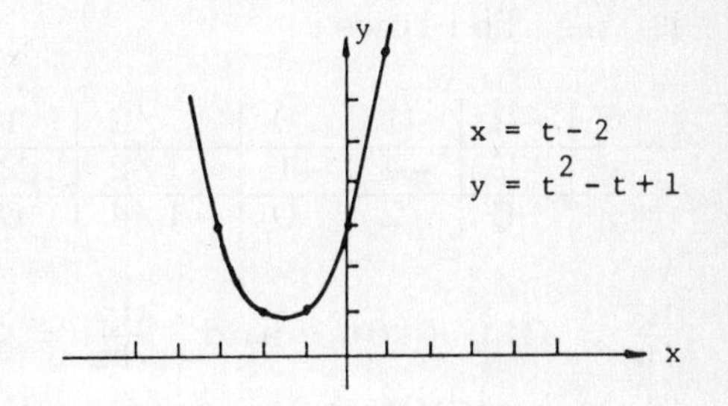

t	-1	0	1	2	3
x	-3	-2	-1	0	1
y	3	1	1	3	7

Substitution of $t = x + 2$ into the parametric equation for y gives

$y = (x + 2)^2 - (x + 2) + 1$, or simplifying algebraically, $y = x^2 + 3x + 3$. This is a Cartesian equation of the parabola sketched in the figure above.

11. Completing the square gives the equation $(x - 1)^2 + y^2 = 1$ which we recognize as an equation of a unit circle centered at $(1,0)$ (see the figure at the right). Hence, $x - 1 = 1 \cos s$ and $y - 0 = 1 \sin s$ or,

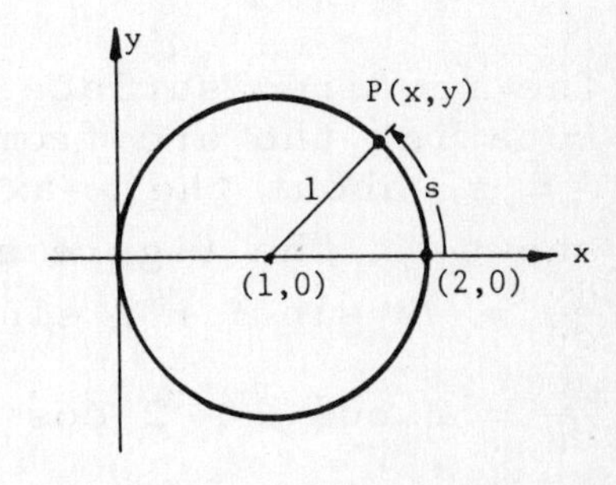

$$x = 1 + \cos s \quad \text{and} \quad y = \sin s$$

are parametric equations for the circle in terms of the arc length parameter s.

12. (a) $\frac{dx}{dt} = 2t$ and $\frac{dy}{dt} = 1$, so that $dx = 2t\,dt$ and $dy = dt$.

(b) $\frac{dy}{dx} = \frac{1}{2t}$ so that $\frac{d^2y}{dx^2} = \frac{dy'/dt}{dx/dt} = \frac{-1/2t^2}{2t} = -\frac{1}{4t^3}$

(c) When $t = 1$, $x = 0$, $y = 2$, and $\frac{dy}{dx} = \frac{1}{2}$; thus $y - 2 = \frac{1}{2}(x - 0)$ or $2y - x = 4$ is an equation of the tangent line.

13. $\frac{dx}{dt} = 3$ and $\frac{dy}{dt} = 2t - 1$, so that $\frac{dy}{dx} = \frac{dy/dt}{dx/dt} = \frac{1}{3}(2t - 1)$. When $t = \frac{1}{2}$, $\frac{dy}{dx} = 0$ which occurs at the point $(x,y) = \left(\frac{1}{2}, -\frac{1}{4}\right)$. To eliminate the parameter t, use the parametric expression for x, $t = \frac{1}{3}(x + 1)$, and substitute into the parametric

expression for y giving

$$y = \frac{1}{9}(x + 1)^2 - \frac{1}{3}(x + 1)$$

$$= \frac{1}{9}(x^2 - x - 2).$$

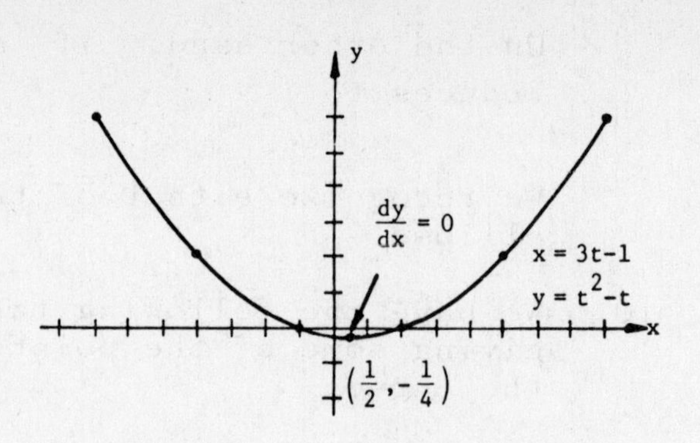

A table of coordinate values is as follows:

t	-2	-1	0	1/2	1	2	3
x	-7	-4	-1	1/2	2	5	8
y	6	2	0	-1/4	0	2	6

14. $\frac{dx}{dt} = 3t^2 + 6t$ and $\frac{dy}{dt} = 3t^2 - 6t$ so that

$$\left(\frac{dx}{dt}\right)^2 + \left(\frac{dy}{dt}\right)^2 = (9t^4 + 36t^3 + 36t^2) + (9t^4 - 36t^3 + 36t^2)$$

$$= 18t^2(t^2 + 4).$$

Thus, the arc length is given by

$$s = \int_0^2 3\sqrt{2}\ t\sqrt{t^2 + 4}\ dt = 3\sqrt{2}\left(\frac{1}{3}\right)\left(t^2 + 4\right)^{3/2}\Big]_0^2$$

$$= 8(4 - \sqrt{2}) \approx 20.7 \text{ units.}$$

15. The required surface is obtained by rotating the arc from $\theta = 0$ to $\theta = \pi$ about the x-axis. The arc is shown in the figure at the right.

$$\frac{dx}{d\theta} = -2 \sin \theta + 2 \sin 2\theta,$$

$$\frac{dy}{d\theta} = 2 \cos \theta - 2 \cos 2\theta \quad \text{so that}$$

$$\left(\frac{dx}{d\theta}\right)^2 + \left(\frac{dy}{d\theta}\right)^2$$

$$= \left(4 \sin^2\theta - 8 \sin\theta \sin 2\theta + 4 \sin^2 2\theta\right) + \left(4 \cos^2\theta - 8 \cos\theta \cos 2\theta + 4 \cos^2 2\theta\right)$$

$$= 8(1 - \sin\theta \sin 2\theta - \cos\theta \cos 2\theta)$$

$$= 8[1 - \cos(2\theta - \theta)]$$

$$= 8(1 - \cos\theta)$$

Therefore, the surface area is given by

$$S = \int_0^{\pi} 2\pi(2 \sin\theta - \sin 2\theta) \cdot 2\sqrt{2}\sqrt{1 - \cos\theta}\ d\theta$$

$$= \int_0^{\pi} 8\sqrt{2}\ \pi \sin\theta(1 - \cos\theta)^{3/2}\ d\theta, \quad \text{with } \sin 2\theta = 2 \sin\theta \cos\theta$$

$$= \frac{16\sqrt{2}}{5}\ \pi(1 - \cos\theta)^{5/2}\Big]_0^{\pi} = \frac{128\pi}{5} \approx 80.4 \text{ square units.}$$

16. (a) $x = -6 \cos \frac{\pi}{4} = -6 \cdot \frac{\sqrt{2}}{2} = -3\sqrt{2}$; $y = -6 \sin \frac{\pi}{4} = -3\sqrt{2}$

(b) $x = 1 \cos\left(-\frac{5\pi}{6}\right) = \cos \frac{5\pi}{6} = -\frac{\sqrt{3}}{2}$;

$y = 1 \sin\left(-\frac{5\pi}{6}\right) = -\sin \frac{5\pi}{6} = -\frac{1}{2}$

(c) $x = -2 \cos\left(-\frac{7\pi}{12}\right) = -2 \cos \frac{7\pi}{12} = \frac{\sqrt{2}}{2}(\sqrt{3} - 1) \approx 0.518$

$y = -2 \sin\left(-\frac{7\pi}{12}\right) = 2 \sin \frac{7\pi}{12} = \frac{\sqrt{2}}{2}(\sqrt{3} + 1) \approx 1.932$

17. (a) $\pm\sqrt{x^2 + y^2} = 5$ or $x^2 + y^2 = 25$

(b) $y = -x$

(c) Equivalently, $r \sin \theta = -5$, or $y = -5$

18. $r = 2 \cos 4\theta$ is symmetric about the x-axis, the y-axis, and the origin.
The graph is the eight-leafed rose sketched at the right.

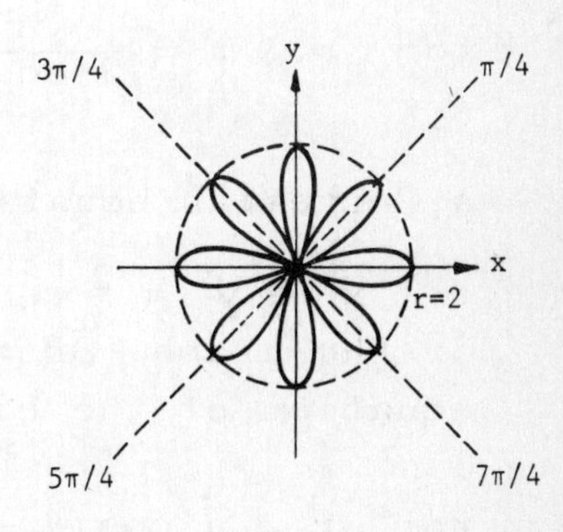

19. $r^2 = -\sin 2\theta$ is symmetric about the origin since the equation is unchanged when r is replaced by $-r$. Notice that $-\sin 2\theta$ must be nonnegative. If θ is restricted to the interval $[0, 2\pi]$, then $-\sin 2\theta \geq 0$ if and only if θ is in $[\frac{\pi}{2}, \pi]$ or $[\frac{3\pi}{2}, 2\pi]$. Using the following table and symmetry we obtain the graph sketched at the right:

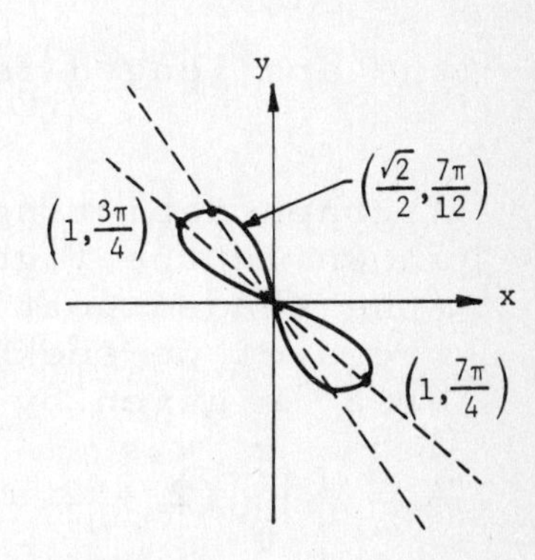

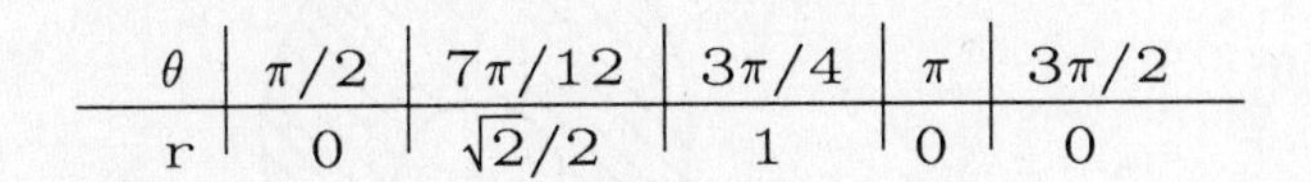

θ	$\pi/2$	$7\pi/12$	$3\pi/4$	π	$3\pi/2$
r	0	$\sqrt{2}/2$	1	0	0

The curve is a lemniscate.

20. Since $r = 5\cos\theta + 5\sin\theta$, for $r \neq 0$ ($r = 0$ is on the graph at $\theta = \frac{3\pi}{4}$), $r^2 = 5r\cos\theta + 5r\sin\theta$, or $x^2 + y^2 = 5x + 5y$. Then, completing the squares in the x and y terms gives

$$\left(x - \frac{5}{2}\right)^2 + \left(y - \frac{5}{2}\right)^2 = \frac{25}{2}.$$

This is a circle with center $\left(\frac{5}{2}, \frac{5}{2}\right)$ and radius $\frac{5}{\sqrt{2}}$.

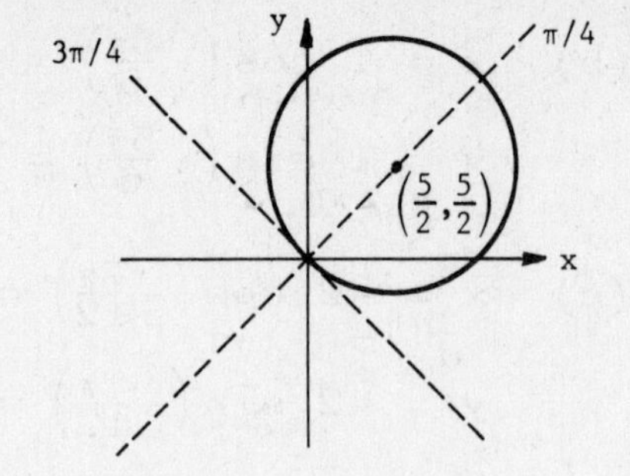

21. A Cartesian equation of the line is given by $y + 3 = -2(x - 1)$ or $2x + y = -1$. Thus, $2r\cos\theta + r\sin\theta = -1$, or solving for r, $r = \frac{-1}{2\cos\theta + \sin\theta}$.

22. A Cartesian equation of the circle is given by $\left(x - \frac{1}{3}\right)^2 + y^2 = \frac{1}{9}$ or $x^2 + y^2 = \frac{2}{3}x$. Thus, $3r^2 = 2r\cos\theta$, or since $r = 0$ lies on the graph $3r = 2\cos\theta$ when $\theta = \frac{\pi}{2}$, the latter gives a polar equation of the circle.

23. For $r = a\cos(\theta + b)$, $\frac{dr}{d\theta} = -a\sin(\theta + b)$ so that $r^2 + \left(\frac{dr}{d\theta}\right)^2 = a^2\cos^2(\theta + b) + a^2\sin^2(\theta + b) = a^2$. Therefore the arc length is given by the integral $s = \int_0^\pi \sqrt{a^2}\, d\theta = |a|\pi$.

24. A graph depicting the region is shown in the figure at the right (the shaded portion represents the area we seek). Thus the area is given by

$$A = \frac{1}{2}\int_0^\pi (2 + 2\sin\theta)^2\, d\theta - \frac{1}{2}\int_0^\pi (2\sin\theta)^2\, d\theta$$

$$= \frac{1}{2}\int_0^\pi (4 + 8\sin\theta)\, d\theta$$

$$= 2(\theta - 2\cos\theta)\Big]_0^\pi$$

$$= 2\pi + 8 \approx 14.28.$$

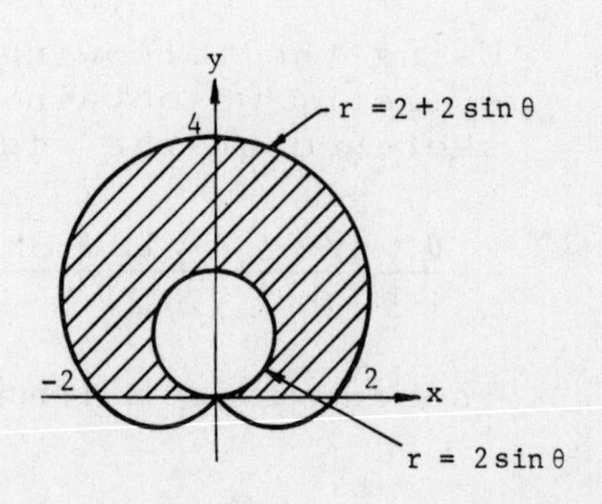

25. A sketch of the surface is shown in the figure at the right. An element of arc length ds generates a portion of surface area

$$dS = 2\pi x\ dx.$$

Now, $\frac{ds}{d\theta} = \sqrt{r^2 + \left(\frac{dr}{d\theta}\right)^2}$

$$= \sqrt{(1 + 2\cos\theta + \cos^2\theta) + \sin^2\theta} = \sqrt{2}\ \sqrt{1 + \cos\theta}$$

Hence, $dS = 2\pi x\ ds = 2\pi r\cos\theta\ ds = 2\sqrt{2}\pi(1 + \cos\theta)^{3/2}\cos\theta\ d\theta$. Therefore, the total surface area generated is given by the integral

$$S = 2\sqrt{2}\pi \int_0^{\pi/2} (1 + \cos\theta)^{3/2} \cos\theta\, d\theta.$$

(The definite integral can be evaluated by using the identity $\cos^2\frac{\theta}{2} = 1 + \cos\theta$, but it is tedious to carry out the calculations. Using Simpson's rule with $n = 12$, an approximate value to the integral is 39.31. This calculation was made using a calculator.)

NOTES.